EXPERIMENTAL DESIGN, STATISTICAL MODELS, AND GENETIC STATISTICS

STATISTICS: Textbooks and Monographs

A SERIES EDITED BY

D. B. OWEN, Coordinating Editor
Department of Statistics
Southern Methodist University
Dallas, Texas

Volume 1: The Generalized Jackknife Statistic, *H. L. Gray and W. R. Schucany*

Volume 2: Multivariate Analysis, *Anant M. Kshirsagar*

Volume 3: Statistics and Society, *Walter T. Federer*

Volume 4: Multivariate Analysis: A Selected and Abstracted Bibliography, 1957-1972, *Kocherlakota Subrahmaniam and Kathleen Subrahmaniam* (out of print)

Volume 5: Design of Experiments: A Realistic Approach, *Virgil L. Anderson and Robert A. McLean*

Volume 6: Statistical and Mathematical Aspects of Pollution Problems, *John W. Pratt*

Volume 7: Introduction to Probability and Statistics (in two parts) Part I: Probability; Part II: Statistics, *Narayan C. Giri*

Volume 8: Statistical Theory of the Analysis of Experimental Designs, *J. Ogawa*

Volume 9: Statistical Techniques in Simulation (in two parts), *Jack P. C. Kleijnen*

Volume 10: Data Quality Control and Editing, *Joseph I. Naus*

Volume 11: Cost of Living Index Numbers: Practice, Precision, and Theory, *Kali S. Banerjee*

Volume 12: Weighing Designs: For Chemistry, Medicine, Economics, Operations Research, Statistics, *Kali S. Banerjee*

Volume 13: The Search for Oil: Some Statistical Methods and Techniques, *edited by D. B. Owen*

Volume 14: Sample Size Choice: Charts for Experiments with Linear Models, *Robert E. Odeh and Martin Fox*

Volume 15: Statistical Methods for Engineers and Scientists, *Robert M. Bethea, Benjamin S. Duran, and Thomas L. Boullion*

Volume 16: Statistical Quality Control Methods, *Irving W. Burr*

Volume 17: On the History of Statistics and Probability, *edited by D. B. Owen*

Volume 18: Econometrics, *Peter Schmidt*

Volume 19: Sufficient Statistics: Selected Contributions, *Vasant S. Huzurbazar (edited by Anant M. Kshirsagar)*

Volume 20: Handbook of Statistical Distributions, *Jagdish K. Patel, C. H. Kapadia, and D. B. Owen*

Volume 21: Case Studies in Sample Design, *A. C. Rosander*

Volume 22: Pocket Book of Statistical Tables, *compiled by R. E. Odeh, D. B. Owen, Z. W. Birnbaum, and L. Fisher*

Volume 23: The Information in Contingency Tables, *D. V. Gokhale and Solomon Kullback*

Vol. 24: Statistical Analysis of Reliability and Life-Testing Models: Theory and Methods, *Lee J. Bain*
Vol. 25: Elementary Statistical Quality Control, *Irving W. Burr*
Vol. 26: An Introduction to Probability and Statistics Using BASIC, *Richard A. Groeneveld*
Vol. 27: Basic Applied Statistics, *B. L. Raktoe and J. J. Hubert*
Vol. 28: A Primer in Probability, *Kathleen Subrahmaniam*
Vol. 29: Random Processes: A First Look, *R. Syski*
Vol. 30: Regression Methods: A Tool for Data Analysis, *Rudolf J. Freund and Paul D. Minton*
Vol. 31: Randomization Tests, *Eugene S. Edgington*
Vol. 32: Tables for Normal Tolerance Limits, Sampling Plans, and Screening, *Robert E. Odeh and D. B. Owen*
Vol. 33: Statistical Computing, *William J. Kennedy, Jr. and James E. Gentle*
Vol. 34: Regression Analysis and Its Application: A Data-Oriented Approach, *Richard F. Gunst and Robert L. Mason*
Vol. 35: Scientific Strategies to Save Your Life, *I. D. J. Bross*
Vol. 36: Statistics in the Pharmaceutical Industry, *edited by C. Ralph Buncher and Jia-Yeong Tsay*
Vol. 37: Sampling from a Finite Population, *J. Hájek*
Vol. 38: Statistical Modeling Techniques, *S. S. Shapiro*
Vol. 39: Statistical Theory and Inference in Research, *T. A. Bancroft and C.-P. Han*
Vol. 40: Handbook of the Normal Distribution, *Jagdish K. Patel and Campbell B. Read*
Vol. 41: Recent Advances in Regression Methods, *Hrishikesh D. Vinod and Aman Ullah*
Vol. 42: Acceptance Sampling in Quality Control, *Edward G. Schilling*
Vol. 43: The Randomized Clinical Trial and Therapeutic Decisions, *edited by Niels Tygstrup, John M. Lachin, and Erik Juhl*
Vol. 44: Regression Analysis of Survival Data in Cancer Chemotherapy, *Walter H. Carter, Jr., Galen L. Wampler, and Donald M. Stablein*
Vol. 45: A Course in Linear Models, *Anant M. Kshirsagar*
Vol. 46: Clinical Trials: Issues and Approaches, *edited by Stanley H. Shapiro and Thomas H. Louis*
Vol. 47: Statistical Analysis of DNA Sequence Data, *edited by B. S. Weir*
Vol. 48: Nonlinear Regression Modelling: A Unified Practical Approach, *David A. Ratkowsky*
Vol. 49: Attribute Sampling Plans, Tables of Tests and Confidence Limits for Proportions, *Robert E. Odeh and D. B. Owen*
Vol. 50: Experimental Design, Statistical Models, and Genetic Statistics, *edited by Klaus Hinkelmann*

OTHER VOLUMES IN PREPARATION

EXPERIMENTAL DESIGN, STATISTICAL MODELS, AND GENETIC STATISTICS

Essays in Honor of Oscar Kempthorne

edited by

Klaus Hinkelmann

Virginia Polytechnic Institute
and State University
Blacksburg, Virginia

MARCEL DEKKER, INC. New York and Basel

Library of Congress Cataloging in Publication Data
Main entry under title:

Experimental design, statistical models, and genetic statistics.

(Statistics, textbooks and monographs ; v. 50)
Includes index.
1. Experimental design--Addresses, essays, lectures.
2. Statistical models (Statistics)--Addresses, essays, lectures. 3. Genetics--Statistical methods--Addresses, essays, lectures. 4. Kempthorne, Oscar. I. Kempthorne, Oscar. II. Hinkelmann, Klaus. III. Series.
QA279.E95 1984 519.5 83-26195
ISBN 0-8247-7151-6

MARCEL DEKKER, INC.
270 Madison Avenue, New York, New York 10016

Current printing (last digit):
10 9 8 7 6 5 4 3 2 1

PRINTED IN THE UNITED STATES OF AMERICA

PREFACE

Over the past 40 years Oscar Kempthorne has made many important contributions to statistics. His name is associated with the areas of experimental design and genetic statistics, his books *Design and Analysis of Experiments* and *Introduction to Genetic Statistics* having become classics in these areas. Topics such as randomization theory, factorial experiments, quantitative genetics bear his stamp. For many years he has been engaged extensively in questions and philosophy of statistical inference. He has written widely on all these subjects, and many more; he has spoken about them in the classroom, at colloquia and at conferences all over the world. In doing so, and in doing it in his style, he has generated many discussions and debates, always in the interest of improving our knowledge and understanding of statistics and its foundations.

Many students have come to study under him at Iowa State University. I was fortunate to have been one of them, and those years bring back fond memories, even though they were not always easy. However in times of despair Oscar Kempthorne, just as a father, would convince us that it was worthwhile to continue. Even if his words were not always complimentary we knew that we could count on him, that he was sincere, and that he acted in our best interest. Above all, we have benefited in so many ways professionally from him through his teaching, formal and informal, and the many conversations with him. We can perhaps never thank him enough for that.

It is in this spirit that former students, colleagues and friends have collaborated to produce this Festschrift in honor of Oscar Kempthorne on the occasion of his 65th birthday. It is a small tribute, but it comes with the best wishes for many more productive years for the benefit of the entire statistics community.

I have to apologize to many who might have liked to contribute to this volume, but in the interest of space, selections had to be made. I would like to thank Herbert A. David for his encouragement to undertake this task, the contributors for their cooperation, and Karen Anderson for the expert typing of the manuscript.

Klaus Hinkelmann

CONTENTS

115,666

CONTRIBUTORS

T. A. Bancroft Iowa State University, Ames, Iowa

David Birkes Oregon State University, Corvallis, Oregon

N. R. Bohidar Merck, Sharp, and Dohme Research Laboratories, West Point, Pennsylvania

George E. P. Box University of Wisconsin, Madison, Wisconsin

John Brocklebank SAS Institute, Cary, North Carolina

R. N. Curnow University of Reading, Reading, United Kingdom

H. A. David Iowa State University, Ames, Iowa

Ted H. Emigh North Carolina State University, Raleigh, North Carolina

Walter T. Federer Cornell University, Ithaca, New York

J. Leroy Folks Oklahoma State University, Stillwater, Oklahoma

K. Ruben Gabriel University of Rochester, Rochester, New York

David Gheva University of Rochester, Rochester, New York

F. G. Giesbrecht North Carolina State University, Raleigh, North Carolina

Dewey L. Harris Agricultural Research Service, U.S. Department of Agriculture and Purdue University, West Lafayette, Indiana

David A. Harville Iowa State University, Ames, Iowa

William G. Hill University of Edinburgh, Edinburgh, Scotland

Klaus Hinkelmann Virginia Polytechnic Institute and State University, Blacksburg, Virginia

Hiromitsu Kanemasu World Bank, Washington, D.C.

H. J. Khamis Wright State University, Dayton, Ohio

John P. Mandeli Virginia Commonwealth University, Richmond, Virginia

Karin Meyer University of Edinburgh, Edinburgh, Scotland

Jochen Müller University of Pittsburgh, Pittsburgh, Pennsylvania

Edward Pollak Iowa State University, Ames, Iowa

C. Radhakrishna Rao University of Pittsburgh, Pittsburgh, Pennsylvania

Stephen S. Rich* Agricultural Research Service, U.S. Department of Agriculture and Purdue University, West Lafayette, Indiana

Basilio A. Rojas M. Centro de Estadística y Cálculo, Chapingo, México

C. Z. Roux Animal and Dairy Science Research Institute, Irene, Republic of South Africa

Justus Seely Oregon State University, Corvallis, Oregon

Bimal Kumar Sinha University of Pittsburgh, Pittsburgh, Pennsylvania

Jaya Srivastava Colorado State University, Fort Collins, Colorado

Robin Thompson University of Edinburgh, Edinburgh, Scotland

Chuan-Ting Wang† Agricultural Research Service, U.S. Department of Agriculture and Purdue University, West Lafayette, Indiana

---

Current affiliations:

*University of Minnesota, Minneapolis, Minnesota.

†Lincoln, Nebraska.

PART I

CONTRIBUTIONS TO IOWA STATE UNIVERSITY AND BEYOND

Chapter 1

THE YEARS 1950-1972

T.A. Bancroft
Iowa State University
Ames, Iowa

I am pleased, of course, to give an account of my association with Oscar Kempthorne during the 22 year period I served as Director of the Statistical Laboratory and Head of the Department of Statistics at Iowa State.

Oscar joined the faculty of the Iowa State Statistical Laboratory (statistical center) in 1947. While I came to Iowa State first in 1941, I was away for about three years, rejoining the faculty in 1949 and became the second "permanent" Director and first "permanent" Head in 1950.

The Department of Statistics, officially established in 1947, actually evolved from a sub-program of courses in statistics primarily in the Mathematics Department, taught for the most part by faculty members with joint appointments with the Statistical Laboratory. After 1947 the faculty of the evolving complete statistical center (Statistical Laboratory and Department of Statistics) were officed together in areas of the, then, Service Building. In 1947, then, the roots of the Iowa State Statistical Center were firmly planted and officially recognized by the University (see Bancroft, 1982).

Many serious problems remained to be solved during the early part of the 1950-1972 period which were of great importance in making possible the contributions in teaching, research, and consulting of Kempthorne as well as other members of the statistical center. Three of these requiring immediate attention were: (i) replacement of resigned key faculty members in statistical theory (Alexander Mood and George Brown), (ii) arrangements for increasing the small amount of space, then available in the Service Building and (iii) updating of the computing hardware from conventional IBM business machines and desk computers to a high speed digital computer (see Bancroft, 1983).

The Iowa State statistical center, after the establishing of the Department in 1947, became committed to statistics as a separate discipline (one of the mathematical sciences) as well as an important part of scientific methodology (a method of obtaining new knowledge in many scientific areas) (see Bancroft, 1983). In view of its early and continuing close association with agriculture and related biological sciences, it was necessary, of course, that the statistical center continually maintain a strong position in the theory and methods of experimental design as well as retaining its established leadership position in consulting and cooperative research in experimental design applications.

Beginning with his joining Iowa State in 1947, Kempthorne rapidly attained a leadership position in consulting and cooperative research in this area. His leadership in these areas was of vital importance, in particular, in backup assistance to other faculty members engaged in the teaching of applied courses involving the methods of experimental design applications and related statistical consulting and cooperative research.

In addition, Kempthorne's leadership in experimental design and analysis, both in teaching and research, complemented the statistical center's leadership position in the methods of survey sampling. Incidentally, the long term strength in survey sampling methods and operational work was to be strengthened, as regards survey sampling theory, when H.O. Hartley joined the Iowa State statistical center in 1953 (see Bancroft, 1978). Recruitment of able faculty members in advanced general theory and methods and multivariate analysis in the early 1950's provided considerable depth for training in statistics, in particular, for advanced graduate students majoring in experimental design with Kempthorne as well as other professors directing Ph.D. programs in other areas.

In addition to his outstanding leadership in teaching, research, and the guidance of graduate students in experimental design and analysis, Kempthorne was equally effective in the area of genetic statistics. Both leadership positions were, of course, of great importance in the agricultural and related biological programs of plant and animal sciences at Iowa State. The latter, in particular, was of great importance in plant and animal breeding.

With the acquisition of additional space in the Service Building and the building of the new Addition it became possible to office Kempthorne, his associates, and graduate students in the areas of experimental design and genetic statistics in a suite of offices in a convenient location on the ground floor and provide a special secretary.

Kempthorne's leadership positions at Iowa State were soon recognized nationally and internationally. His recognized excellence in research resulted in many off-campus research grants and contracts providing financial support for some graduate students and part time salaries for certain faculty members. Also many able graduate students were attracted to enroll in statistics at Iowa State and arrange for graduate programs under Kempthorne's guidance.

The curriculum vitae for the 1950-1972 period shows that Kempthorne published some 88 papers singly or co-authored and 7 books singly, co-authored or edited. In addition, during this period, he directed some 30 Ph.D. and 14 M.S. theses. It is of interest to note that subsequent to graduation many of these graduates distinguished themselves in academic, government or industrial positions.

As would be expected, Kempthorne's contributions were recognized by statistical and scientific societies and universities. He was named a Distinguished Professor of Sciences and Humanities at Iowa State in 1964 and granted a Sc.D. by Cambridge University in 1960. In addition he served as a president of the Biometric Society, ENAR; chairman of Section U (Statistics) of AAAS; chairman of the R.A. Fisher Memorial Lectures for Statistical Societies; and in many other leadership activities.

On a personal note, I appreciate very much Oscar Kempthorne's decision to devote, during the 1950-72 period and until now, his outstanding productive scholarly career primarily to the programs of the statistical center at Iowa State. With at least some additional time in the future off from official academic duties, opportunity should be available for spending additional time with his wife Val and his children.

REFERENCES

Bancroft, T.A. (1978). Greetings to HOH for 1977. In: *Contributions to Survey Sampling and Applied Statistics, Papers in Honor of H.O. Hartley*, H.A. David (ed.). XVII-XIX, New York: Academic Press.

Bancroft, T.A. (1978). Roots of the Iowa State University Statistical Center: 1914-1950. Iowa State Journal of Research 57: 3-10.

Bancroft, T.A. (assisted by Margaret G. Kirwin) (1983). Highlights of Some Expansion Years of the Iowa State Statistical Laboratory (statistical center): 1947-1972. *Fiftieth Anniversary of the Iowa State Statistical Laboratory*, Ames: I.S.U. Press.

Kempthorne, O. (1952). *The Design and Analysis of Experiments*. New York: John Wiley.

Kempthorne, O., T.A. Bancroft, J.W. Gowen and J.L. Lush (eds.) (1954). *Statistics and Mathematics in Biology*. Ames: Iowa State University Press.

Kempthorne, O. (1957). *An Introduction to Genetic Statistics*. New York: John Wiley.

Kempthorne, O. (ed.)(1960). *Biometrical Genetics*. New York: Pergamon Press.

Kempthorne, O. and J.L. Folks (1971). *Probability, Statistics, and Data Analysis*. Ames: Iowa State University Press.

Pollak, E., O. Kempthorne and T.B. Bailey, Jr. (eds.)(1977). *Proceedings of the International Conference on Quantitative Genetics*. Ames: Iowa State University Press.

Chapter 2

THE YEARS 1972-1984

H.A. David
Iowa State University
Ames, Iowa

Oscar Kempthorne made his name early with the publication in 1952 of the widely used Wiley text *The Design and Analysis of Experiments*. Since then he has consolidated his position as a leading statistician by numerous research publications and several more books. As a result, it is my impression that many statisticians will be surprised that he is turning only 65. In fact, it is a very vigorous 65, as I will try to establish by looking at his activities over the past ten years or so.

First, some personal remarks. When I accepted, with alacrity, the opportunity to head up Statistics at Iowa State I naturally had some apprehensions about Kemp's feelings towards me. I say "naturally" because anyone who has been around knows that Kemp is not only a distinguished statistician but also one who does not hesitate to express strong views. As it soon turned out, I need not have worried. His strong views stem from a very deep concern about the proper interpretation and application of statistical ideas. He gets irritated by those propounding, as brand new, ideas which can be found in earlier work; the cumulative achievements of the field of statistics are important to him. On the local scene

Oscar Kempthorne provides not only intellectual leadership, as one would expect, but also outstanding devotion to the good of the entire department. His interest in all facets of the department's activities is insatiable. Kemp became a friend to turn to for advice and who is ready to step in whenever his help is needed. He continually astonishes by finding time to encourage a young colleague, to talk at length to visitors, and to quiz every prospective faculty member. He is generous with his time and in other ways; the warm hospitality frequently extended by him and his wife Val adds much to the department. Kemp's wide knowledge is invaluable at seminar time when he is likely to break the ice by a searching question or a clarifying comment. In all these activities he manages to combine friendliness with a slightly provocative manner or perhaps it is the other way around. Since his comments are always constructive, unless he feels a wrong notion must be scotched, his approach has earned him the affection as well as the respect of all who have come into contact with him.

What then has been the main thrust of Oscar Kempthorne's work since 1972? In his teaching his enthusiasm for introducing graduate students to the theory and application of linear models has been unflagging; he is constantly revising his class notes. I was happy to learn that this major effort will also benefit wider audiences in a planned two-volume revision of the 1952 text, never updated before. The project is joint with his former student, Klaus Hinkelmann. Kempthorne has also lectured frequently on advanced aspects of the linear model and on experimental design. He is a challenging and entertaining teacher, with an empathy for students no doubt enhanced by watching his three children go through the student stages.

During this period Kempthorne has published some 30 papers, nearly all concerned with some aspect of statistical inference. Typically he has tackled the big issues critically reviewing the pertinent literature and throwing light on subjects still in a controversial state. A good illustration is his *Biometrics* Invited

Paper (1978) on the interpretation of nature-nurture data. He leaves us in no uncertainty on where he stands:

> The idea that there are racial-genetic differences in mental abilities and behavioral traits of humans is, at best, no more than idle speculation.

A conclusion stated in a paper on fixed and mixed models (1975b) seems to apply to much of his more recent work:

> This essay does not give tight answers to any of the basic questions. Those who have a desperate need for tight answers should go to a theology.

Another example is his paper (1976) on tests of significance and tests of hypothesis. The latter is a decision procedure requiring the choice of a fixed significance level whereas the former is an inference procedure, the level of significance corresponding to a particular test of significance being a statistic. This distinction (simplified here) was made earlier in the book by Kempthorne and Folks (1971). He has never hesitated, at an appropriate time, to repeat expounding an important but not sufficiently well-known concept.

The logic underlying randomization has been a long-time interest of Kempthorne's, e.g., (1975a). In 1976 he made the incisive observation

> ... Fisher rejected as irrelevant any sort of repeated sample sampling principle in some classes of situations. Yet, in contrast, Fisher's exposition of the force of experiment randomization is based, as far as I can see, on a total acceptance of the principle. This ambivalence is a deep mystery, I believe, which receives quite inadequate discussion.

This brings us to the tremendous influence of Fisher's ideas on Kempthorne's work. With his dual interests in statistics and quantitative genetics Kempthorne is in a particularly good position to appreciate Fisher's contribution. Indeed, most of Kempthorne's papers have references to Fisher's work. He is clearly an admirer but not an uncritical one. The master is consulted on any issue where he has spoken but is then frequently accused of obscurity!

My remarks on some of Kempthorne's research over the past ten years are not, of course, meant to be an appraisal but are merely intended to convey the flavor of his writings. There is no space here to include the many peppery asides which Kempthorne uses as seasoning. Needless to say, he is a much sought after speaker. In a 1978 visit to Australia he presented eleven seminars but in his characteristically unsparing way he did not just repeat the same few talks in different States. The titles are illuminating: Foundations of Statistical Thinking and Reasoning (Knibbs lecture); Population Genetics; Mixed Linear Models; Randomization: The Scientific World Needs It; Analysis of Variance: Linear and Non-Linear; The Teaching of Statistics (twice), R.A. Fisher on Tests of Significance; Algebraic Structures in Experimental Design; IQ and Heredity: A Deep Statistical Problem; Varieties of Inference.

There may be an impression that Kempthorne is not a committee man. Nothing could be further from the truth, at least on home territory where I have been able to observe how he achieves great effectiveness by a combination of thoroughness, tenacity, resourcefulness, and the courage to think big. Kemp was the moving spirit and did much of the work, including fund raising, for the major 1976 International Conference on Quantitative Genetics and also for the 1979 Regional Conference featuring J.F.C. Kingman on Population Genetics. He was a key member of the Program Committee for the 50th Anniversary Conference of the Statistical Laboratory held in 1983. But also more mundane matters such as nominations of colleagues for promotion and awards are handled with great thoughtfulness.

National recognition of Oscar Kempthorne has continued over the past ten years. He has served in numerous editorial capacities and also as Chair for Section U (Statistics) of AAAS and the Fisher Memorial Lecture Committee of the Statistical Societies. It is my impression, however, that he was most touched by an award many might have thought eclipsed by earlier honors, namely the Iowa State University Alumni Achievement Faculty Citation for long, outstanding, and inspiring service on the University Staff.

Kemp once remarked to me: It is important to put on a good show. We in Statistics at ISU count ourselves fortunate that he has been a leader in our midst for over 35 years and hope he will continue to contribute and put on a good show for many more years to come.

REFERENCES

Kempthorne, O. and L.J. Folks (1971). *Probability, Statistics, and Data Analysis.* Ames: Iowa State University Press.

Kempthorne, O. (1975a). Inference from Experiments and Randomization. In: *A Survey of Statistical Design and Linear Models.* J.N. Srivastava (ed.). Amsterdam: North Holland, 303-31.

Kempthorne, O. (1975b). Fixed and Mixed Models in the Analysis of Variance. *Biometrics 31,* 473-86.

Kempthorne, O. (1976). Of What Use Are Tests of Significance and Tests of Hypothesis? *Commun. Statist. - Theory Meth. A5,* 763-77.

Kempthorne, O. (1978). Logical, Epistemological, and Statistical Aspects of Nature-Nurture Data Interpretation. *Biometrics 34,* 1-23.

PART II

DESIGN AND ANALYSIS OF EXPERIMENTS

Chapter 3

USE OF RANDOMIZATION IN EXPERIMENTAL RESEARCH

J. Leroy Folks
Oklahoma State University
Stillwater, Oklahoma

I. INTRODUCTION

I still remember my first encounter with Oscar Kempthorne on a statistical topic. As a beginning doctoral student at Iowa State University, I made the comment to him that a two-factor experiment and a randomized block design resulted in the same model. I learned very quickly that my education was not complete - that there was a fundamental difference between these two situations resulting from the randomization.

In years to come I was to realize the depth of his thinking about inference and, in particular, about the role of randomization in experimental inference. Professor Kempthorne has stated and restated the role of randomization in the foundation of experimental inference. When I was invited to write a paper for this Festschrift, I could not get the randomization idea out of my mind. So I offer one more summary of this important topic, hoping to add some small bit to this ongoing dialogue.

II. ORIGINS OF EXPERIMENTAL RANDOMIZATION

The idea of a random sample may be a fairly recent one in human history. Carnap (1962) refers to Venn and C.S. Peirce for original formulations of the concept. Although Venn gives much discussion of randomness he does not give an explicit formulation of a random sample from a population. Peirce, on the other hand, gives in a number of his writings the definition of a random sample and its role in inductive inference. Peirce (1883) states,

> The first premise of a scientific inference is that certain things (in the case of induction) or certain characters (in the case of hypothesis) constitute a fairly chosen sample of the class of things or the run of characters from which they have been drawn. The rule requires that the sample should be drawn at random and independently from the whole lot sampled.

It is also clear that a random sample was, for Peirce, made possible only by a physical act. Peirce (1896) wrote,

> A sample is a random one, provided it is drawn by such machinery, artificial or physiological, that in the long run any one individual of the whole lot would get taken as often as any other.

The conviction with which Peirce believed that randomization should be physically performed is indicated by the great care with which he and Joseph Jastrow randomized the order of treatments in an experiment to test the threshold theory of sensory perception. This experiment, performed in 1884, is summarized by Stigler (1978). Stigler says,

> The Peirce-Jastrow experiment is the first of which I am aware where the experimentation was performed according to a precise, mathematically sound randomization scheme!

As Stigler notes, however, randomization of treatments did not become an accepted part of statistical thought until promoted by Fisher in the twentieth century.

Fisher was at Rothamsted from 1919 until 1933 and in his early years there he began to think about the role of randomization in experimental design. Fisher (1950) says,

> From about 1923 onwards the Statistical Department at Rothamsted had been much concerned with the precision of field experiments in agriculture, and with modifications in their design, having the dual aim of increasing the precision and of providing a valid estimate of error.

Fisher (1926) stated explicitly the case for randomization in his paper on field trials and in many subsequent publications. Today the principle of randomization is the accepted norm in experimentation and is put forward matter-of-factly in many recent books. For example, Box, Hunter, and Hunter (1978) say,

> Fisher has shown how we can extricate ourselves from this seemingly insurmountable difficulty. By introducing randomization as part of the physical conduct of the experiment itself, we can validate our inferential procedures, whatever the form of the unknown disturbances.

III. WHAT IS THE RANDOMIZATION THEORY OF EXPERIMENTAL INFERENCE?

I believe that this theory can be stated as follows:

> The act of physically assigning treatments to experimental units at random
>
> 1. tends to improve the experiment regardless of what the analysis may be, and
> 2. justifies an analysis based upon probability models.

These two principles are stated by Cox (1958) as follows:

> To return to a less statistical description, the positive advantages of randomization are assurances
>
> a. that in a large experiment it is very unlikely that the estimated treatment effects will be appreciably in error...
> b. that the random error of the estimated treatment effects can be measured and their level of statistical significance examined, ...

Fisher was quite emphatic in maintaining that randomization provided a physical basis for the analysis. In *The Design of Experiments*, first published in 1935, one of the section headings is "Randomization; the Physical Basis of the Validity of the Test." In the 1926 field trials paper he wrote,

> One way of making sure that a valid estimate of error will be obtained is to arrange the plots deliberately at random, so that no distinction can creep in between pairs of plots treated alike and pairs treated differently; in such a case an estimate of error, derived in the usual way from the variations of sets of plots treated alike, may be applied to test the significance of the observed difference between the averages of plots treated differently.

As already stated, randomization has been accepted widely in the scientific community but some doubts remain. Many people have questioned whether an experiment can really be improved by assigning treatments at random. Also despite Fisher's assurances that randomization provides a physical basis for the validity of a test, he was obscure on how it does this. As I wrote (Folks, 1981),

> How it does this is not clear. Perhaps it is psychological; it is not purely mathematical.

From a totally different viewpoint, that of Bayesian analysis, Lindley and Novick (1981) conclude that randomization may be desirable:

> In order to make reasonably sure that our design does not confound the effects of T and A, we may assign treatments at random, that is, independent of A. This does not ensure lack of confounding but reduces its possibility to an acceptable level.

Harville (1975) argues that neither of the randomization principles cited above are valid. He says, referring to the alleged benefits of randomization,

> In what follows, it will be my contention that none of these arguments should be taken seriously. I claim that some of the benefits attributed to the use of randomization are an illusion, while the remainder can be obtained just as well, if not better, through the optimal-design approach.

In order to carry the discussion further, we need to state more fully what we mean when we say that physically assigning treatments to experimental units justifies an analysis based on probability models. Given a set of data from a randomized block design and having calculated a treatment difference, it is only natural to wonder how seriously we are to consider the calculated number. This contemplation leads directly to consideration of a conceptual population of repetitions. Attempts to model such a population have proceeded along two lines: (1) consideration of the observations as a sample of a conceptual but finite population which would be generated if every treatment could be assigned to every experimental unit, and (2) consideration of the observations as a sample from conceptual and vaguely defined normal populations with means depending upon treatments, blocks, etc.

Kempthorne and several of his colleagues (Wilk and Kempthorne, 1955 and Zyskind, 1962), have done extensive work on the finite randomization model and these models are widely presented. Scheffé (1959) also gives a summary of randomization models.

Fisher (1935) presented the randomization test, or permutation test, as many prefer to say. Given a set of experimental data, we can test the hypothesis of equal treatment effects by considering all possible assignments of treatments to the same experimental units and assuming the responses would be the same, calculate a test statistic for each such permutation. Then in the usual situation, the observed significance level is the fraction of cases yielding a statistic more extreme than that observed.

This thought process may be described as embedding the observed set of experimental results in a reference set or a conceptual set of repetitions. What is the proper set of repetitions? Conventional wisdom says that the proper set of repetitions is the set which would be generated by repeating the experiment in the way that the obtained one was actually conducted. However, there is certainly no reason that we can not look at other reference sets. For example, data from a randomized block design can be embedded in a conceptual set of repetitions which would result from a completely randomized design. Box et al. (1978) open their book by embedding a set of experimental data in several different reference sets.

This lack of uniqueness of a reference set has appeared to be an insurmountable stumbling block to some. Basu (1980) says,

> The author concludes that the Fisher randomization test is not logically viable.

In a discussion of the Basu paper, Lane offers the following criticism of the randomization test:

> The randomization test addresses the question: Might the two diets really be equally effective and the apparent superiority of the improved diet be attributed to chance variability? The success of the test depends on the relevance of the interpretation it requires for the notions of "equally effective diets" and "chance variability." According to the Fisherian foundation of the test, two treatments can be considered equally effective only if they would each elicit exactly the same response from each experimental unit. The experimenter may, however, be interested in a weaker notion of equality between two treatments: Their distributions for the responses over the experimental units (or over all potential recipients of the treatments) should coincide. For example, a physician may not believe that each cancer patient faces the same prospect for a cure from radiotherapy as from chemotherapy, but he or she still might want to entertain the hypothesis that the overall success rates of the two treatments might be the same. The randomization test would be of no help to the physician.

Despite such criticisms, the randomization test is considered worth performing by many individuals and its appeal appears to be strengthened if randomization has actually been performed. There has been a journal devoted to randomization for several years and randomization test procedures are available as part of statistical software packages.

IV. NATURE OF STATISTICAL INFERENCE

Some describe statistical inference as decision making. Others do not. Although there were a few occasions where it seems that Fisher was ambivalent, in the main he did not regard statistical inference as decision making. Rather, for the most part he argued against this viewpoint. In his last book (1956) he stated:

> .... it would still be true that the Natural Sciences can only be successfully conducted by responsible and independent thinkers applying their minds and imaginations to the detailed interpretation of verifiable observations. The idea that this responsibility can be delegated to a vast computer programmed with decision functions belongs to the fantasy of circles rather remote from scientific research.

Kempthorne has long maintained the distinction between inference and decision making. For example, see Kempthorne (1955, 1966, 1979) and Kempthorne and Folks (1971). Of course, in a certain sense, all human thought processes can be formulated as decision making. But is it helpful to do so? I believe that it is more helpful to regard statistical inference as a process of searching for hypotheses rather than as a process for selecting a hypothesis. Is the EDA work of Tukey and others statistical inference? I believe so. Is it decision making? Only in the loosest sort of interpretation of the term. A distinction in vogue is that between descriptive statistics and inferential statistics based upon random sampling and probability models. But from where do we get the assumption of random sampling and probability models. Kempthorne (1979) commented upon this:

> It may be argued by some that without an assumption of random sampling, nothing is possible. I am inclined to agree with this view, but make what is, I believe, a quite different interpretation. In many studies the only way to make any sort of inference is to determine by data examination a probability distribution or a family of probability distributions that is such that the data are like a random sample. The inclusion of the word 'like' gives a strongly different attitude to the conclusions one reaches.

To elaborate a bit upon Kempthorne's idea, what are we doing when we calculate statistics? What is the meaning of "parameter estimates" calculated from real data with four decimal places (for example), bounded above and below, with numerous ties when the background model is that of normality? The answer is "These observations are somewhat 'like' what I would get if I simulated the normal model using the 'parameter estimates'." I agree with Kempthorne that the word 'like' gives a different meaning to the statement. It also makes it clear that the distinction between descriptive and inferential statistics is one of detail only, not of logical substance. Thus parameter estimates, intervals, and outcomes of tests are data constructs only - descriptive statistics which utilize probability distributions.

V. TESTS OF SIGNIFICANCE

Fisher presented many examples of significance tests and wrote extensively about them. However, he did not give a tightly worded definition of a test of significance. Neyman and Pearson (1933) formulated a test as a decision rule for accepting or rejecting a null hypothesis. Although it is recognized that this formulation is too rigid, the Neyman-Pearson theory is usually presented in statistical inference. However, it is clear that Fisher had something else in mind. On several occasions he argued against ever accepting the null hypothesis. In the *Design of Experiments* he wrote,

> In relation to the test of significance, we may say that a phenomenon is experimentally demonstrable when we know how to conduct an experiment which will rarely fail to give us a statistically significant result.

In *Statistical Methods and Scientific Inference,* he wrote,

> To a practical man, also, who rejects a hypothesis, it is, of course, a matter of indifference with what probability he might be led to accept the hypothesis falsely, for in his case he is not accepting it.

It is also clear that Fisher regarded the significance level, or observed significance level, or P-value, as a statistic which provided a measure of the evidence in the data against the null hypothesis. In the 1956 book he wrote,

> There is the logical disjunction: Either an intrinsically improbable event will occur, or, the prediction will not be verified. The psychological resistance has been, I think wrongly, ascribed to the fact that the event in question has, in the proper sense of the Theory of Probability, the low probability assigned to it, rather than to the fact, very near in this case, that the correctness of the assertion would entail an event of this low probability. The probability statement is a sufficient, but not a necessary, condition for disbelief in this degree.

Statistical tests from the Fisher viewpoint have been studied by quite a number of workers (see Kempthorne and Doerfler, 1969,

Kempthorne and Folks, 1971, Stone, 1969, Cox and Hinkley, 1974, and Lambert, 1982). Although the Bayesian argument is that the calculation of an observed significance level should not be done, the practice is well established and virtually every statistical package gives such P-values routinely.

It is also interesting that the inversion of a significance test often proceeds in a very direct way to give an interval of values for the parameter of interest. The interpretation of such intervals, although very clear, does not seem to be completely satisfactory to those accustomed to Neyman confidence interval statements. We say simply that the parameter values in such an interval are in agreement with (in consonance with) the data in the sense that the observed significance level is large. Kempthorne and Doerfler (1969) studied such intervals arising from randomization tests. Kempthorne and Folks (1971) called such intervals consonance intervals hoping to emphasize the interpretation. Easterling (1976) combined the idea of goodness of fit and interval estimation assuming a model with unknown parameters to obtain a consonance interval for the parameters of a class of probability distributions.

VI. NORMAL MODEL

Fisher, in *The Design of Experiments* states,

> In these discussions it seems to have escaped recognition that the physical act of randomization, which, as has been shown, is necessary for the validity of any test of significance, affords the means, in respect of any particular body of data, of examining the wider hypothesis in which no normality of distribution is implied.

I have already discussed the idea that randomization is thought to validate the randomization test. But Fisher did not stop there. From his other writings it seems that he was saying that randomization justifies the normality assumption. Hinkley (1980) says clearly,

> As I see it, the purpose of randomization in the design of agricultural field experiments was to help ensure the validity of normal-theory analysis.

Although several people, including Kempthorne and his colleagues have studied how well the randomization test is approximated by the normal theory test, it remains unclear how randomization justifies the normality assumption. Kempthorne (1966) concluded that,

> The use of randomization does not, therefore, ensure the validity of the normal theory test, unless the normal theory test is a good approximation to the randomization test.

Kempthorne (1955) had written more fully on this point earlier.

> The important aspect of the validity of an inference is the validity of the assumptions on which the inference is based ... The making of assumptions of normality and then applying the battery of mathematical statistical tests is not a satisfactory basis for experimental inference, because the extent to which the reliability of an inference depends on the assumptions made in the analysis is usually unknown.

There remain other obscurities in linking randomization theory to the normal model theory. For example, the sixth edition of *Statistical Methods* by Snedecor and Cochran (1967) repeats formulas previously given for the efficiency of one design relative to another. These formulas are based upon a randomization model - not upon a normal model.

Another obscurity relates to the unbiasedness of designs in the Yates sense. A well established practice is to base the choice of an error term for an F test upon the expected mean squares. When the null hypothesis is true, the expected value of numerator and denominator should be equal. The obscurity arises because randomization models do not give the same expected mean squares as models based on normality.

VII. REPRESENTATIVENESS

Students sometimes come into my classes having been taught that a random sample should be a representative sample. I mean representative in the sense that the sample is like the population or is a typical cross-section of the population. This is not what random means and I was prompted to state (Folks, 1981) that,

The sample is a random sample, not a representative sample.

Of course, any sample is representative in the sense that it is a proxy or stand-in for the population.

I believe it to be a common practice that unusual looking random assignments are discarded and another randomization performed. Lindley and Novick (1981) state,

> In practice scientists do not allocate completely at random; instead they obtain a random allocation from the mechanism and then inspect it for any unusual features before using it. Thus, if in the random selection of a Latin square, one in which the treatments lay down the diagonal was obtained, it would be discarded and a new allocation selected. In other words, the scientist always thinks about the proposed allocation before using it; which is essentially the argument here - use an allocation which you think is free from confounding.

Fisher (1926) had thought to some extent about the problem that a random assignment may not "look random" and wrote,

> Most experimenters on carrying out a random assignment of plots will be shocked to find how far from equally the plots distribute themselves; three or four plots of the same variety, for instance, may fall together at the corner where four blocks meet. This feeling affords some measure of the extent to which estimates of error are vitiated by systematic regular arrangements, for, as we have seen, if the experimenter rejects the arrangement arrived at by chance as altogether "too bad," or in other ways "cooks" the arrangement to suit his preconceived ideas, he will either (and most probably) increase the standard error as estimated from the yields; ...

The problem is that if one believes in randomization, assignments should not be rejected arbitrarily but some assignments look so bad to us that we feel we must "toss them out." Because of this problem, a number of writers have studied subsets of randomizations which preserve the properties of the full set (Sutter, Zyskind, and Kempthorne, 1963, Grundy and Healy, 1950, and Youden, 1956).

VIII. CAUSALITY

Kempthorne (1966) in distinguishing between experimental inference and non-experimental inference stated,

> I believe the distinction between the two types of situation is real, because one can talk about causality in a tight way only when intervention by an organism is possible. One can "infer" causality in a nonexperimental situation only by hypothesizing the possibility of intervention. Even in the case of an effect being produced by an act of human intervention, the only causal inference possible is that the whole act of intervention produced the effect.

The difficulty of giving a consistent definition of causality is recognized: the subject has been discussed at great length in philosophical literature. Nevertheless, it must be recognized that many experiments are conducted to determine causes (admittedly in a loose sense). The results of some such experiments become public knowledge. In everyday terminology, the application of chemical fertilizer will "cause" an increase in yield, the use of polio vaccine will "cause" a lower incidence of polio, etc. We, of course, are familiar with Fisher's argument that smoking does not necessarily cause lung cancer and we have to admit the validity of his argument. Nevertheless, we cannot deny the almost universal opinion that smoking causes an increase in lung cancer.

Why is it important to discuss causality? Kerlinger (1964), in a widely used book on behavioral research says,

> Now, about prediction and control. It can be said that scientists do not really have to be concerned with explanation and understanding. Only prediction and control are necessary. Proponents of this point of view would say that the adequacy of a theory is its predictive power. If by using the theory we are able to predict successfully, then the theory is confirmed and this is enough. We need not necessarily look for further underlying explanations. Since we can predict reliably we can control because control is deducible from prediction.

Toda (1977) describes the situation concerning people's desire to discover causes as follows:

> I am afraid that some readers may have resented my way of discussing causality so far; I have done it as if the existence of causal relations is a fact. No one can really <u>prove</u> the existence of causality, as, strictly speaking, the number of potentially causally related variables is infinite, and no control can be perfect. However, this existence has been

> unproven in almost the same sense as no one has ever really proved the existence of atoms. Human beings apparently have constantly been on the alert to discover causal relations, and, for reasons unknown, tentative causal relations begin to shape themselves up into a consistent system of natural laws.

Rubin (1974, 1978) unabashedly discusses causation and gives the following definition:

> The causal effect of one treatment relative to another for a particular experimental unit is the difference between the result if the unit had been exposed to the first treatment and the result if, instead, the unit had been exposed to the second treatment.

Rubin (1978) goes on to say,

> Classical randomized designs stand out as especially appealing assignment mechanisms designed to make inference for causal effects straightforward by limiting the sensitivity of a valid Bayesian analysis.

IX. DISCUSSION

In this chapter, I have tried to summarize some of the ideas of the randomization theory of experimental inference and some of the arguments for and against this theory. The most intriguing question for me always has been, "Exactly what do we gain by actually carrying out the assignment of treatments to experimental units?"

Beforehand, the arguments that we will tend to avoid harmful confounding seem quite compelling (see Cox, 1958, and Lindley and Novick, 1981). After the experiment is performed, however, can we really benefit in our analysis from knowing that the assignment of treatments was made at random?

The randomization test is merely a consideration of some test statistic calculated from the data at hand for a set of permutations of treatment labels. The same set of permutations can be considered regardless of whether the data in hand was obtained through a random assignment of treatments. Statements of unbiasedness refer to averages over a conceptual set of repetitions but surely we can consider these same averages regardless of how the experiment was actually conducted.

The argument usually advanced is that if we do not randomize we would not be entitled to use the conceptual population of repetitions. To this I advance the following counter argument. Suppose I list all possible randomizations (possibly with some blocking restrictions). I could then simply use the first one on the list as my assignment of treatments. If I were to run the experiment again I would use the second; if again the third, etc. In such a way I would generate exactly the same conceptual population of repetitions without using random assignment of treatments at any point.

Still deep within me, I have the feeling that the interpretation is clearer, the conclusions are stronger and the analysis has greater validity if treatments have actually been assigned at random. But why? I do not know and none of the explanations which have been advanced are totally satisfactory to me.

REFERENCES

Basu, D. (1980). Randomization analysis of experimental data: the Fisher randomization test. *J. Am. Stat. Assoc. 75,* 575-595.

Box, G.E.P., W.G. Hunter, and J.S. Hunter (1978). *Statistics for Experimenters.* New York: John Wiley.

Cox, D.R. (1958). *Planning of Experiments.* New York: John Wiley.

Cox, D.R. and David Hinkley (1974). *Theoretical Statistics.* London: Chapman and Hall.

Dempster, A.P. and M. Schatzoff (1965). Expected significance level as a sensitivity index for test statistics. *J. Am. Stat. Assoc. 60,* 420-436.

Easterling, Robert G. (1976). Goodness of fit and parameter estimation, *Technometrics 18,* 1-14.

Fisher, R.A. (1926). The arrangement of field trials. *J. Mining Agricul. 33,* 503-513.

Fisher, R.A. (1935). *The Design of Experiments.* Edinburgh: Oliver and Boyd.

Fisher, R.A. (1950). Author's note to Paper 17 in *Contributions to Mathematical Statistics.* New York: John Wiley.

Fisher, R.A. (1956). *Statistical Methods and Scientific Research.* Edinburgh: Oliver and Boyd.

Folks, J. Leroy (1981). *Ideas of Statistics.* New York: John Wiley.

Grundy, P.M. and M.J.R. Healy (1950). Restricted randomization and quasi-Latin squares. *J. Roy. Statist. Soc. B 12,* 286-291.

Harville, David A. (1975). Experimental randomization: Who needs it? *Amer. Stat. 29,* 27-51.

Hinkley, David (1980). Discussion of Basu's paper. *J. Am. Stat. Assoc. 75,* 575-595.

Kempthorne, O. (1955). The randomization theory of experimental inference. *J. Am. Stat. Assoc. 50,* 946-967.

Kempthorne, O. (1966). Some aspects of experimental inference. *J. Am. Stat. Assoc. 61,* 11-34.

Kempthorne, O. (1979). Sampling inference, experimental inference, and observation inference. *Sankhya 40,* 115-145.

Kempthorne, O. and T.E. Doerfler (1969). The behavior of some significance tests under experimental randomization. *Biometrika 56,* 231-248.

Kempthorne, O. and L. Folks (1971). *Probability, Statistics, and Data Analysis.* Ames: Iowa State University Press.

Kerlinger, Fred N. (1964). *Foundations of Behavioral Research.* New York: Holt, Rinehart, and Winston.

Lambert, D. (1982). Qualitative robustness of tests. *J. Am. Stat. Assoc. 77,* 352-357.

Lane, David (1980). Discussion of Basu's paper. *J. Am. Stat. Assoc. 75,* 575-595.

Lindley, Dennis (1980). Discussion of Basu's paper. *J. Am. Stat. Assoc. 75,* 575-595.

Lindley, Dennis A. and Melvin R. Novick (1981). The role of exchangeability in inference. *Ann. Stat. 9,* 45-48.

Neyman, J. and E.S. Pearson (1933). On the problem of the most efficient tests of statistical hypothesis. *Philos. Trans. R. Soc. A 231,* 289-337.

Peirce, C.S. (1883). *The Johns Hopkins Studies in Logic.* Boston: Little, Brown, and Co.

Peirce, C.S. (1896). Lessons from the History of Science, written around 1896 in *Collected Papers of Charles Sanders Peirce*, edited by A. Burks, Cambridge: Harvard University Press.

Rubin, Donald B. (1974). Estimating causal effects of treatments in randomized and non-randomized studies. *J. Educ. Psych. 66,* 688-701.

Rubin, Donald B. (1978). Bayesian inference for causal effects: the role of randomization. *Ann. Stat. 6,* 34-58.

Scheffé, Henry (1959). *The Analysis of Variance*. New York: John Wiley.

Snedecor, G.W. and W.G. Cochran (1967). *Statistical Methods*. Ames: Iowa State University Press.

Stigler, Stephen M. (1978). Mathematical statistics in the early stages. *Ann. Stat. 6,* 239-265.

Stone, M. (1969). The role of significance testing: some data with a message. *Biometrika 56,* 485-493.

Sutter, C.J., G. Zyskind, and O. Kempthorne (1963). *Some Aspects of Constrained Randomization*. ARL Report 63-18. Wright-Patterson AFB, Ohio.

Wilk, M.B. and O. Kempthorne (1955). Fixed, mixed, and random models in the analysis of variance. *J. Am. Stat. Assoc. 50,* 1144-1167.

Youden, W.J. (1956). *Randomization and Experimentation*. Washington: National Bureau of Standards.

Zyskind, George (1962). On structure, relation, sigma, and expectation of mean squares. *Sankhya 24,* 115-118.

Chapter 4

CONFIDENCE INTERVALS UNDER EXPERIMENTAL RANDOMIZATION

R.N. Curnow
University of Reading
Reading, United Kingdom

I. INTRODUCTION

In 1959 I spent six months at Ames as a Harkness Fellow of the Commonwealth Fund. During that time Oscar Kempthorne spared me much more of his time and patience than I, as a raw 'post-doc', had any right to expect. Our shared Cornish origins or, more specifically, my wife's ability to cook Cornish pasties may have had something to do with Kempthorne's generosity of time!

Kempthorne's emphasis on randomization tests as the basic method for analyzing experimental data has been an important corrective to the more generally held view that analyses can safely be based on the assumption that the 'errors' are a sample of independent values from some hypothetical single distribution of infinite size, generally a Normal one. The justification for this latter assumption is rarely clear and even more rarely discussed. There is little that is essentially new in the use of randomization tests and Kempthorne has always emphasized the earlier work of Fisher, Pitman and Welch.

Most statisticians are unaware of the way in which statistical inference is, and has to be, embedded in the wider philosophy of scientific method. Kempthorne's interest in and understanding of the general philosophy of science has, I am sure, led him to emphasize the limited nature of the statistical inferences that can be made from an experiment, even a properly randomized one. The inference from the randomization test can only be to the experimental units used in the experiment. The inference that must then be made from the experimental units to the wider population of relevant units of which they are hopefully representative, is not one that can be made statistically valid simply by assuming, without justification, that the experimental units are a random uncorrelated sample from the wider population of units. The inference has to be made but is not a statistical one. The argument has to be based on knowledge of the subject-matter involved not on the use of unsupportable assumptions about the origin, and hence the uncorrelated representativeness, of the experimental units.

Unfortunately, randomization analyses generally result only in the calculation of a significance level related to some null hypothesis. Hypothesis testing, equivalent to a decision based on the 'rounding-off' of possible significance levels to one of the three commonly used levels ($P = 0.05$, 0.01 and 0.001), has been rightly criticized by Kempthorne on many occasions (Kempthorne, 1976). I would go further and criticize the calculation of the exact significance levels except at a very preliminary stage in the analysis. The null hypothesis is nearly always known to be untrue and so why should it be so central to the analysis? The significance level attempts to summarize the size of any effects and their precision in one measure and hence confuses the two. On the other hand, confidence intervals keep the two separate and allow a proper assessment of the probable range of values for the effect. Kempthorne and Doerfler (1969) appear to support this view, for they remark that "The procedure of significance testing reaches its full utility only when the significance test can be inverted to obtain sets of values of the parameter under tests which give the possible significance levels". Confidence intervals have not often been mentioned in the context of randomization analysis, presumably because they are tedious to calculate. They also require the choice of a confidence level or a set of confidence levels. This choice is an inevitable consequence of the variability in the data and so has to be faced.

The emphasis on testing null hypotheses is probably one reason why the extrapolation from the experimental units to a wider population of units has so often been uncritically accepted. A zero effect may be thought to be more repeatable than an effect of any particular non-zero magnitude. Unfortunately a real zero effect is very unlikely and hence discussion of its likely repeatability irrelevant. Many scientists have rejected statistical methods for the wrong reasons. Some good scientists have rejected them because we talk about null hypotheses that are unbelievable and samples from infinite distributions that do not exist.

In this paper, randomization confidence intervals will be calculated for simulated data from the same set of distributions as was used by Kempthorne and Doerfler (1969) in their paper entitled 'The behaviour of some significance tests under experimental randomization'. A paired comparison experiment will be studied. It will be shown that the confidence intervals derived from the randomization are often similar to those based on the more usual assumptions of Normality and hence the main conclusion will concern not the method of analysis but the nature of the inferences that can be made.

II. PAIRED COMPARISON EXPERIMENT

Consider an experiment consisting of eight pairs of experimental units. One unit of each pair is allocated, at random, to treatment A and the other is then allocated to treatment B. $d_1, d_2, \ldots, d_8$ are the observed differences between the experimental units in each pair allocated to treatments A and B. The difference is always treatment A minus treatment B. The randomization, or distribution-free, method to test the hypothesis that treatments A and B are identical in effect would be to compare

$$\bar{d} = \frac{1}{8} \sum_{i=1}^{8} d_i$$

with the $2^8 = 256$ possible means that could have been obtained from the randomization of the units within pairs of units to the treatments,

$$\frac{1}{8} \sum_{i=1}^{8} (\pm\, d_i)$$

Using a two-sided test, if the observed $\bar{d}$ is either the smallest or largest of the 256 possible means then the treatment effects differ significantly at the 2/256 = 0.78% level. This is the nearest we can get to a 1% significance level. Since 12/256 = 4.69%, the nearest to a two-tailed test at the 5% significance level would be based on seeing if the observed mean is among the six largest or six smallest possible means.

Assuming only that the treatment and experimental unit effects are additive, a confidence interval for the difference, Δ, between the effects of treatments A and B can be calculated. To obtain a confidence interval at the (1 - 2n/256) confidence level we need to calculate the values of Δ that can be subtracted from the observed differences d_i so that the observed $\bar{d} - \Delta$ is either the nth largest or nth smallest of the 256 possible means

$$\frac{1}{8} \Sigma \pm (d_i - \Delta)$$

These two Δ values are the limits of a (1 - 2n/256) confidence interval. Kempthorne and Doerfler (1969) have shown, in an Appendix, that this procedure does lead to a unique continuous interval.

To obtain the 254/256 = 99.22% confidence level, $\bar{d} - \Delta$ has to be the largest or smallest among all the possible means,

$$\frac{1}{8} \Sigma \pm (d_i - \Delta)$$

Clearly this requires $\Delta = \underset{i}{\text{Min}}\ d_i$ and $\Delta = \underset{i}{\text{Max}}\ d_i$.

To obtain an interval at about the 95% level of confidence we take n = 6 to achieve a confidence level of 244/256 = 95.31%. Calculating the values of Δ that make $\bar{d} - \Delta$ the sixth largest or sixth smallest among the 256 possible means is not straightforward. We used the following algorithm, which is related to methods described by Green (1977) in the context of randomization tests.

For a given Δ, rank the values of $|d_i - \Delta|$ and call the jth largest of these D_j. Set $S_j = 1$ if the sign of $d_i - \Delta$ corresponding to D_j is negative and $S_j = 0$ if the sign is positive. Then the observed mean

$$\frac{1}{8} \Sigma(1 - 2S_j)D_j$$

is among the sixth largest if and only if

$$\begin{aligned}
& \Sigma S_j = 0 \\
\text{or} \quad & \Sigma S_j = 1 \text{ and } S_6 + S_7 + S_8 = 1 \\
\text{or} \quad & \Sigma S_j = 1 \text{ and } S_5 = 1 \text{ and } D_5 < D_6 + D_8 \\
\text{or} \quad & \Sigma S_j = 1 \text{ and } S_4 = 1 \text{ and } D_4 < D_7 + D_8 \\
\text{or} \quad & \Sigma S_j = 2 \text{ and } S_7 = S_8 = 1 \text{ and } D_4 > D_7 + D_8 \\
\text{or} \quad & \Sigma S_j = 2 \text{ and } S_6 = S_8 = 1 \text{ and } D_5 > D_6 + D_8
\end{aligned}$$

where Σ always means sum over j = 1,2,...,8.

A similar algorithm based on $|-d_i - \Delta|$ can be used to check if, for a given Δ, the observed mean is among the six smallest means. An iterative search, using the above conditions, for the largest and smallest values of Δ that make $(\bar{d} - \Delta)$ the sixth largest or sixth smallest among the 256 possible means

$$\frac{1}{8} \Sigma \pm (d_i - \Delta)$$

completes the process.

We need to generate values of the d_i so that we can study the properties of the confidence intervals generated by the randomization argument and compare them with the more usual confidence intervals based on Normal theory. We shall do this by sampling values from seven of the eight distributions used by Kempthorne and Doerfler (1969). The distribution omitted is the "right-triangular" distribution which is asymmetric and therefore not relevant. The distributions are given in Table 4.1. They have, without loss of generality, all been standardized to have zero mean and unit variance. Standard Normal values, y, and values, u, uniformly distributed over [0,1] were used to generate the samples as indicated in Table 4.1.

A standard analysis would calculate the 95% confidence interval for the treatment effect as

$$\bar{d} \pm t_7 \, s_d/\sqrt{8}$$

where t_7 is the value appropriate to the confidence level of the t-distribution with seven degrees of freedom and s_d is the standard

TABLE 4.1

Distribution	Generation
1. Normal	y
2. Uniform	$\sqrt{12}(u-\frac{1}{2})$
3. M shaped	$\pm [\sqrt{(2u)}]$, $\pm$ prob = $\frac{1}{2}$
4. Triangular	$\pm \sqrt{6}(1-\sqrt{u})$, $\pm$ prob = $\frac{1}{2}$
5. 90% Normal, 10% Uniform	N(0,1), prob=0.9; $\sqrt{12}\ [u-\frac{1}{2}]$, prob=0.1
6. 70% Normal, 30% Uniform	N(0,1), prob=0.7; $\sqrt{12}\ [u-\frac{1}{2}]$, prob=0.3
7. Bi-rectangular	$\pm [\frac{1}{2}(\sqrt{(11/3)}-1)+u]$, $\pm$ prob = $\frac{1}{2}$

deviation of the eight differences, d_i. Numerical integration showed that the t-values for confidence levels of 254/256 = 99.22% and 244/256 = 95.31% are 3.68 and 2.41, respectively.

Individual randomization confidence intervals will not be symmetric but there will be no bias in the asymmetry. If L and U are the lower and upper limits respectively then the standardized half length of the confidence interval to be compared with the value of t_7 in the Normal theory will be

$$R = \frac{1}{2}\left[\frac{|U-\bar{d}| + |\bar{d}-L|}{s_d/\sqrt{8}}\right]$$

Four hundred samples of eight observations were taken from each distribution. Sufficient decimals were used to avoid problems with ties in the data. The average values of R are shown in Table 4.2. The Normal theory confidence intervals will only be applicable when the sample of values is from an infinite Normal distribution. Table

TABLE 4.2

Standardized Half-length R of Randomization Confidence Intervals Averaged over 400 Samples

Distribution	Confidence Level	
	99.22%	95.31%
1. Normal	4.15	2.44
2. Uniform	3.91	2.50
3. M shaped	3.57	2.55
4. Triangular	4.13	2.45
5. 90% Normal, 10% Uniform	4.15	2.45
6. 70% Normal, 30% Uniform	4.13	2.45
7. Bi-rectangular	3.56	2.54
Using Normal theory	3.68	2.41
Approx. S.E.'s of averages (400 samples)	0.02	0.01

TABLE 4.3

Number of Normal Theory Intervals, out of 400, Not Including True Value

Distribution	Confidence Level	
	99.22%	95.31%
1. Normal	5	17
2. Uniform	2	23
3. M shaped	6	20
4. Triangular	2	20
5. 90% Normal, 10% Uniform	3	17
6. 70% Normal, 30% Uniform	6	13
7. Bi-rectangular	9	20
Theoretical value	3.1	18.8

4.3 shows the number of the Normal theory confidence intervals, out of the 400, which did not contain the true value. The theoretical expected values are 3.1 and 18.8.

Theoretical expected values cannot be calculated for the 99.22% values of R in Table 4.2 because R is not the mean half-range in a sample of size 8 but the mean standardized half-range. The standardization increased accuracy and allowed the values obtained to be compared directly with the two-sided 0.78% point of the t-distribution on 7 d.f., $t_7 = 3.68$. Figure 4.1 shows the histograms of the unstandardized half-lengths of the 99.22% confidence intervals for the Normal distribution using randomization theory and Normal theory. The randomization theory values are now unstandardized half-ranges and the histogram does have approximately the right mean and standard deviation, namely, 1.42 and 0.41, respectively (Biometrika Tables, 1970). The Normal theory values are $t_7(0.78\%) \times s_d/\sqrt{8} = 130\ s_d$ and the histogram is therefore the empirical distribution of s_d multiplied by a constant and has approximately the right mean and standard deviation, 1.25 and 0.34, respectively (Biometrika Tables, 1970).

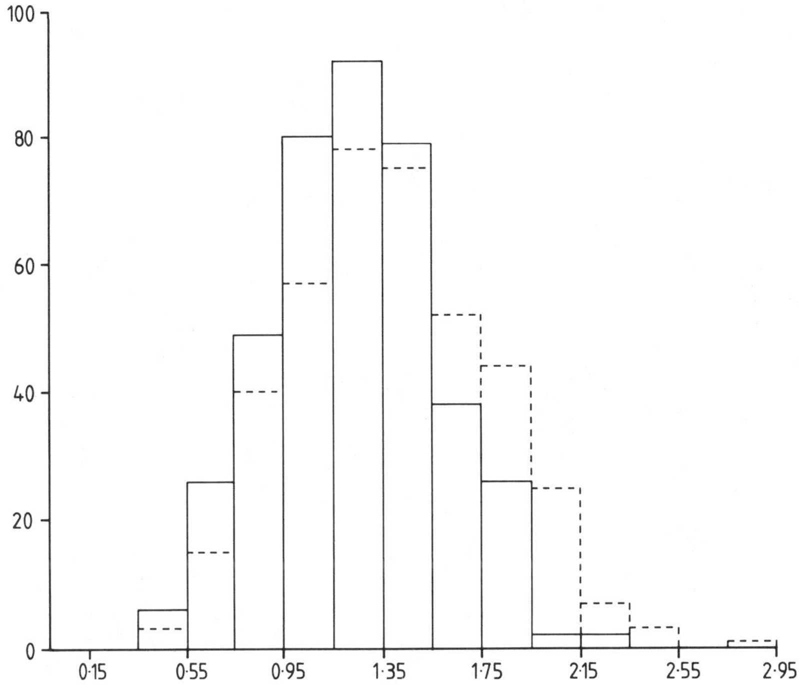

Figure 4.1 Histogram of Half-Length of Confidence Intervals. Confidence level = $\frac{254}{256}$ = 99.22%. 400 simulations.

__________ "Normal" theory

---------- Randomization

Figure 4.2 shows the histograms for the half-lengths of the 95.31% confidence intervals. There are no theoretical values available for the randomization histogram. The mean and standard deviation of the simulated results were 0.82 and 0.23, respectively. For the Normal theory values, the histogram is the histogram of $s_d/\sqrt{8}$ now multiplied by $t_7(4.69\%) = 2.41$. The theoretical mean and standard deviation for this histogram are 0.82 and 0.22, respectively.

Forsythe and Hartigan (1970) discuss the use of subsets, random or 'balanced', of the possible means to derive confidence intervals for the population mean. They derived some asymptotic re-

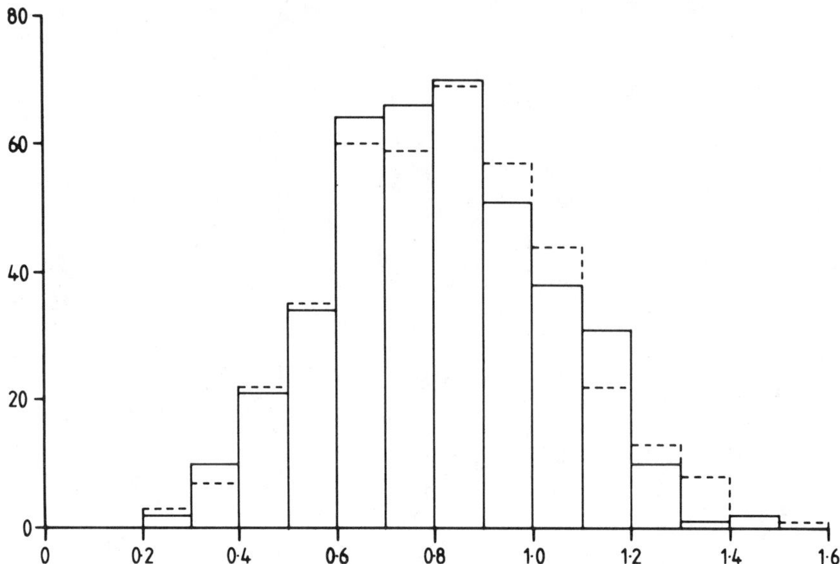

Figure 4.2 Histogram of Half-Lengths of Confidence Intervals. Confidence level = $\frac{244}{256}$ = 95.31%. 400 simulations.
__________ "Normal" theory
---------- Randomization

sults and compared the confidence intervals generated by the use of the subsets of means with those generated by the t-distribution for various sample sizes when the Normal distribution applies. When all possible means are used, as in this paper, the confidence intervals were not much longer than those based on t-values except for very small sample sizes.

Recent papers by Robinson (1982) and Wei (1982) describe asymptotic approximations to confidence intervals based on permutation tests and give references to earlier work. The approximations are often good but the work involved in calculating them is often of the same magnitude as the work needed to calculate randomization confidence intervals by the type of numerical algorithm presented in this paper.

III. CONCLUSIONS

Table 4.2 and Figures 4.1 and 4.2 show that, in our particular case, confidence intervals calculated on the basis of randomization distributions are rarely much longer than those calculated using Normal theory. Table 4.3 shows that the Normal theory confidence intervals contained the true value about the right proportion of times. All this parallels general findings about the power of randomization tests and is probably of wide application. The use of confidence intervals derived from Normality assumptions will rarely be misleading. The importance of randomization tests and randomization confidence intervals therefore lies not in the actual analysis of data but in highlighting the avoidance of unjustifiable assumptions and the restrictions this imposes on the nature of the inferences that can be drawn from experimental data.

ACKNOWLEDGMENT

I am grateful to K.H. Freeman for assistance with the computer simulations and the numerical integrations.

REFERENCES

Biometrika Tables for Statisticians (1970). Volume 1. 3rd Ed. E.S. Pearson and H.O. Hartley (eds.). Cambridge University Press.

Forsythe, A. and Hartigan, J.A. (1970). Efficiency of confidence intervals generated by repeated sub-sample calculations. *Biometrika 57*, 629-639.

Green, B.F. (1977). A practical interactive program for randomization tests of location. *Amer. Stat. 31*, 37-39.

Kempthorne, O. (1976). Of what use are tests of significance and tests of hypothesis? *Commun. Statist. Theor. Math. A5 (8)*, 763-777.

Kempthorne, O. and Doerfler, T.E. (1969). The behaviour of some significance tests under experimental randomization. *Biometrika 56*, 231-248.

Robinson, J. (1982). Saddlepoint approximations for permutation tests and confidence intervals. *J. R. Statist. Soc. (B) 44*, 91-101.

Wei, L.J. (1982). Asymptotically distribution-free simultaneous confidence region of treatment differences in a randomized complete block design. *J. R. Statist. Soc. (B) 44*, 201-208.

Chapter 5

COMPLETE SETS OF ORTHOGONAL F-SQUARES OF PRIME POWER ORDER WITH DIFFERING NUMBERS OF SYMBOLS

John P. Mandeli
Virginia Commonwealth University
Richmond, Virginia

Walter T. Federer
Cornell University
Ithaca, New York

I. INTRODUCTION AND DEFINITIONS

An F-square has been defined by Hedayat (1969) and Hedayat and Seiden (1970) as follows:

Definition 1.1. Let $A = [Q_{ij}]$ be an n×n matrix and let $\Sigma = \{A_1, A_2, \ldots, A_m\}$ be the ordered set of m distinct elements or symbols of A. In addition, suppose that for each $k = 1,2,\ldots,m$, A_k appears exactly λ_k times ($\lambda_k \geq 1$) in each row and column of A. Then A will be called a frequency square or, more concisely, an F-square, on Σ of order n and frequency vector $(\lambda_1, \lambda_2, \ldots, \lambda_m)$. The notation we use to denote this F-square is $F(n; \lambda_1, \lambda_2, \ldots, \lambda_m)$. Note that $\lambda_1 + \lambda_2 + \ldots + \lambda_m = n$ and that when $\lambda_k = 1$ for all k and m = n, a latin square of order n results.

As with latin squares, one may consider orthogonality of a pair of F-squares of the same order. The above cited authors have given the following definition to cover this situation:

Definition 1.2. Given an F-square $F_1(n;\ \lambda_1,\lambda_2,\ldots,\lambda_k)$ and an F-square $F_2(n;\ u_1,u_2,\ldots,u_t)$, we say F_2 is an orthogonal mate for F_1 (and write $F_2 \perp F_1$), if upon superposition of F_2 on F_1, A_i appears $\lambda_i u_j$ times with B_j. Note that when $\lambda_i = 1 = u_j$ for all i and j and $\sum_{i=1}^{k}\lambda_i = n = \sum_{j=1}^{t}u_j$, we have the familiar definition of the orthogonality of two latin squares of order n.

The definition of a set of orthogonal F-squares is given as:

Definition 1.3. Let $\{F_1,F_2,\ldots,F_t\}$ be a set of two or more F-squares of order $n \geq 3$. The F-squares in this set are called orthogonal, and we refer to $\{F_1,F_2,\ldots,F_t\}$ as an orthogonal set, provided that F_i and F_j are orthogonal for each $i \neq j$. If $F_1,F_2,\ldots,F_t$ are all latin squares of order n then a set of t mutually orthogonal latin squares results and is denotes as OL(n,t).

If a complete set of orthogonal latin squares of order n exists, then t = n - 1 and the set is denoted as OL(n,n-1). If a complete set of orthogonal F-squares of order n exists, the number will depend upon the number of symbols in each F-square. This leads to the following definition which is a generalization of the one given by Hedayat and Seiden (1970):

Definition 1.4. A complete set of t orthogonal F-squares of order n is denoted as CSOFS(n,t), where $t = \sum_{i=2}^{n} N_i$, N_i is the number of $F(n;\ \lambda_1,\lambda_2,\ldots,\lambda_i)$-squares in the set (i.e., N_i is the number of squares with i distinct elements), $\sum_{h=1}^{i}\lambda_h = n$, and $\sum_{i=2}^{n} N_i(i-1) = (n-1)^2$.

The fact that $\sum_{i=2}^{n} N_i(i-1) = (n-1)^2$ in order to have a CSOFS follows directly from analysis of variance theory and from factorial theory in that the interaction of two n-level factors has $(n-1)^2$ degrees of freedom and from the fact that only interaction degrees of freedom are available to construct F-squares. For each $F(n;\ \lambda_1,\lambda_2,\ldots,\lambda_i)$-square, there are (i-1) degrees of freedom associated with the i distinct symbols of an F-square, there are N_i F-squares containing i symbols, and hence $(n-1)^2 = \sum_{i=2}^{n} N_i(i-1)$.

II. ONE-TO-ONE CORRESPONDENCE BETWEEN FACTORIAL EFFECTS AND ORTHOGONAL LATIN SQUARES AND ORTHOGONAL F-SQUARES

From results in Bose (1938) and (1946), we may write the following theorem on the construction of the complete set of orthogonal latin squares from a symmetrical factorial experiment.

Theorem 2.1. Let $n = S^m$ where S is a prime number and m is a positive integer. The complete set of orthogonal latin squares of order n, i.e., the OL(n,n-1) set, can be constructed from a $(S^m)^2$ symmetrical factorial experiment, i.e., a symmetrical factorial experiment with 2 factors each at S^m levels.

A similar theorem for F-squares due to Hedayat, Raghavarao, and Seiden (1975), is:

Theorem 2.2. Let S be a prime number and m be a positive integer. For any integer p, that is a divisor of m, the complete set of $(S^m-1)^2/(S^p-1)$ orthogonal $F(S^m; S^{m-p}, S^{m-p},\ldots,S^{m-p})$-squares can be constructed from a $(S^p)^{2m/p}$ symmetrical factorial experiment, i.e., a symmetrical factorial experiment with 2m/p factors each at S^p levels.

III. DECOMPOSING LATIN SQUARES INTO F-SQUARES

We consider the following decomposition theorems.

Theorem 3.1. Each latin square in the set of orthogonal latin squares, $OL(S^m, S^m-1)$, can be decomposed into $(S^m-1)/(S-1)$ orthogonal $F(S^m; S^{m-1}, S^{m-1},\ldots,S^{m-1})$-squares, and the entire $OL(S^m, S^m-1)$ set can be decomposed into $(S^m-1)^2/(S-1)$ orthogonal $F(S^m; S^{m-1}, S^{m-1},\ldots,S^{m-1})$-squares.

Proof. Consider a $(S)^{2m}$ symmetrical factorial experiment, i.e., a symmetrical factorial experiment with 2m factors each at S levels. Taking p = 1 in theorem 2.2, we can construct from the unconfounded-with-rows-and-columns pencils $Q_1, Q_2, \ldots, Q_k$ where $k = (S^m-1)^2/(S-1)$, the complete set of k orthogonal $F(S^m; S^{m-1}, S^{m-1}, \ldots, S^{m-1})$-squares. It is also true from theorem 2.1, since $S^{2m} = (S^m)^2$, that we may construct the complete set of $n - 1 = S^m - 1$ orthogonal latin squares of order $n = S^m$, i.e., the OL(n,n-1) set, from the unconfounded-with-rows-and-columns pencils $\tilde{Q}_1, \tilde{Q}_2, \ldots, \tilde{Q}_{n-1}$ in a symmetrical factorial experiment with 2 factors, each at S^m levels. Since it is true that each pencil $\tilde{Q}_i$ is made up of $(S^m-1)/(S-1)$ orthogonal Q_j's, we have that each latin square of order $n = S^m$ is made up of or decomposes into $(S^m-1)/(S-1)$ orthogonal $F(S^m; S^{m-1}, S^{m-1}, \ldots, S^{m-1})$-squares. Hence we have that the entire $OL(S^m, S^m-1)$ set decomposes into $(S^m-1)^2/(S-1)$ orthogonal $F(S^m; S^{m-1}, S^{m-1}, \ldots, S^{m-1})$-squares. □

The second decomposition theorem is a generalization of theorem 3.1.

Theorem 3.2. If p is a divisor of m, then each latin square in the set of orthogonal latin squares, $OL(S^m, S^m-1)$, can be decomposed into $(S^m-1)/(S^p-1)$ orthogonal $F(S^m; S^{m-p}, S^{m-p}, \ldots, S^{m-p})$-squares, and the entire $OL(S^m, S^{m-1})$ set can be decomposed into $(S^m-1)^2/(S^p-1)$ orthogonal $F(S^m; S^{m-p}, S^{m-p}, \ldots, S^{m-p})$-squares.

Proof. Consider a $(S^m)^2$ symmetrical factorial experiment, i.e., a symmetrical factorial experiment with 2 factors each at S^m levels. By theorem 2.1, we can construct from the unconfounded-with-rows-and-columns pencils $\tilde{Q}_1, \tilde{Q}_2, \ldots, \tilde{Q}_{n-1}$, where $n = S^m$, the complete set of n - 1 orthogonal latin squares of order n, i.e., the OL(n,n-1) set. It is also true from theorem 2.2, since $S^{2m} = (S^p)^{2m/p}$, that we may construct the complete set of k orthogonal $F(S^m; S^{m-p}, S^{m-p}, \ldots, S^{m-p})$-squares, where $k = (S^m-1)^2/(S^p-1)$, from the unconfounded-with-rows-and-columns pencils $Q_1, Q_2, \ldots, Q_k$ in a symmetrical factorial experiment with 2m/p factors each at S^p levels. Since it is true that each pencil $\tilde{Q}_i$ is made up of $(S^m-1)/(S^p-1)$ orthogonal Q_j's, we have that each latin square of order $n = S^m$ is made up of, or decomposed into, $(S^m-1)/(S^p-1)$ orthogonal $F(S^m; S^{m-p}, S^{m-p}, \ldots, S^{m-p})$-squares. Hence we have that the entire $OL(S^m,$

S^m-1) set decomposes into $(S^m-1)^2/(S^p-1)$ orthogonal $F(S^m; S^{m-p}, S^{m-p}, \ldots, S^{m-p})$-squares. □

A third decomposition theorem illustrates how each latin square in an $OL(S^m, S^m-1)$ set may be decomposed into F-squares with two different numbers of symbols.

Theorem 3.3. Each latin square in the set of orthogonal latin squares, $OL(S^m, S^m-1)$, can be decomposed into one $F(S^m; S, S, \ldots, S)$ plus S^{m-1} $F(S^m; S^{m-1}, S^{m-1}, \ldots, S^{m-1})$ orthogonal F-squares of order S^m, and the entire $OL(S^m, S^m-1)$ set can be decomposed into $(S^m-1)F(S^m; S, S, \ldots, S)$ plus $S^{m-1}(S^m-1)F(S^m; S^{m-1}, S^{m-1}, \ldots, S^{m-1})$ orthogonal F-squares of order S^m.

Proof. Consider a $(S)^{2m}$ symmetrical factorial experiment, i.e., a symmetrical factorial experiment with 2m factors each at S levels. Taking p = 1 in theorem 2.2 we can construct from the unconfounded-with-rows-and-columns pencils $Q_1, Q_2, \ldots, Q_k$ where $k = (S^m-1)^2/(S-1)$, the complete set of k orthogonal $F(S^m; S^{m-1}, S^{m-1}, \ldots, S^{m-1})$-squares. There exist (S^m-1) sets of $(S^{m-1}-1)/(S-1)$ Q_i's that form (S^m-1) pencils each with $S^{m-1}-1$ degrees of freedom. We can use these to form (S^m-1) orthogonal $F(S^m; S, S, \ldots, S)$-squares. Hence we have formed a set of $(S^m-1)F(S^m; S, S, \ldots, S)$ plus

$$(S^m-1)^2/(S-1) - (S^m-1)[(S^{m-1}-1)/(S-1)] = S^{m-1}(S^m-1)$$

$F(S^m; S^{m-1}, S^{m-1}, \ldots, S^{m-1})$ orthogonal F-squares of order S^m.

It is also true from theorem 2.1, since $S^{2m} = (S^m)^2$, that we may construct the complete set of $n - 1 = S^m - 1$ orthogonal latin squares of order $n = S^m$, i.e., the $OL(n, n-1)$ set, from the unconfounded-with-rows-and-columns pencils $\tilde{Q}_1, \tilde{Q}_2, \ldots, \tilde{Q}_{n-1}$ in a symmetrical factorial experiment with 2 factors each at S^m levels. Since it is true that each pencil $\tilde{Q}_i$ is made up of $(S^m-1)/(S-1)$ orthogonal Q_j's, we have that each latin square of order $n = S^m$ is made up of or decomposes into (S^m-1) $F(S^m; S, S, \ldots, S)$ plus $S^{m-1}(S^m-1)$ $F(S^m;$

$S^{m-1}, S^{m-1},\ldots,S^{m-1}$) orthogonal F-squares of order S^m; we have that the entire OL(S^m, S^{m-1}) set decomposes into (S^m-1) F(S^m; S,S,...,S) plus $S^{m-1}(S^m-1)$ F(S^m; $S^{m-1}, S^{m-1},\ldots,S^{m-1}$) orthogonal F-squares of order S^m. □

IV. COMPLETE SETS OF ORTHOGONAL F-SQUARES WITH DIFFERING NUMBERS OF SYMBOLS

We show in this section, how to construct complete sets of orthogonal F-squares of order $n = S^m$, where the F-squares in the sets have differing numbers of symbols, instead of a constant number. This is useful to experimenters who have differing numbers of treatments from square to square.

Theorem 4.1. There exist and one can construct complete sets of orthogonal F-squares of order $n = S^m$ where the F-squares are of varying types. In particular, any complete set of orthogonal F-squares of order S^m can contain F(S^m; $S^{m-p}, S^{m-p},\ldots,S^{m-p}$)-squares for any integer p, $1 \leq p \leq m$, that divides m, and F(S^m; S,S,...,S)-squares.

Proof. Take the OL(S^m, S^m-1) set. We can decompose as many latin squares as we wish from the OL set into F(S^m; $S^{m-p}, S^{m-p},\ldots, S^{m-p}$)-squares for integers p that divide m, including p = 1 and p = m by theorem 3.2. We can also decompose as many latin squares as we wish into F(S^m; S,S,...,S)-squares and F(S^m; $S^{m-1}, S^{m-1},\ldots, S^{m-1}$)-squares by theorem 3.3. □

Example 4.1. There exists a complete set of orthogonal F-squares of order $n = 2^2 = 4$ consisting of

(a) 9F(4; 2,2)-squares

or

(b) 6F(4; 2,2)-squares and 1F(4; 1,1,1,1)-square

or

(c) 3F(4; 2,2)-squares and 2F(4; 1,1,1,1)-squares

or

(d) 3F(4; 1,1,1,1)-squares.

These four complete sets of orthogonal F-squares of order four are obtained from the OL(4, 3) set. We get set (a) by decomposing all 3 latin squares in the OL set into F(4; 2,2)-squares by theorem 3.1, set (b) is gotten by decomposing 2 latin squares in the OL set into F(4; 2,2)-squares and leaving the third latin square undecomposed, set (c) is gotten by decomposing 1 latin square in the OL set into F(4; 2,2)-squares and leaving the other two latin squares undecomposed, and set (d) is obtained by leaving all 3 latin squares in the OL set undecomposed. (One can also use a latin square of order 4 with no orthogonal mate, decompose it into 3F(4; 2,2)-squares and then can find 6 other F(4; 2,2) squares to construct sets (a) or (b).)

Example 4.2. One of the many complete sets of orthogonal F-squares of order $n = 2^6 = 64$ that exists, consists of

10(63) + 9(32) = 918 F(64; 32,32)-squares,
12(21) = 252 F(64; 16,16,16,16)-squares,
21(9) = 189 F(64; 8,8,...,8)-squares,
9(1) = 9 F(64; 2,2,...,2)-squares, and
11(1) = 11 F(64; 1,1,...,1)-squares

This set is obtained from the OL(64, 63) set by decomposing 10 latin squares into F(64; 32,32)-squares by theorem 3.2, decomposing 12 latin squares into F(64; 16,16,16,16)-squares by theorem 3.2, decomposing 21 latin squares into F(64; 8,8,...,8)-squares by theorem 3.2, decomposing 9 latin squares into F(64; 2,2,...,2)-squares and F(64; 32,32,32)-squares by theorem 3.3, and leaving 11 latin squares undecomposed.

V. EXAMPLE

As an example consider the $OL(2^3, 2^3-1) = OL(8, 7)$ set. We may relate the complete set of 7 orthogonal latin squares of order 8 to a $2^{2(3)} = 2^6$ factorial treatment design. Let the six main effects be A, B, C, D, E, and F each at two levels 0 and 1. We set up an 8×8 square consisting of the $2^6 = 64$ treatment combinations, confounding three main effects and their interactions with rows and three main effects and their interactions with columns. Without loss of generality let us confound main effects A, B, C and their interactions AB, AC, BC, ABC with rows and let us confound main effects D, E, F and their interactions DE, DF, EF, DEF with columns. Then we have the square in Figure 5.1. We obtain the following analysis of variance table relating F-squares and latin squares to the effects in the 2^6 factorial treatment design:

	Source of variation	d.f.		
	CFM	1		
	ROWS	7		
	A		1	
	B		1	
	AB		1	
	C		1	
	AC		1	
	BC		1	
	ABC		1	
	COLUMNS	7		
	D		1	
	E		1	
	DE		1	
	F		1	
	DF		1	
	EF		1	
	DEF		1	
	LATIN SQUARE NUMBER ONE TREATMENTS	7		
$F_1(8;2,2,2,2)$ treatments	AD $= F_1(8;4,4)$ treatments		1	3
	BE $= F_2(8;4,4)$ treatments		1	
	ABDE $= F_3(8;4,4)$ treatments		1	
	CF $= F_4(8;4,4)$ treatments		1	
	ACDF $= F_5(8;4,4)$ treatments		1	

	Source of variation	d.f.		
	BCEF = $F_6(8;4,4)$ treatments		1	
	ABCDEF = $F_7(8;4,4)$ treatments		1	
	LATIN SQUARE NUMBER TWO TREATMENTS	7		
	ADEF = $F_8(8;4,4)$ treatments		1	
$F_2(8;2,2,2,2)$ treatments	BD = $F_9(8;4,4)$ treatments		1	3
	ABEF = $F_{10}(8;4,4)$ treatments		1	
	CDE = $F_{11}(8;4,4)$ treatments		1	
	ACF = $F_{12}(8;4,4)$ treatments		1	
	BCE = $F_{13}(8;4,4)$ treatments		1	
	ABCDF = $F_{14}(8;4,4)$ treatments		1	
	LATIN SQUARE NUMBER THREE TREATMENTS	7		
	AEF = $F_{15}(8;4,4)$ treatments		1	
$F_3(8;2,2,2,2)$ treatments	BCF = $F_{16}(8;4,4)$ treatments		1	3
	ABCE = $F_{17}(8;4,4)$ treatments		1	
	ABDF = $F_{18}(8;4,4)$ treatments		1	
	BDE = $F_{19}(8;4,4)$ treatments		1	
	ACD = $F_{20}(8;4,4)$ treatments		1	
	CDEF = $F_{21}(8;4,4)$ treatments		1	
	LATIN SQUARE NUMBER FOUR TREATMENTS	7		
	ADF = $F_{22}(8;4,4)$ treatments		1	
$F_4(8;2,2,2,2)$ treatments	ABCF = $F_{23}(8;4,4)$ treatments		1	3
	BCD = $F_{24}(8;4,4)$ treatments		1	
	ABE = $F_{25}(8;4,4)$ treatments		1	
	BDEF = $F_{26}(8;4,4)$ treatments		1	
	CEF = $F_{27}(8;4,4)$ treatments		1	
	ACDE = $F_{28}(8;4,4)$ treatments		1	
	LATIN SQUARE NUMBER FIVE TREATMENTS	7		
	AF = $F_{29}(8;4,4)$ treatments		1	
$F_5(8;2,2,2,2)$ treatments	BDF = $F_{30}(8;4,4)$ treatments		1	3
	ABD = $F_{31}(8;4,4)$ treatments		1	
	CE = $F_{32}(8;4,4)$ treatments		1	
	ACEF = $F_{33}(8;4,4)$ treatments		1	
	BCDEF = $F_{34}(8;4,4)$ treatments		1	
	ABCDE = $F_{35}(8;4,4)$ treatments		1	

	Source of variation	d.f.	
	LATIN SQUARE NUMBER SIX TREATMENTS	7	
$F_6(8;2,2,2,2)$ treatments	ADE = $F_{36}(8;4,4)$ treatments	1	3
	BF = $F_{37}(8;4,4)$ treatments	1	
	ABDEF = $F_{38}(8;4,4)$ treatments	1	
	BCDF = $F_{39}(8;4,4)$ treatments	1	
	ABCEF = $F_{40}(8;4,4)$ treatments	1	
	CD = $F_{41}(8;4,4)$ treatments	1	
	ACE = $F_{42}(8;4,4)$ treatments	1	
	LATIN SQUARE NUMBER SEVEN TREATMENTS	7	
$F_7(8;2,2,2,2)$ treatments	BEF = $F_{43}(8;4,4)$ treatments	1	3
	BCDE = $F_{44}(8;4,4)$ treatments	1	
	CDF = $F_{45}(8;4,4)$ treatments	1	
	AE = $F_{46}(8;4,4)$ treatments	1	
	ABF = $F_{47}(8;4,4)$ treatments	1	
	ABCD = $F_{48}(8;4,4)$ treatments	1	
	ACDEF = $F_{49}(8;4,4)$ treatments	1	
	TOTAL	64	

To construct latin square number one from effects AD, BE, ABDE, CF, ACDF, BCEF, and ABCDEF we let the symbols I,II,...,VIII in the latin square be represented as shown in Figure 5.2. We now take the 8×8 square of the 2^6 treatment combinations (Figure 5.1) and put our "treatments" I,II,...,VIII in the appropriate cells. We then get the following 8×8 latin square:

I	V	III	VII	II	VI	IV	VIII
V	I	VII	III	VI	II	VIII	IV
III	VII	I	V	IV	VIII	II	VI
VII	III	V	I	VIII	IV	VI	II
II	VI	IV	VIII	I	V	III	VII
VI	II	VIII	IV	V	I	VII	III
IV	VIII	II	VI	III	VII	I	V
VIII	IV	VI	II	VII	III	V	I

	Columns							
Rows	1	2	3	4	5	6	7	8
1	000000	000100	000010	000110	000001	000101	000011	000111
2	100000	100100	100010	100110	100001	100101	100011	100111
3	010000	010100	010010	010110	010001	010101	010011	010111
4	110000	110100	110010	110110	110001	110101	110011	110111
5	001000	001100	001010	001110	001001	001101	001011	001111
6	101000	101100	101010	101110	101001	101101	101011	101111
7	011000	011100	011010	011110	011001	011101	011011	011111
8	111000	111100	111010	111110	111001	111101	111011	111111

Figure 5.1

The remaining six latin squares are constructed in the same manner from their corresponding set of seven single degree of freedom effects in the analysis of variance table. The seven latin squares of order 8 constructed in this manner are pairwise orthogonal. Hence we have constructed the OL(8, 7) set from the analysis of variance of the 2^6 factorial treatment design.

Each single degree of freedom effect can in turn be used to construct an F(8; 4,4)-square by theorem 2.2. To construct the F-square $F_1(8; 4,4)$ from the AD effect in the analysis of variance table we let the symbols α and β in the $F_1(8; 4,4)$-square be represented as follows:

Level of Effect	Combinations (see Figure 5.2)
$(AD)_0$	I, II, III, IV = α
$(AD)_1$	V, VI, VII, VIII = β

We now take the 8×8 square of the 2^6 treatment combinations (Figure 5.1) and put our "treatments" α and β in the appropriate cells. Or alternatively, we could take the previously constructed latin square number one and replace "treatments" I, II, III, and IV by "treatment"

Level of Effect	Combinations	
$(AD)_0, (BE)_0, (CF)_0, (ABDE)_0, (ACDF)_0, (BCEF)_0, (ABCDEF)_0$	000000, 100100, 010010, 110110, 001001, 101101, 011011, 111111 =	I
$(AD)_0, (BE)_0, (CF)_1, (ABDE)_0, (ACDF)_1, (BCEF)_1, (ABCDEF)_1$	001000, 101100, 011010, 111110, 000001, 100101, 010011, 110111 =	II
$(AD)_0, (BE)_1, (CF)_0, (ABDE)_1, (ACDF)_0, (BCEF)_1, (ABDCEF)_1$	010000, 110100, 000010, 100110, 011001, 111101, 001011, 101111 =	III
$(AD)_0, (BE)_1, (CF)_1, (ABDE)_1, (ACDF)_1, (BCEF)_0, (ABCDEF)_0$	011000, 111100, 001010, 101110, 010001, 110101, 000011, 100111 =	IV
$(AD)_1, (BE)_0, (CF)_0, (ABDE)_1, (ACDF)_1, (BCEF)_0, (ABCDEF)_1$	100000, 000100, 110010, 010110, 101001, 001101, 111011, 011111 =	V
$(AD)_1, (BE)_0, (CF)_1, (ABDE)_1, (ACDF)_0, (BCEF)_1, (ABCDEF)_0$	101000, 001100, 111010, 011110, 100001, 000101, 110011, 010111 =	VI
$(AD)_1, (BE)_1, (CF)_0, (ABDE)_0, (ACDF)_1, (BCEF)_1, (ABCDEF)_0$	110000, 010100, 100010, 000110, 111001, 011101, 101011, 001111 =	VII
$(AD)_1, (BE)_1, (CF)_1, (ABDE)_0, (ACDF)_0, (BCEF)_0, (ABCDEF)_1$	111000, 011100, 101010, 001110, 110001, 010101, 100011, 000111 =	VIII

Figure 5.2

α and "treatments" V, VI, VII, and VIII by "treatment" β. In either case we get the following $F_1(8;\ 4,4)$-square:

α	β	α	β	α	β	α	β
β	α	β	α	β	α	β	α
α	β	α	β	α	β	α	β
β	α	β	α	β	α	β	α
α	β	α	β	α	β	α	β
β	α	β	α	β	α	β	α
α	β	α	β	α	β	α	β
β	α	β	α	β	α	β	α

The remaining forty-eight F(8; 4,4)-squares are constructed in the same manner from their corresponding single degree of freedom effect in the analysis of variance table. The forty-nine F(8; 4,4)-squares constructed in this manner are pairwise orthogonal. Hence each latin square in the OL(8, 7) set decomposes into 7 orthogonal F(8; 4,4)-squares and the entire OL(8, 7) set decomposes into 49 orthogonal F(8; 4,4)-squares.

In the preceding analysis of variance table, under Latin Square Number One Treatments, we see that the set of three effects AD, BE, and ABDE is closed under multiplication and hence can be used to construct an $F_1(8;\ 2,2,2,2)$-square. To construct this $F_1(8;\ 2,2,2,2)$-square we let the symbols W, X, Y, and Z in the $F_1(8;\ 2,2,2,2)$-square be represented as follows:

Level of Effect	Combinations (see Figure 5.2)
$(AD)_0, (BE)_0, (ABDE)_0$	I, II = W
$(AD)_0, (BE)_1, (ABDE)_1$	III, IV = X
$(AD)_1, (BE)_0, (ABDE)_1$	V, VI = Y
$(AD)_1, (BE)_1, (ABDE)_0$	VII, VIII = Z

We now take the 8×8 square of the 2^6 treatment combinations (Figure 5.1) and put our "treatments" W, X, Y, and Z in the appropriate cells. Or alternatively, we could take the previously constructed latin square number one and replace "treatments" I and II by "treatment" W, "treatments" III and IV by "treatment" X, "treatments" V and VI by Y, and "treatments" VII and VIII by Z. In either case we get the following F_1(8; 2,2,2,2)-square:

W	Y	X	Z	W	Y	X	Z
Y	W	Z	X	Y	W	Z	X
X	Z	W	Y	X	Z	W	Y
Z	X	Y	W	Z	X	Y	W
W	Y	X	Z	W	Y	X	Z
Y	W	Z	X	Y	W	Z	X
X	Z	W	Y	X	Z	W	Y
Z	X	Y	W	Z	X	Y	W

Note that the set of seven effects corresponding to each latin square has such a subset of three effects that it is closed under multiplication. Hence we see that each latin square in the OL(8, 7) set decomposes into one F(8; 2,2,2,2) and four F(8; 4,4)-squares. And so we can say that the entire OL(8,7) set decomposes into seven F(8; 2,2,2,2)-squares and twenty-eight F(8; 4,4)-squares. This is a direct application of theorem 3.3.

REFERENCES

Bose, R.C. (1938). On the application of the properties of Galois fields to the problem of construction of hyper-Graeco-Latin squares. *Sankhya 3,* 323-338.

Bose, R.C. (1947). Mathematical theory of symmetrical factorial design. *Sankhya 8,* 107-166.

Federer, W.T. (1977). On the existence and construction of a complete set of orthogonal F(4t; 2t,27)-squares design. *Ann. Statist. 5,* 561-564.

Hedayat, A. (1969). On the theory of the existence, non-existence, and the construction of mutually orthogonal F-squares and latin squares. Ph.D. thesis. Biometrics Unit, Cornell University, Ithaca, N.Y.

Hedayat, A., Raghavarao, D., and Seiden, E. (1975). Further contributions to the theory of F-square designs. *Ann. Statist. 3*, 712-716.

Hedayat, A. and Seiden, E. (1970). F-square and orthogonal F-square design: A generalization of latin squares and orthogonal latin squares design. *Ann. Math. Statist. 41*, 2035-2044.

Chapter 6

GENERALIZATION OF THE RECTANGULAR LATTICE

Basilio A. Rojas M.
Centro de Estadística y Cálculo
Chapingo, México

I. INTRODUCTION

Randomized incomplete block designs were proposed as a way to attain more homogeneity among experimental units and increase the efficiency in the comparison of treatments. Yates (1936) introduced the lattice designs for k^2 and k^3 treatments in blocks of size k. Harshbarger (1947) developed the rectangular lattices for k(k+1) treatments in blocks of k experimental units. Both designs are resolvable since the incomplete blocks are assembled in groups such that each group constitutes a complete replication of the treatments. This feature is important for then the designs can be also analyzed as a complete block design. Full details of the methods of construction and analysis of these lattice designs are given by Kempthorne (1952).

The partially balanced incomplete block designs proposed by Bose and Nair (1939) require in general more than four replications of the treatments and are not resolvable in many cases. David (1967) constructed cyclic and resolvable designs with the restriction that k (the block size) must equal either r (the number of replicates) or a multiple of r. Patterson and Williams (1976) stated that both the lack of rectangular lattices for a number of treatments different from k(k+1) and the specific agronomic researcher's needs led them to develop a new class of resolvable designs with b blocks of equal size and kb treatments. They also derived designs with unequal block sizes.

A method is presented here for constructing and analyzing an extension of the classical rectangular lattice k(k+1) to the more general case k(k+m), where t = k(k+m) treatments are allocated in r replications, each with k+m incomplete blocks of size k, m being an integer larger than one. The statistical efficiency of the new designs is studied and their precision relative to randomized complete blocks is examined on two uniformity trials. The theoretical and empirical results obtained are indicative of the merits of these new designs, which extend appreciably the number of possibilities for the experimenter to select from. Within the range from 4 to 100 treatments there are only 17 lattices and standard rectangular lattices, whereas with this new family of designs there are 71 cases k(k+m), $m \geq 0$.

II. CONSTRUCTION OF DESIGNS

A. Construction

The construction of the generalized rectangular lattice k(k+m), $m \geq 1$, requires the following steps:

1. Select a set of f mutually orthogonal Latin squares of side k+m. If $(k+m) = p^n$, where p is prime, then there are k+m-1 mutually orthogonal Latin squares. For a simple rectangular lattice one Latin square is required; for a triple rectangular lattice a basic set of two orthogonal squares is necessary, etc. A single Latin square exists for every k+m, therefore, a simple rectangular lattice k(k+m) can always be constructed.

2. From each of the orthogonal Latin squares take k rows.

3. Form a rectangular array with k rows and k+m columns. Each of the k(k+m) cells will have the respective f languages of the set of orthogonal Latin squares.

4. The t = k(k+m) treatments are numbered so that the cells of the first row are 1,2,...,k+m; the cells of the second row are k+m+1,...,2(k+m); and so on up to the k-th row. This type of ordering combined with the standard numbering of the languages in the set of orthogonal Latin squares has the property of simplifying the solution of the normal equations.

5. The first basic replicate of the rectangular lattice contains as incomplete blocks the columns of the rectangular array. The successive basic replicates have as incomplete blocks the treatments which have the same letter within each language.

An example of the rectangular lattice 2×4 follows to illustrate the above procedure. There are three orthogonal Latin squares. Table 6.1 shows the rectangular array 2×4 with the orthogonal languages in parentheses and the treatments numbered in each cell. From the array a quadruple rectangular lattice 2×4 is constructed.

It can be seen that the method of constructing the design assures that a pair of treatments occurs at most once in a block.

B. Randomization

The randomization consists of three steps:

1. Randomize the incomplete blocks separately and independently within each replication.

2. Randomize the treatments separately and independently within each incomplete block.

3. Allot the treatments to the treatment numbers at random. If the original numbers are changed keep track of the cor-

TABLE 6.1
Construction of a 2×4 Quadruple
Rectangular Lattice Array

(0,0,0)	1	(1,1,1)	2	(2,2,2)	3	(3,3,3)	4
(1,2,3)	5	(0,3,2)	6	(3,0,1)	7	(2,1,0)	8

Replicate

1		2		3		4	
Same Column		Same 1st letter		Same 2nd letter		Same 3rd letter	
1	5	1	6	1	7	1	8
2	6	2	5	2	8	2	7
3	7	3	8	3	5	3	6
4	8	4	7	4	6	4	5

respondence among them in order to simplify the structure of the normal equations.

III. STATISTICAL ANALYSIS

A. Model

The statistical model is the following

$$y_{ijq} = \mu + \tau_i + \beta_j + \pi_q + e_{ijq} \tag{1}$$

where μ, τ_i, and π_q represent the fixed effects of the mean, the treatment and the replicate, respectively. For the intra-block analysis β_j is a fixed block effect. For the inter-block analysis β_j is a random variable, normally and independently distributed with zero mean and variance σ_b^2 and also independent of e_{ijq}. For both analyses e_{ijq} is the intra-block error, assumed normally and independently distributed with mean zero and variance σ^2. y_{ijq} is the observation made at plot ijq.

B. Notation

The number of treatments is k(k+m). The basic set of replications is λ which is repeated s times. The total number of replications in the design is $r = \lambda s$. For example, a simple lattice replicated three times has $\lambda = 2$, $s = 3$ and $r = 6$. The total number of blocks is $b = r(k+m)$. The number of experimental units or plots is $n = bk = rk(k+m)$.

With the n observations y_{ijq} the following values are required

$$\begin{aligned}
&\text{Treatment } i \text{ total} && T_i = \sum_{j,q} y_{ijq} \\
&\text{Block } j \text{ total} && B_j = \sum_{i} y_{ijq} \\
&\text{Replicate } q \text{ total} && R_q = \sum_{i,j} y_{ijq} \\
&\text{Grand total} && G = \sum_{i,j,q} y_{ijq}
\end{aligned} \tag{2}$$

The statistical analysis to be presented is in matrix algebra. Computer programs for matrix operations are now common procedures available to experimenters. R is a diagonal matrix $t \times t$ with r as the common element. N is the incidence matrix $t \times b$ of the design, such that $N = (n_{ij})$, $n_{ij} = 1$ if treatment i appears in block j, $n_{ij} = 0$ if treatment i is not included in block j. B is a column vector with B_j as elements, $j = 1,2,\ldots,b$. T is a column vector with T_i as elements, $i = 1,2,\ldots,t$. H is a matrix $t \times t$ with all elements equal to one. Finally C is the contrasts matrix $(t-1) \times t$, defined as follows

$$C = \begin{bmatrix} 1 & -1 & 0 & 0 & \ldots\ldots & 0 \\ 1 & 0 & -1 & 0 & \ldots\ldots & 0 \\ 1 & 0 & 0 & -1 & \ldots\ldots & 0 \\ \cdot & \cdot & \cdot & \cdot & \cdots & \cdot \\ 1 & 0 & 0 & 0 & \ldots\ldots & -1 \end{bmatrix} \tag{3}$$

C. Statistical Analysis

The statistical analysis for the generalized rectangular lattices follows the method for a two-way table with unequal frequencies, see Kempthorne (1952).

1. Intra-block Analysis. The intra-block analysis is also referred to as analysis without recovery of inter-block information. The basic purposes of this analysis are the estimation of the intra-block error variance σ^2 and the determination of the efficiency factor of the rectangular lattice. Table 6.2 presents the required analysis of variance. The sums of squares for replicates, blocks within replicates ignoring treatment effects, and total are calculated in the standard form. SST(A), sum of squares for treatments adjusted for block effects, is obtained as follows, see John (1971). Let

$$A = R - \frac{NN'}{k} \quad \text{and} \quad Q = T - \frac{NB}{k} \tag{4a}$$

TABLE 6.2

Analysis of Variance. Intra-block Error

Source	d.f.	S.S.
Replicates	r-1	$\Sigma R_q^2/t - G^2/n$
Blocks within reps (ignoring treatments)	b-r	$\Sigma B_j^2/k - \Sigma R_q^2/t$
Treatments (adjusted)	t-1	SST(A), formula (6)
Intra-block error	n-b-t+1	SSE (by subtraction)
Total	n-1	$\Sigma y_{ijq}^2 - G^2/n$

Define

$$P = A + H \tag{4b}$$

The vector column $\hat{T}$ of intra-block treatment estimates is

$$\hat{T} = P^{-1}Q \tag{5}$$

and, finally,

$$SST(A) = Q'\hat{T} \tag{6}$$

2. Efficiency Factor. Let $g\sigma^2$ be the average variance of the differences between all possible pairs of treatments in the rectangular lattice. If the experiment had been in randomized complete blocks the variance of the difference between any two treatments is $2\sigma_c^2/r$, where σ_c^2 stands for the error variance in the complete block design. The relative efficiency of the intra-block estimates of the rectangular lattice with respect to the complete block design is then $(2\sigma_c^2/r)/g\sigma^2$ which has to be larger than one if the lattice is to be preferred; therefore,

$$\frac{\sigma^2}{\sigma_c^2} < \frac{2}{rg} = \text{E.F.} \tag{7}$$

where E.F. stands for efficiency factor. The intra-block analysis of the rectangular lattice is more efficient than the analysis of the corresponding complete block analysis only if the ratio of the respective error variances is less than the efficiency factor. This disadvantage is removed with the recovery of inter-block information.

It can be shown that the coefficient g in (7) is

$$g = \text{tr } CP^{-1}C'/(t-1)$$

where tr stands for trace. Now, if $P^{-1} = (p^{ij})$, then

$$g = 2(tp^{11} - \sum_{j=1}^{t} p^{1j})/(t-1) \tag{8}$$

3. Recovery of Inter-block Information. Model (1) with both β_j and e_{ijq} random normal independent variates with means zero and variances σ^2 and σ_b^2, respectively, makes the random variables y_{ijq} correlated within each block and simple least squares cannot be applied. Instead, the method of maximum likelihood is used, which is equivalent to the minimization of a weighted sum of squares. Let

$$M = W_1 P + \frac{W_2}{k} NN'$$
$$L = W_1 Q + \frac{W_2 N}{k} (B - \frac{G}{b} \underset{\sim}{1}) \tag{9}$$

where P and Q were defined in (4). G is the grand total stated in (2). $\underset{\sim}{1}$ is a column vector with all elements equal to one. W_1 and W_2 are the weights, where

$$W_1 = \frac{1}{\sigma^2}, \; W_2 = \frac{1}{\sigma^2 + k\sigma_b^2} \tag{10}$$

The column vector of treatment estimates $\overset{*}{T}$, with recovery of inter-block information, is

$$\overset{*}{T} = M^{-1} L \tag{11}$$

It can be seen that $W_2 \leq W_1$. When σ_b^2 is very large the blocks are very heterogeneous, W_2 approaches zero and the blocks do not afford valuable information on the treatment effects and $\overset{*}{T} \to \hat{T}$, that is, formula (11) approaches (5). On the other hand, when $\sigma_b^2 = 0$ the blocks are as homogeneous as the plots within the blocks, in other words, all plots in a replicate have the same variance σ^2, then $W_1 = W_2$ and formula for $\overset{*}{T}$ simplifies to that of a complete block design with r replicates.

4. Estimation of Weights W_1 and W_2. The estimate w_1 of W_1 is

$$w_1 = \frac{1}{\hat{\sigma}^2} \tag{12}$$

where $\hat{\sigma}^2$ is the mean square of intra-block error, calculated from Table 6.2.

From Table 6.3,

$$\hat{\sigma}_b^2 = \frac{r}{k(r-1)} \text{(MSB-MSE)}$$

and the estimate w_2 of W_2 is

$$w_2 = 1/(\hat{\sigma}^2 + k\hat{\sigma}_b^2) \tag{13}$$

Estimated weights w_1 and w_2 are put in place of W_1 and W_2 in expressions (9) to obtain $\overset{*}{T}$ given by (11).

5. Average Variance of Treatment Differences. Let V be the average variance of the difference between any two treatment estimates with recovery of inter-block information. It can be shown that

TABLE 6.3

Analysis of Variance. Inter-block Error

Source	S.S.	M.S.	E(M.S.)
Replicates	From Table 6.2		
Treatments (unadjusted)	$\sum_i T_i^2/r - G^2/n$		
Blocks within replicates (adjusted)	By subtraction	MSB	$\sigma^2+[k(r-1)/r]\sigma_b^2$
Intra-block error	From Table 6.2	MSE	σ^2
Total	From Table 6.2		

$$V = tr(CM^{-1}C')/(t-1)$$

where tr stands for trace. If $M^{-1} = (m^{ij})$ then

$$V = 2(tm^{11} - \sum_{i=1}^{t} m^{1j})/(t-1) \quad (14)$$

The relative efficiency R.E. of the rectangular lattice with recovery of inter-block information with respect to the corresponding complete block design is

$$R.E. = 2\ MSE(CB)/rV \quad (15)$$

where MSE(CB) is the mean square error as if the treatments had been distributed at random within each of the r replicates (randomized complete blocks).

6. Matrices P and M. These matrices have been defined in expressions (4) and (9). Both are of dimension $t \times t$. The structure of P is simple and that of M follows a similar pattern. The rectangular lattice design starts with a basic set of λ replications, the basic set is then repeated s times. The total number r of replicates is then $r = \lambda s$. The design has $t = k(k+m)$ treatments in incomplete blocks of size k. Row 1, corresponding to treatment 1, of matrix P has the following elements:

1. The diagonal term is $1 + r(k-1)/k$.
2. The elements corresponding to other treatments appearing together with treatment 1 in the incomplete blocks are $(k-\lambda)/k$.
3. The elements corresponding to treatments which do not appear with treatment 1 in the incomplete blocks are 1.

All other rows are formed in a similar manner. Take, for example a simple rectangular lattice 2×4 replicated twice and consider as basic set the first two replicates shown in Table 6.1. We have $\lambda = 2$, $s = 2$, $r = 4$, $k = 2$. Matrix P is

$$P = \begin{bmatrix} 3 & 1 & 1 & 1 & 0 & 0 & 1 & 1 \\ 1 & 3 & 1 & 1 & 0 & 0 & 1 & 1 \\ 1 & 1 & 3 & 1 & 1 & 1 & 0 & 0 \\ 1 & 1 & 1 & 3 & 1 & 1 & 0 & 0 \\ 0 & 0 & 1 & 1 & 3 & 1 & 1 & 1 \\ 0 & 0 & 1 & 1 & 1 & 3 & 1 & 1 \\ 1 & 1 & 0 & 0 & 1 & 1 & 3 & 1 \\ 1 & 1 & 0 & 0 & 1 & 1 & 1 & 3 \end{bmatrix}$$

It can be seen that in general matrices P and M are centrosymmetric if the treatments have been numbered as explained in section II.A, paragraph 4.

This important property holds for lattices k^2 and rectangular lattices $k(k+m)$ and has a bearing on facilitating the solutions of the normal equations, as will be shown.

Muir (1960), pages 364-368, describes and defines the properties of the centrosymmetric matrices. A square matrix F of dimension t is centrosymmetric if the r-th row, $r \leq t$, when reversed forms the $(t+1-r)$-th row. Let F be partitioned as follows:

$$F = \begin{bmatrix} F_{11} & F_{12} \\ F_{21} & F_{22} \end{bmatrix}$$

If $t = 2q$ then all matrices F_{ij} are of size $q \times q$. If $t = 2q+1$ then F_{11} and F_{22} are of sizes $q \times q$ and $(q+1) \times (q+1)$, respectively. F can be transformed into matrix D, such that

$$D = \begin{bmatrix} D_{11} & 0 \\ D_{21} & D_{22} \end{bmatrix}, \; D_{11} \text{ of size } q \times q$$

To obtain the above transformation the following row and column elementary operations are performed in succession on F.

1. Add row i to row t+1-i, i = 1,2,...,q for t = 2q and t = 2q + 1.

2. Subtract column j from column t+1-j, j = t, t-1,...,(t-q+1).

The specified row and column elementary operations are equivalent to

$$D = E_1 F E_2$$

Matrices E_1 and E_2 perform the stated row and column operations, respectively. It can be shown that $E_1^{-1} = E_2$. Therefore, $F^{-1} = E_2 D^{-1} E_1$ that is, by performing the same column and row elementary operations on D^{-1} we obtain F^{-1}.

The inverse of D can be shown to be

$$D^{-1} = \begin{bmatrix} D_{11}^{-1} & 0 \\ -D_{22}D_{21}D_{11}^{-1} & D_{22}^{-1} \end{bmatrix}$$

It can then be concluded that a centrosymmetric matrix of size t = 2q or t = 2q+1 can be inverted through the inversion of matrices of sizes q and q+1. This property can be useful for lattices or rectangular lattices when t is large. Matrices D_{11} and D_{22} can in turn be also partitioned, although they may not be centrosymmetric in general, and inverted with a known formula, Graybill (1969). The inversion of F can be reduced in this manner to the inversion of matrices of size around t/4.

IV. EFFICIENCY OF THE GENERALIZED RECTANGULAR LATTICES

Table 6.4 includes 37 lattices and rectangular lattices for different values of k, k+m, t, λ, s and r. k is block size, k+m the number of incomplete blocks in a single replicate; t = k(k+m) is the number of treatments, for $m \geq 0$; λ is the number of replications in the basic set; s is the number of repetitions of the basic set; and $r = \lambda s$ is the total number of replications.

TABLE 6.4

Efficiencies of Lattices and Rectangular Lattices

k	k+m	t	λ	s	r	E.F.	R.E.(1)	R.E.(2)
3	3	9	4	1	4	0.75	1.00	
3	4	12	2	2	4	0.60		
3	4	12	4	1	4	0.71	1.05	
4	4	16	4	1	4	0.79	1.00	
4	5	20	2	2	4	0.68		
4	5	20	4	1	4	0.77	1.05	1.02
5	5	25	4	1	4	0.82	1.03	1.01
3	9	27	4	1	4	0.64	1.16	1.05
5	6	30	2	2	4	0.73		
4	8	32	4	1	4	0.73	1.10	1.12
6	6	36	2	2	4	0.78		
5	8	40	4	1	4	0.79		
6	7	42	4	1	4	0.83		
5	9	45	4	1	4	0.78	1.12	1.11
7	7	49	4	1	4	0.86		
4	13	52	4	1	4	0.70		
7	8	56	4	1	4	0.85	1.04	1.07
5	12	60	3	1	3	0.75		
8	8	64	4	1	4	0.87		
4	17	68	4	1	4	0.68	1.02	1.23
8	9	72	4	1	4	0.87	1.01	1.28
7	11	77	4	1	4	0.84		
9	9	81	4	1	4	0.88		
5	17	85	4	1	4	0.75		
9	10	90	2	2	4	0.83		
5	19	95	4	1	4	0.75		1.21
10	10	100	4	1	4	0.89		
9	13	117	4	1	4	0.87	1.00	1.08
8	16	128	4	1	4	0.85		
4	41	164	4	1	4	0.56	1.16	1.49
6	29	174	4	1	4	0.78	1.17	1.25
9	20	180	4	1	4	0.86	1.21	1.30
10	19	190	4	1	4	0.87	1.15	1.01
12	17	204	4	1	4	0.90		
8	29	232	4	1	4	0.84	1.06	1.23
13	19	247	4	1	4	0.91	1.19	1.19
16	16	256	4	1	4	0.93	1.05	1.05

(1): Peanut uniformity trial.

(2): Bromegrass uniformity trial.

For each of the designs the efficiency factor E.F. was calculated with formulas (7) and (8). Some of the designs were superimposed over two uniformity trials; the first was a peanut field and was taken from Robinson et al. (1948); the second uniformity data referred to bromegrass and was obtained from Wassom and Kalton (1953). Relative efficiencies of the lattices over the respective complete block designs were calculated with expressions (14) and (15).

The main conclusions derived from Table 6.4 are the following:

1. The efficiency factor increased, as expected, when $\lambda \to r$.
2. The efficiency factor increased with t.
3. The efficiency factor diminished with k when $t = k(k+m)$ remained essentially constant.
4. The relative efficiencies of the rectangular lattices increased in general for smaller values of k and $t = k(k+m)$ remained essentially constant. This result is of importance to experimenters for it indicates that in field experimental work the smaller the block the more homogeneity among plots within the block was reached.

The author developed the construction and analysis of the generalized rectangular lattices in 1979. Carlos Meza and Jose Luis Martinez collaborated with the author to get Table 6.4. The work was done at the Biometric Unit of the National Institute of Agricultural Research (INIA) in Mexico. The computing work used the facilities of the Statistical and Computing Center (CEC) of the Graduate College in Chapingo, Mexico.

REFERENCES

Bose, R.C., and K.R. Nair (1939). Partially balanced incomplete block designs. *Sankhyā 4*, 337-372.

David, H.A. (1967). Resolvable cyclic designs. *Sankhyā A 29*, 191-198.

Harshbarger, B. (1947). Rectangular lattices. Virginia Agr. Exp. Sta. Mem. 1.

John, Peter W.M. (1971). *Statistical Design and Analysis of Experiments*. New York: Macmillan.

Kempthorne, O. (1952). *The Design and Analysis of Experiments*. New York: John Wiley.

Muir, Thomas (1960). *A Treatise on the Theory of Determinants*. New York: Dover Publications.

Patterson, H.D. and E.R. Williams (1976). A new class of resolvable incomplete block designs. *Biometrika 63 1*, 83-92.

Robinson, H.F., J.A. Rigney and P.H. Harvey (1948). Investigations in peanut plot technique. North Carolina Agr. Exp. Sta., Tech. Bull. 86.

Wassom, C.E. and R.R. Kalton (1953). Estimation of optimum plot size using data from bromegrass uniformity trials. Iowa Agr. Exp. Sta. Res. Bull. 296.

Yates, F. (1936). A new method of arranging variety trials involving a large number of varieties. *J. Agr. Sci. 26*, 424-455.

Chapter 7

PARAMETRIZATIONS AND RESOLUTION IV

Justus Seely and David Birkes
Oregon State University
Corvallis, Oregon

I. INTRODUCTION

This paper is concerned with the role of parametrizations in the area of even resolution designs. It presents two examples in Section V which show that the notion of resolution IV depends not only on the data set, but also on the choice of parametrization. Another example in the same section shows that the Margolin (1969) lower bound is not true for all types of parametrizations. The paper attempts to answer questions raised by these examples.

Our discussion is primarily from a linear model point of view. Sections II-IV are introductory. Section VI considers Margolin's lower bound. It gives a condition on main effect parametrization matrices (defined in Section IV) that is sufficient to insure the validity of the bound. Section VIII investigates relationships between different parametrizations for the same data set. It is found that when the ranges of corresponding main effect parametrization matrices in two different parametrizations are the same, then one parametrization is of resolution IV if and only if the other one is. Section VIII studies factorial models with the usual constraints and shows that such a model has the resolution IV property if and only if the property holds for unconstrained models whose main effect parametrization matrices have columns that sum to zero.

II. LINEAR MODEL PREREQUISITES

The discussion in the following sections is presented from a linear model viewpoint. To facilitate the discussion, let us briefly review some estimability facts about partitioned linear models. First, for notation, we let $\underline{R}(H)$, $\underline{r}(H)$, and H' denote the range, rank, and transpose of a matrix H, and we let $g^{\perp}$ denote the set of all vectors that are orthogonal to a vector g.

Suppose that Y is an n×1 random vector with expectation $E(Y) = X\pi$ where π is a vector of unknown parameters. The *regression space* of the model is $\underline{R}(X)$. Now suppose E(Y) is partitioned into the form

$$E(Y) = X_1\pi_1 + X_2\pi_2 = X\pi \tag{II.1}$$

Let $df(\pi_2)$ denote the dimension of the vector space of estimable parametric functions of the form $g'\pi_2$. Some facts we need are:

$$df(\pi_2) = \underline{r}(X) - \underline{r}(X_1) \tag{II.2a}$$

The parametric vector π_2 is estimable if and only if

$\underline{R}(X_1) \cap \underline{R}(X_2) = \{0\}$ and X_2 has full column rank (II.2b)

Justification for the statements above can be found, for example, in Section 3 of Seely and Birkes (1980).

We need the additional concept of a reparametrization (or parametrization) for E(Y). We say that $E(Y) = W\phi$ is a *reparametrization* for E(Y) provided that the regression space is the same as in the original model, that is, $\underline{R}(W) = \underline{R}(X)$.

The number $df(\pi_2)$ also has an interpretation in terms of degrees of freedom. In particular, from (II.2a) it is clear that $df(\pi_2)$ is the degrees of freedom associated with the familiar reduction sum of squares $R(\pi_2|\pi_1)$.

III. PARAMETRIZATION MATRICES AND HADAMARD PRODUCTS

To investigate the resolution IV property in classification models, it is necessary to precisely define main effects and interaction effects. One popular means of doing this (e.g., see Webb, 1968, and Margolin, 1969) is to first parametrize the main effects and then take Hadamard products of the main effect column vectors of the design matrix to form the interaction column vectors. This is the method we adopt. Another method of defining main effects and interaction effects is given in Chapter 4 of Raktoe, Hedayat, and Federer (1981).

To develop the main effect parametrizations in which we are interested, it is sufficient to consider reparametrizations for the one-way classification model

$$E(Y_{ip}) = \mu + \alpha_i$$

where $i = 1,\ldots,t$ and $p = 1,\ldots,n_i$. In matrix notation we have

$$\text{M1:} \qquad E(Y) = 1_n\mu + A\alpha = X\pi$$

where Y is the $n\times 1$ vector of observations, $\alpha = (\alpha_1,\ldots,\alpha_t)'$, and 1_n, A are defined in the obvious way.

Our interest is in full (column) rank parametrizations of the form

$$\text{M1}_f\text{:} \qquad E(Y) = 1_n\mu^f + A^f\alpha^f = X^f\pi^f$$

where A^f is an $n \times (t-1)$ matrix. Because $\underline{R}(X) = \underline{R}(A)$, it follows that M1_f is a parametrization for E(Y) if and only if A^f satisfies

$$\underline{R}(A) = \underline{R}(1_n, A^f) \qquad \text{(III.1)}$$

One way to form such an A^f matrix is the following: Let A_d be any $t \times (t-1)$ matrix such that $(1_t, A_d)$ is nonsingular. Then $A^f = AA_d$

is an $n \times (t-1)$ matrix satisfying (III.1). In fact, any $n \times (t-1)$ matrix A^f satisfying (III.1) can be formed in this fashion. Forming an A^f matrix via such an A_d matrix is similar in spirit to the Raktoe et al. (1981) approach for defining main effects and is easy to understand. In particular, the ith row of the A_d matrix spells out the coefficients of the α^f vector in $M1_f$ for observations in the ith group; and the matrix A simply replicates these rows according to the number of observations in the ith group.

Definition III.1. For a factor with t levels we say that any $t \times (t-1)$ matrix L is a *main effect parametrization matrix* provided that $(1_t, L)$ is nonsingular.

Main effect parametrization matrices allow one to conveniently define design matrix column vectors for main effects in factorial experiments. Some common examples are given next.

Example III.1. Suppose a factor has $t = 3$ levels and assume x_1, x_2, x_3 are three distinct real numbers. Each of the following matrices is a valid main effect parametrization matrix:

$$L_p = \begin{bmatrix} x_1 & x_1^2 \\ x_2 & x_2^2 \\ x_3 & x_3^2 \end{bmatrix} \quad L_{op} = \begin{bmatrix} -1 & 1 \\ 0 & -2 \\ 1 & 1 \end{bmatrix} \quad L_h = \begin{bmatrix} 1 & 1 \\ -1 & 1 \\ 0 & -2 \end{bmatrix} \quad L_c = \begin{bmatrix} 1 & 0 \\ 0 & 1 \\ -1 & -1 \end{bmatrix}$$

The L_p (p = polynomial) matrix might be used when the levels of the factor correspond to a quantitative variable at levels x_1, x_2, x_3. The L_{op} (op = orthogonal polynomial) matrix might be used in place of L_p when the levels x_1, x_2, x_3 are equally spaced. The L_h (h = Helmert) matrix is used by Anderson and Thomas (1979) in defining main effects. The last matrix L_c (c = constraints) is a very natural matrix to use when model M1 is assumed to have the side constraints $\Sigma_i \alpha_i = 0$.

To define the design column vectors for interaction effects, we use the notion of Hadamard products as defined by Margolin (1969).

The *Hadamard product* of two k×1 vectors $g = (g_1,\ldots,g_k)'$ and $h = (h_1,\ldots,h_k)'$ is the k×1 vector $g \odot h = (g_1h_1,\ldots,g_kh_k)'$. It is convenient to extend the notion of a Hadamard product to matrices. The Hadamard product of a k×p matrix $G = (G_1,\ldots,G_p)$ and a k×q matrix $H = (H_1,\ldots,H_q)$ is the k×pq matrix $G \odot H = (G_1 \odot H_1, G_1 \odot H_2,\ldots, G_p \odot H_q)$. Some useful properties of the Hadamard product are:

$$G \odot (H+K) = G \odot H + G \odot K \tag{III.2a}$$

$$\text{If } \underline{R}(G) \subset \underline{R}(P), \text{ then } \underline{R}(G \odot H) \subset \underline{R}(P \odot H) \tag{III.2b}$$

$$GL \odot HM = (G \odot H)(L \otimes M) \text{ where } \otimes \text{ denotes the Kronecker product} \tag{III.2c}$$

Properties (III.2a) and (III.2b) are straightforward to verify and property (III.2c) can be found in Khatri and Rao (1968, p. 169).

IV. RESOLUTION IV PARAMETRIZATIONS

A k-factor factorial design is said to be of resolution IV provided that all main effects are estimable with respect to the factorial model that assumes all three-factor and higher order interactions are negligible. This rather loosely stated principle has been formalized in various places in the literature (e.g., see the first paragraph of the previous section). We give here a brief introduction which we believe is consistent with the published literature.

For notational convenience, we concentrate on the two-factor experiment. The treatment of more than two factors is similar and only briefly commented upon.

Consider observations $\{Y_{ijp}\}$ from a two-factor experiment whose expectations include two-factor interactions. The basic parametrization from which we build is

$$E(Y_{ijp}) = \mu + \alpha_i + \beta_j + \theta_{ij}$$

where $i = 1,\ldots,a$, $j = 1,\ldots,b$, and $p = 1,\ldots,n_{ij}$ with the understanding that $n_{ij} = 0$ implies there are no observations with ij as the first two subscripts. Let n denote the total number of observations and let Y denote the $n\times 1$ vector of the Y_{ijp}. Express E(Y) in matrix form as

$$M2: \quad E(Y) = 1_n\mu + A\alpha + B\beta + T\theta = X\pi$$

For convenience we assume that the components of the parameter vectors α, β, and θ are all in lexicographic order. This ensures that $T = A \circledcirc B$.

Let A_d and B_d be parametrization matrices for factors α and β, respectively. Set $A^f = AA_d$ and $B^f = BB_d$. Using (III.1) and (III.2b), one can show

$$\underline{R}(A \circledcirc B) = \underline{R}[(1_n,A^f) \circledcirc (1_n,B^f)] = \underline{R}(1_n,A^f,B^f,A^f \circledcirc B^f) \qquad \text{(IV.1)}$$

Set $T^f = A^f \circledcirc B^f$. Because $\underline{R}(X) = \underline{R}(T)$ and $T = A \circledcirc B$, it follows that

$$M2_f: \quad E(Y) = 1_n\mu^f + A^f\alpha^f + B^f\beta^f + T^f\theta^f = X^f\pi^f$$

is a parametrization for E(Y). We say that Model $M2_f$ is of *resolution IV* if and only if the main effects α^f and β^f are estimable.

The generalization of the above ideas to more than two factors is essentially straightforward. We will, however, briefly discuss the three-factor case. Consider a collection of observations $\{Y_{ijkp}\}$ whose expectations include all two-factor interactions, parametrized as

$$E(Y_{ijkp}) = \mu + \alpha_i + \beta_j + \gamma_k + \theta_{ij} + \eta_{ik} + \omega_{jk} ,$$

where i, j, k, and p vary from 1 to a, b, c, and n_{ijk}, respectively. Let Y be the observation vector and express E(Y) in matrix form as

$$\text{M3:} \quad E(Y) = 1_n\mu + A\alpha + B\beta + C\gamma + T\theta + U\eta + V\omega = X\pi$$

where the parameter vectors are all lexicographically ordered. Now form A^f, B^f and C^f from parametrization matrices A_d, B_d and C_d. Next form the two-factor interaction Hadamard product matrices T^f, U^f and V^f. By noting that $\underline{R}(X) = \underline{R}(T,U,V)$, and by applying (IV.1) to $T = A \odot B$, $U = A \odot C$ and $V = B \odot C$, it can be seen that

$$\text{M3}_f\text{:} \quad E(Y) = 1_n\mu^f + A^f\alpha^f + B^f\beta^f + C^f\gamma^f + T^f\theta^f + U^f\eta^f + V^f\omega^f = X^f\pi^f$$

is a parametrization for E(Y). Then model M3_f is said to be of resolution IV provided that the main effects α^f, β^f, and γ^f are all estimable.

V. EXAMPLES

Below we give three examples to illustrate some points mentioned in the introduction. For the first two examples we suppose a 3^3 experiment with the following n = 17 observations:

111	211	121	321	231	331
112	312	322	132	232	
213	313	123	223	133	333

Here ijk denotes an observation with its three factors at levels i, j, and k, respectively. Also, we suppose the levels of each factor are determined by quantitative variables taking values 0, 1 and 2. The assumed model in both examples is M3_f defined in Section IV.

Example V.1. For a situation like the present setting, Raktoe et al. seem to prefer main effect parametrization matrices whose coefficients come from orthogonal polynomial tables. That is, use $A_d = B_d = C_d = L_{op}$ to form model M3_f where L_{op} is defined in Example III.1. With these parametrization matrices, it can be checked using

(II.2b) that model $M3_f$ is of resolution IV. (An alternative and somewhat easier method for checking the resolution IV property in this example is given in Section VIII.)

Example V.2. Instead of using the parametrization matrices of Example V.1, one might prefer to use orthogonal polynomials that take into account the unequal frequencies of the levels of each factor. The x values 0, 1, 2 for the three levels of each factor occur with frequencies 6, 5, 6, respectively. If one constructs orthogonal polynomials taking the unequal frequencies into account, then

$$A_d = B_d = C_d = \begin{bmatrix} 1 & -5 \\ 0 & 12 \\ -1 & -5 \end{bmatrix}$$

Unlike the previous example, these parametrization matrices have the property that the columns of $(1_n, H^f)$ are all orthogonal to one another for H = A, B, and C. Using these parametrization matrices, it can be checked that the column vector

$$g = (0\ 34\ -2\ -34\ 2\ 0\ 0\ 0\ -34\ 34\ 0\ 289\ -17\ 17\ 3\ -289\ 17\ -17\ -3)'$$

satisfies $X^f g = 0$. Since g has nonzero coefficients for some main effect column vectors and some interaction effect column vectors, it follows that the disjoint intersection condition in (II.2b) with $X_2 = (A^f, B^f, C^f)$ is not true. Hence model $M3_f$ is not of resolution IV.

The two examples above illustrate an aspect of the resolution IV property that oftentimes appears to be neglected in the literature. In particular, two very reasonable 'orthogonal polynomial' type parametrizations are employed on the same data set and yet lead to seemingly contradictory statements with respect to the resolution IV property. The problem is, of course, that the notion of resolution IV depends on the main effect parametrizations that are employed. Thus, when one wants to use a resolution IV design in an experiment, particular attention should be paid to the definitions of the main effects. If you select a design from the literature where the resolution IV property was verified using a particular parametrization, there is no guarantee that the resolution IV property will carry over to a different parametrization.

Our last example illustrates that not only does the resolution IV property itself depend on the parametrization, but so does the Margolin lower bound on the minimal number of runs needed to obtain resolution IV (see Section VI).

Example V.3. Consider a 3×3 factorial experiment with (see model M2) $n_{33} = 0$ and all other n_{ij} nonzero. Assume the levels of each factor are determined by quantitative variables taking values 0, 1, and 2. Instead of using 'orthogonal polynomial' type parametrization matrices, suppose one decides to simply use the raw values 0, 1, 2, leading to $A_d = B_d = L_p$ where L_p is defined in Example III.1. With these parametrization matrices, it is easy to check using (II.2b) that model $M2_f$ is of resolution IV. (In fact, if $n_{22} = n_{23} = n_{32} = n_{33} = 0$ and all other n_{ij} are nonzero, the resulting model $M2_f$ will still be of resolution IV.)

We found Example V.3 to be somewhat surprising because a corollary to Margolin's lower bound result (1969, p. 518) is that in an a×b factorial arrangement one needs a minimum of ab observations to obtain the resolution IV property.

VI. MARGOLIN'S LOWER BOUND

An example was given in the previous section that was of resolution IV and had fewer runs than Margolin's lower bound. Margolin (1969) used an orthogonal polynomial parametrization (like the one in Example V.2) in deriving his lower bound and stated that the proof follows for other 'full rank' parametrizations because one can always transform to orthogonal polynomials. However, as Examples V.1 and V.2 illustrate, the resolution IV property is not necessarily invariant under such a transformation. In the present section we give a sufficient condition which insures the validity of Margolin's lower bound.

Suppose $L = (L_1,\ldots,L_s)$ is a parametrization matrix for a factor having $t = s+1$ levels. Because $(1_t,L)$ is nonsingular, it follows that $\underline{R}(L \odot L) \subset \underline{R}(1_t,L)$. Thus, for each $p,q = 1,\ldots,s$ there exists a unique real number d_{pq} and a unique vector $x_{pq} \in \underline{R}(L)$ such that

$$L_p \odot L_q = d_{pq}1_t + x_{pq} \tag{VI.1}$$

Let D denote the $s \times s$ matrix whose elements are the numbers d_{pq}. We say that L is *regular* provided that D is nonsingular.

Lemma VI.1. Let L be a regular main effect parametrization matrix for a factor having t levels, let A be an $n \times t$ matrix such that each row has a single 1 and the rest zeros, and let F be an arbitrary $n \times q$ matrix. Set $H = AL$. If (H,F) has full column rank and if $R(H,F) \cap R(1_n,H \odot F) = \{0\}$, then $(1_n,H,F,H \odot F)$ has full column rank.

Proof. (This is based on the proof of Theorem 1 in Margolin, 1969.) It suffices to show the columns of $(1_n,H \odot F)$ are linearly independent. Suppose $(1_n,H \odot F)u = 0$. We must show $u = 0$. Partition u so that $(1_n,H \odot F)u = u_0 1_n + \Sigma_{j=1}^{s}(H_j \odot F)u_j$ (where $s = t-1$, $u_0 \in R^1$, $u_j \in R^q$). For any column H_p of H, $0 = H_p \odot (u_0 1_n + \Sigma_{j=1}^{s}(H_j \odot F)u_j) = u_0 H_p + \Sigma_{j=1}^{s} H_p \odot (H_j \odot F)u_j$. Note $(H_j \odot F)u_j = H_j \odot (Fu_j)$, $H_p \odot (H_j \odot Fu_j) = (H_p \odot H_j) \odot Fu_j$, and $H_p \odot H_j = AL_p \odot AL_j = A(L_p \odot L_j) = A(d_{pj}1_t + x_{pj}) = d_{pj}1_n + Ax_{pj}$, using (VI.1). Thus, $0 = u_0 H_p + \Sigma_{j=1}^{s} d_{pj}Fu_j + \Sigma_{j=1}^{s} Ax_{pj} \odot Fu_j$. The last summation is in $\underline{R}(H \odot F)$ because $Ax_{pj} \in \underline{R}(AL) = \underline{R}(H)$. By our hypotheses, we must have $u_0 H_p = 0$, which implies $u_0 = 0$, and we must have $\Sigma_{j=1}^{s} d_{pj}F_{uj} = 0$ (for all p), which says $FUD = 0$, where $U = (u_1,\ldots, u_s)$. Since D is nonsingular and F has full column rank, U must be 0. Therefore $u = 0$. □

The lemma above can be used to establish the Margolin lower bound. Consider a k-factor factorial experiment. Let A_d, B_d,... be the main effect parametrization matrices leading to A^f, B^f,... . Consider the model

$$Mk_f: \quad E(Y) = 1_n\mu^f + A^f\alpha^f + F\delta + W\phi + G\eta$$

where F consists of all main effect matrices except A^f, $W = A^f \circledcirc F$ consists of all Hadamard product matrices involving A^f, and G consists of all the remaining Hadamard product matrices. Suppose model Mk_f is of resolution IV. Using (II.2b) with $X_2 = (A^f,F)$, we can conclude that X_2 has full column rank and that $\underline{R}(X_2) \cap \underline{R}(1_n,W) = \{0\}$. If A_d is regular, then Lemma VI.1 with $L = A_d$ implies $(1_n, A^f, F, W)$ has full column rank. In particular, if n denotes the number of runs in the experiment, if A^f has (a-1) columns and F has m = (b-1) + (c-1) + ... columns, then

$$n \geq 1 + (a-1) + m + (a-1)m = a(m+1) \tag{VI.2}$$

And this is precisely Margolin's lower bound when a is the maximum number of levels. Thus, whenever a factor having the maximum number of levels in the experiment has a main effect matrix A^f formed from a regular main effect parametrization matrix A_d, we can be assured that Margolin's lower bound is valid.

Example VI.1. Suppose a main effect parametrization matrix L is such that all columns of $A^f = AL$ sum to zero, such as in Example V.2. Noting that $\Lambda = A'A$ is diagonal and $A1_t = 1_n$, we have

$$L_p'\Lambda L_q = 1_t'\Lambda(L_p \circledcirc L_q) = d_{pq}1_t'A'A1_t + 1_t'A'Ax_{pq} = nd_{pq}$$

Thus $D = n^{-1}L'A'AL$, which is nonsingular. So L is regular.

Example VI.2. Suppose L is a parametrization matrix whose columns sum to zero, such as L_{op}, L_h, and L_c in Example III.1. Arguing as in Example VI.1 with $\Lambda = I_t$, one can show $D = t^{-1}L'L$, which is nonsingular. So L is regular.

Example VI.3. Suppose a factor has t levels determined by a quantitative variable x taking values $x_1,\ldots,x_t$. Set $s = t - 1$ and let L be the t×s matrix whose ij element is x_i^j (for t = 3, see L_p in Example III.1). It can be verified that L is a main effect parametrization matrix if and only if the x_i are all distinct. For $p + q \leq s$, the vector $L_p \odot L_q$ is a column of L, so $d_{pq} = 0$. Thus the determinant of D, except possibly for a sign change, is given by the product of the reverse diagonal elements d_{pq}, $p + q = t$. These elements are all identical and are nonzero if and only if all of the x_i are nonzero. So, L is a regular main effect parametrization matrix if and only if $x_1,\ldots,x_t$ are distinct and nonzero. This helps explain Example V.3. In fact, any main effect parametrization matrix L with a zero row cannot be regular.

VII. RELATIONSHIPS BETWEEN DIFFERENT PARAMETRIZATIONS

Consider model $M3_f$ of Section IV. Additionally, assume A_d^*, B_d^*, C_d^* are another set of parametrization matrices leading to

$$M3_f^*: \quad E(Y) = 1_n\mu_* + A_*^f\alpha_*^f + \ldots + V_*^f\omega_*^f = X_*^f\pi_*^f$$

where each symbol is defined like its counterpart in $M3_f$. The present section is devoted to determining conditions under which $df(\pi_2^f) = df(\pi_{2*}^f)$ for various choices of π_2. Extensions to more than three factors are straightforward.

Throughout we take π_2 to be one of the symbols μ, $\alpha,\ldots,$ or ω; and we assume $M3_f$ and $M3_f^*$ are partitioned as

$$E(Y) = X_1^f\pi_1^f + X_2^f\pi_2^f \quad \text{and} \quad E(Y) = X_{1*}^f\pi_{1*}^f + X_{2*}^f\pi_{2*}^f$$

where all symbols in the partitions are defined to be consistent with the choice of π_2. When π_2 is an interaction symbol, it is generally known that $\underline{R}(X_1^f) = \underline{R}(X_{1*}^f)$. (A proof can easily be constructed. For example, if $\pi_2 = \eta$, then apply (IV.1) to $A \odot B$ and to $B \odot C$.) From (II.2a) we obtain

$$df(\pi_2^f) = df(\pi_{2*}^f) \quad \text{for} \quad \pi_2 = \theta,\ \eta,\ \text{or}\ \omega \qquad \text{(VII.1)}$$

However, as Examples V.1 and V.2 illustrate, this statement is not true for main effects without additional assumptions.

Lemma VII.1. Suppose $\underline{R}(L_d) = \underline{R}(L_d^*)$ for L = A, B, and C. Then $\underline{R}(X_1^f) = \underline{R}(X_{1*}^f)$ and $\underline{R}(X_2^f) = R(X_{2*}^f)$ for $\pi_2 = \mu, \alpha, \ldots,$ or ω.

Proof. The lemma will follow if we show $\underline{R}(X_2^f) = \underline{R}(X_{2*}^f)$ for each choice of π_2. For $\pi_2 = \mu$, α, β, or γ the result is easy to show. Now suppose $\pi_2 = \eta$. Using (III.2b) with $\underline{R}(A^f) = \underline{R}(A_*^f)$ and $\underline{R}(C^f) = \underline{R}(C_*^f)$ we see that $U^f = A^f \odot C^f$ and $U_*^f = A_*^f \odot C_*^f$ have the same range. A similar proof works for $\pi_2 = \theta$ or ω. □

Using this lemma and (II.2a), we obtain the corollary

$$\text{If } \underline{R}(L_d) = \underline{R}(L_d^*) \text{ for L = A, B, and C, then} \qquad \text{(VII.2)}$$
$$df(\pi_2^f) = df(\pi_{2*}^f) \text{ for } \pi_2 = \mu, \alpha, \ldots, \text{ and } \omega$$

There are several useful observations which follow from (VII.1) and (VII.2). If π_2 is an interaction symbol, then π_2^f is estimable if and only if π_{2*}^f is estimable. Under the conditions of (VII.2) this is also true for $\pi_2 = \mu$, α, β, or γ. And this implies that model $M3_f$ is of resolution IV if and only if model $M3_f^*$ is of resolution IV. The conditions in (VII.2) can also be related to the discussion on Margolin's lower bound.

Lemma VII.2. Suppose L and L_* are two main effect parametrization matrices with $\underline{R}(L) = \underline{R}(L_*)$. Then L is regular if and only if L_* is regular.

Proof. We can write $L = L_*G$ for some nonsingular G. Using (III.2c), we obtain $L \odot L = (L_* \odot L_*)(G \otimes G)$, which yields $D = G'D_*G$. Hence D is nonsingular if and only if D_* is nonsingular. □

Remark VII.1. The validity of (VII.1) and (VII.2) depended only on $\underline{r}(X_1^f) = \underline{r}(X_{1*}^f)$. The stronger equality $\underline{R}(X_1^f) = \underline{R}(X_{1*}^f)$ which was proved further implies the reduction sums of squares equality $R(\pi_2^f|\pi_1^f) = R(\pi_{2*}^f|\pi_{1*}^f)$.

We conclude by examining the range of the three types of main effect parametrization matrices considered in the previous section.

Example VII.1. Suppose L is a parametrization matrix of the type discussed in Example VI.1. Observe that $1_n'A^f = v'L$ where $v' = 1_n'A$ is the vector of frequencies for the levels of the factor. Since the columns of A^f sum to zero, it follows that $\underline{R}(L) = v^\perp$. Hence, all L such that the columns of AL sum to zero have the same range.

Example VII.2. Suppose L is a parametrization matrix whose columns sum to zero. Then $R(L) = 1^\perp_t$. Hence, all such L have the same range.

Example VII.3. Consider the parametrization matrix L of Example VI.3. One might consider making a scale and location change before doing an analysis. Does such a transformation affect $\underline{R}(L)$? A scale change does not, but a location change does, because it adds to the first column of L a multiple of 1_t, which is not in $\underline{R}(L)$.

VIII. CONSTRAINED PARAMETRIZATION

Consider the usual constrained version of model M2 introduced in Section IV. In particular, suppose

$$E(Y_{ijp}) = \mu^c + \alpha_i^c + \beta_j^c + \theta_{ij}^c$$

$$\alpha_.^c = \beta_.^c = 0 \text{ and } \theta_{i.}^c = \theta_{.j}^c = 0 \text{ for all } i, j$$

where i, j, p vary as in model M2. Write the model in matrix form as

$$M2_c: \ E(Y) = 1_n\mu^c + A\alpha^c + B\beta^c + T\theta^c = X\pi^c, \quad \Delta'\pi^c = 0$$

It seems reasonable to say that model $M2_c$ is of resolution IV provided that α^c and β^c are estimable. How does this definition of resolution IV relate to the model $M2_f$ definition?

Before answering this question, we review some terminology and facts from Seely and Birkes (1980, Section 3). Suppose Y is a random vector with expectation $E(Y) = X\pi^c$, $\Delta'\pi^c = 0$. The regression space is $\Omega = \{Xu: \Delta'u = 0\}$. Suppose E(Y) is partitioned as

$$E(Y) = X_1\pi_1^c + X_2\pi_2^c, \ \Delta_1'\pi_1^c = 0 \text{ and } \Delta_2'\pi_2^c = 0 \qquad \text{(VIII.1)}$$

To partition E(Y) in this way it must be true that the π_1^c constraints are independent of the π_2^c constraints. For example, we could partition $M2_c$ with $\pi_2^c = \theta^c$ but not with $\pi_2^c = \theta_{11}^c$. Let $df(\pi_2^c)$ be the dimension of the vector space of estimable functions of the form $g'\pi_2^c$. And set $\Omega_k = \{X_k u: \Delta_k' u = 0\}$ for k = 1,2. Then:

$$df(\pi_2^c) = \dim \Omega - \dim \Omega_1 \qquad \text{(VIII.2a)}$$

π_2^c is estimable if and only if $\Omega_1 \cap \Omega_2 = \{0\}$ and $(X_2', \Delta_2)'$ has full column rank (VIII.2b)

The notion of a reparametrization is now any description $E(Y) = W\phi$, $\Gamma'\phi = 0$, for which $\{Wu: \Gamma'u = 0\} = \Omega$. Notice that by setting $\Delta = 0$ and $\Gamma = 0$ in this paragraph we recover the discussion in Section II.

Now let us return to model $M2_c$. It is known that $\Omega = \underline{R}(X)$. This means models M2, $M2_f$, and $M2_c$ are all parametrizations for E(Y). Next let π_2 be any one of the symbols μ, α, β, or θ. Partition $M2_f$ and $M2_c$ as follows:

$$E(Y) = X_1^f\pi_1^f + X_2^f\pi_2^f \quad \text{and} \quad E(Y) = X_1\pi_1^c + X_2\pi_2^c, \ \Delta_1'\pi_1^c = 0, \ \Delta_2'\pi_2^c = 0$$

Here all symbols are defined to be consistent with the choice of π_2. When $\pi_2 = \theta$ it can be checked that $\Omega_1 = \underline{R}(X_1^f)$. From (VIII.2a) we get

$$df(\theta^f) = df(\theta^c) \tag{VIII.3}$$

Without additional information, it is not possible to state a similar result for other choices of π_2.

Lemma VIII.1. Suppose the columns of the $M2_f$ parametrization matrices A_d and B_d sum to zero. Then $\Omega_1 = \underline{R}(X_1^f)$ and $\Omega_2 = \underline{R}(X_2^f)$ for $\pi_2 = \mu, \alpha, \beta$, or θ.

Proof. The lemma will follow if we show $\Omega_2 = \underline{R}(X_2^f)$ for each choice of π_2. For $\pi_2 = \mu$ this is immediate because $\Delta_2 = 0$ and $X_2 = X_2^f = 1_n$. The case $\pi_2 = \alpha$ is quickly established by noting that $a_\cdot^c = 0$ if and only if $a^c \in \underline{R}(A_d)$. The case $\pi_2 = \beta$ is similar. Now suppose $\pi_2 = \theta$. We must show $\Omega_2 = \{Tu: \Delta_2'u = 0\}$ is equal to $\underline{R}(T^f)$. Use (III.2c) to write $T^f = AA_d \odot BB_d = TG$ where $T = A \odot B$ and $G = A_d \otimes B_d$. Next, observe that Δ_2' can be expressed as the $(a+b) \times ab$ matrix

$$\Delta_2' = \begin{bmatrix} I_a \otimes 1_b' \\ \\ 1_a' \otimes I_b \end{bmatrix}$$

Using standard Kronecker product matrix multiplication rules, we see $\Delta_2'G = 0$. This means $\underline{R}(G) \subset \underline{N}(\Delta_2')$, the null space of Δ_2'. Next, check that $\underline{r}(\Delta_2) = a + b - 1$. Then

$$\dim \underline{N}(\Delta_2') = ab - \underline{r}(\Delta_2) = (a-1)(b-1) = \underline{r}(A_d)\underline{r}(B_d) = \underline{r}(G).$$

Hence $\underline{R}(G) = \underline{N}(\Delta_2')$. This implies $\underline{R}(T^f) = \Omega_2$. □

From (VIII.2a) and Lemma VIII.1 it is straightforward to conclude:

If the columns of A_d and B_d sum to zero,
then $df(\pi_2^f) = df(\pi_2^c)$ for $\pi_2 = \mu, \alpha, \beta$, or θ (VIII.4)

From (VIII.4) one can conclude the corollaries that π_2^f is estimable if and only if π_2^c is estimable and that model $M2_f$ is of resolution IV if and only if α^c and β^c are estimable.

The previous discussion concerning $M2_f$ and $M2_c$ can be extended to Mk_f and Mk_c. In particular: (i) Statement (VIII.3) remains true for any interaction symbol. (ii) Lemma VIII.1 and Statement (VIII.4) with its corollaries generalize provided that all main effect parametrization matrices have the zero sum property. (iii) Because of the range condition proved in Lemma VIII.1 the content of Remark VII.1 carries over to Mk_f and Mk_c. That is, provided all main effect parametrization matrices have the zero sum property, then $R(\pi_2^c|\pi_1^c) = R(\pi_2^f|\pi_1^f)$ where π_2 can be μ, a main effect symbol, or an interaction symbol. [Here the constrained reduction sum of squares is the difference in the regression sums of squares for the full model and the submodel $E(Y) = X_1\pi_1^c$, $\Delta_1'\pi_1^c = 0$.] We conclude with an example.

Example VIII.1. Let us verify the resolution IV claim in Example V.1 using model $M3_c$, which we can do since the parametrization matrices of the example have the zero sum property. Consider the following two linear combinations of the observation:

linear combination	expectation
111 + 231 + 321 - 112 - 232 - 322	$3(\gamma_1^c - \gamma_2^c)$
121 + 211 + 331 - 123 - 213 - 333	$3(\gamma_1^c - \gamma_3^c)$

Thus we see that two independent γ^c-constraints are estimable. Because $\gamma_.^c = 0$ is also estimable, it follows that γ^c is estimable. Hence γ^f is estimable. By interchanging the j and k subscripts in the table, a similar analysis shows that β^f is estimable. By interchanging the i and k subscripts, one finds that α^f is estimable. Hence model $M3_f$ of Example V.1 is of resolution IV.

REFERENCES

Anderson, D.A. and Thomas, A.M. (1979). Resolution IV fractional factorial designs for the general asymmetric factorial. *Commun. Stat. Ser. A 8,* 931-943.

Khatri, C.G. and Rao, C.R. (1968). Solutions to some functional equations and their applications to characterizations of probability distributions. *Sankhyā, Ser. A 30,* 167-180.

Margolin, B.H. (1969). Resolution IV fractional factorial design. *J. R. Stat. Soc. Ser. B 31,* 514-523.

Raktoe, B.L., Hedayat, A. and Federer, W.T. (1981). *Factorial Designs.* New York: Wiley.

Seely, J. and Birkes, D. (1980). Estimability in partitioned linear models. *Ann. Stat. 8,* 399-406.

Webb, S.R. (1968). Non-orthogonal designs of even resolution. *Technometrics 10,* 291-299.

Chapter 8

SENSITIVITY AND REVEALING POWER: TWO FUNDAMENTAL STATISTICAL CRITERIA OTHER THAN OPTIMALITY ARISING IN DISCRETE EXPERIMENTATION*

Jaya Srivastava
Colorado State University
Fort Collins, Colorado

I. INTRODUCTION: THE DEVELOPMENT OF DESIGN THEORY

It certainly gives the author a great pleasure to contribute to this Volume in honor of Professor Kempthorne. His book on experimental design, and the one by Fisher, are the two classics in this subject which inspired her in the late 1950's, and which she still holds in the highest regard. She particularly benefited from his balanced presentation of the subject. The author believes that the subject needs to have a balanced development in the coming decades. To discuss this issue, we first present a brief overview of the subfield of multifactor experiments.

The immense benefits inherent in discrete experimentation were pointed out by Fisher (1960, 1935). The possible existence of "nuisance factors," necessitating the development of confounding theory, was also recognized in those early stages. Several good (and relatively complicated) designs for factors with 2 or 3 levels were ob-

*This work was supported by AFOSR Grant No. 82-0156.

tained by Fisher and Yates (e.g., Yates, 1937). The foundations of a general theory of confounded s^n factorial designs was developed in Bose and Kishen (1940) where s is a prime power (i.e., a prime number or a power of a prime number) through the use of the finite field GF(s) and related finite geometries. The special case s = 2 was independently approached through the use of group theory by Fisher (1942).

As empirical experience accumulated, it was observed that in most experiments, the majority of factorial effects turn out to be negligible. Since the number of treatments to be tried increases rapidly with the number of factors (even if only a single replication is to be used) the important concept of fractional replication (i.e., trying only a subset of the treatments) was put forward (Finney, 1945). It was observed that in some situations almost all 2-factors and higher effects are negligible, leading to the important paper by Plackett and Burman (1946). Rao (1946) considered situations where the information matrix would be diagonal, discovering orthogonal arrays. Bose (1947) made an advancement in the geometric aspects of the confounding theory of symmetrical factorial designs, which later on was connected (Bose, 1961) with coding theory, and which became an important part of combinatorial mathematics on its own. A generalization of the above theory to the asymmetrical case was made in Kishen and Srivastava (1960) who gave general mathematical techniques (as opposed to most of the earlier ones which were based on trial and error) which provide suitable designs in a large number of practical cases. It may be remarked here that a completely general theory (covering all possible situations likely to arise in practice) is not available even now, and even for the symmetrical case. Designs for eliminating heterogeneity in two or more directions are also developed. These included the Latin Squares, and later on Shrikhande Squares, Youden Squares and Rectangles, and finally, the "Generalized Youden Designs" (GYD's) introduced by Kiefer, which we shall refer to as "Kiefer Rectangles." Such row-column, or higher dimensional designs, include also confounded factorial designs. Of course, "large" designs of this type suffer from the drawback that row-column interactions may be present, causing an unknown amount of bias in the results. The lattice square, to some extent, is helpful from this viewpoint, since v treatments (with $v = t^2$) are accommodated in (t×t) [rather than $(t^2 \times t^2)$] squares. A generalization of this idea is the Nested Multidimensional-Block Design, introduced by Srivastava (1981), where a square (or a multidimensional rectangle) can be as small as 2×2 (or 2×2×2, etc.) with v general. In these designs, in each block, heterogeneity can be removed in several directions. Since the number of (used) levels of nuisance factors can be controlled (and hence, kept desirably small) the chance of bias arising out of interactions between nuisance factors is minimal. With adequate usage of recovery of "interblock" information, these designs have the potential of being more useful than the classical incomplete block designs.

In factorial designs, work was pursued in the 1950's and early 1960's by Bose, Bush, Conner & Young, Zelen, Patel, Addelman, etc. This was largely related to orthogonal arrays, both of the general type, and of the "parallel-flats" type (which consists of treatments corresponding to points lying on a set of parallel hyper-planes inside a finite Euclidean space). In 1961 the author (being inspired by the work of Kiefer, 1959) started out in the direction of developing optimal designs for (discrete) factorial experiments. Soon it was realized that this was an extremely difficult undertaking, since important and well-known unsolved mathematical problems were special cases of the general problem at hand. A special case was therefore considered. This consists of the situation where the variance-covariance matrix is "balanced," i.e., it is invariant under a permutation of factor symbols. In order to be able to handle the variance-covariance matrix in the general case and develop a mathematical theory for optimal designs, it was necessary to consider two aspects of the problem. These are the analytical and the combinatorial aspects.

The former consists of obtaining the spectrum of the information matrix in terms of the combinatorial parameters of the set of treatments which correspond to balance in the information matrix. The latter consists of determining the combinatorial structure of the set of treatments in order that the variance-covariance matrix be balanced and the information matrix be "maximized" in some sense.

At this point, attention was restricted to a further special case, namely, designs of resolution d, for d = 3,4,5. (Recall that resolution $(2\ell+1)$ means that all $(\ell+1)$-factor and higher effects are assumed negligible, while the ℓ-factor and lower effects are to be estimated. In designs of resolution 2ℓ, $(\ell-1)$-factor and lower effects are estimated, assuming the $(\ell+1)$-and higher effects to be negligible.) These concepts are found in Box and Hunter (1961), and in the earlier work of Bose and others. For example, the famous "packing problem" of Bose, first encountered in Bose (1947), concerns the problem of developing confounded designs of the s^n type in blocks of size s^k so that no ℓ-factor or lower order effect is confounded. In Bose (1961), this was shown to be identical with the problem of obtaining linear group codes so that every set of ℓ, or a lesser number of errors, can be corrected with certainty. This is also similar to the problem of obtaining a design of resolution $(2\ell+1)$ so that the information matrix is diagonal. Bose translated this problem to the language of finite geometry: In PG(n,s) the finite projective geometry of n dimensions, based on GF(s), find a maximal set of points so that no ℓ of them lie in a hyperspace of dimension less than $(\ell-1)$. For $\ell \geq 3$, situations where a relatively large number of ℓ-factor (besides lower order) effects are nonnegligible, while "almost all" $(\ell+1)$-factor and higher effects are negligible, appear to arise <u>extremely</u> rarely in practice. However, the cases where $\ell \leq 2$ are quite common. A partial theoretical justification for this comes from response surface theory (which corresponds to factorial experimentation with continuous factors). Given any point in the factor space, most response functions arising in scientific investigations are "fairly smooth," at least within a certain neighborhood of the point, and, therefore (by Taylor's Theorem), can be approximated "sufficiently" closely by a first, or at least second, degree polynomial.

This is, indeed, observed in many fields of industry, for example, chemical engineering. The above explains the interest in the study of designs of resolution V and less. Since good resolution III designs were available through the work of Plackett and Burman, the work on orthogonal arrays of strength two, BIB designs, Hadamard matrices, and papers such as Addelman and Kempthorne (1961), the author felt interested in resolution IV and V more, and particularly resolution V because of the much greater mathematical challenge that it involves. Firstly, the concept of balance was linked (Srivastava, 1961) to certain combinatorial arrays, which are a class of (0,1) matrices, and were called "balanced arrays." [Though independently discovered by the author, these were first noted by Chakravarti (1956). We may add, however, that only a few paragraphs of Chakravarti's paper, containing basically the definitions, were devoted to the above topic.] Balanced arrays and the corresponding optimal designs (for 2^m factorials) were studied by the author and her associates Anderson and Chopra in a series of articles, a review of which is made in Srivastava (1976). This theory consists of combinatorial optimization guided by the demands from the analytical angle. The analytical part, on the other hand, consists mainly of dealing with the information through a non-commutative linear associative algebra which arises from the multidimensional, partially balanced association scheme defined by the author (1961). This, in turn, was a generalization of the ordinary partially balanced association scheme defined by Bose and Shimamoto (1954), which was implicitly contained in the fundamental paper on partially balanced designs due to Bose and Nair (1939). Further work on optimal designs was carried on by the Japanese school, consisting of Yamamoto and his associates (e.g., Yamamoto et al., 1975) such as Nishii, Shirakura, Kuwada, etc. Their work led to optimal designs of resolution VII. The analytical part, which was formidable, was facilitated by the use of ideal theory, while the combinatorial part was taken care of basically by the (present) author's earlier results.

All through the 1960's the author felt uneasy over the basic assumption behind the work on (fractional) factorial designs, namely, that the 3-factor and higher effects are negligible. In many areas where such designs are commonly used, the assumption is not very far from being correct, but certainly it is never exactly true either. In other words, the possibility of an unknown amount of bias in the estimates always exists. Even in the continuous case, it is apparent that, in general, the approximation of a response function by a second degree polynomial would, at best, hold only in suitably small regions of the factor space. The idea that nature usually comes in such a way that even over relatively large parts of the factor space, the above assumption about all the higher order effects being negligible holds, is unfounded and incorrect. In view of the above, the "optimality" of the optimal balanced designs produced in the above work is misleading, since it ignores the almost certain presence of unknown amounts of biases.

It slowly became clear to the author that the above shortcomings were not a monopoly just of the fractional factorial designs. For example, the Latin Square design is no doubt variance-optimal. However, unknown row-column interactions may be present and the (usual) model under which the above optimality is demonstrated may not be valid. An illustration is provided by the data (in parentheses) in Figure 8.1 for the 5×5 Latin Square experiment with five treatments A,...,E.

An appropriate analysis will reveal an interaction between row 3 and column 4, and that E has a higher expected yield than A.

B(221)	D(241)	E(162)	C(232)	A(138)
C(148)	A(124)	B(81)	D(127)	E(64)
D(188)	E(181)	C(152)	A(238)	B(122)
A(141)	B(168)	D(121)	E(141)	C(121)
E(129)	C(167)	A(72)	B(132)	D(90)

Figure 8.1

The presence of "large" parameters, which influence the expected value of observations, and which are not included in the model, is, therefore, an important hazard to the outcome of an experiment. It is clear, for example, that such a hazard is present whenever we assert that our observations obey a given (linear) model, but are unable to defend our assertion to a sufficiently high degree. On the other hand, we recall that all of optimality theory is variance-based, or at the least, based on the "mean square error" (MSE). Note that the emphasis here is on the fact that this MSE is computed under the assumption that the model asserted by us is correct. To the extent we cannot defend well enough the model asserted by us, to that extent our "optimality" concept may be inappropriate as an indicator of "goodness" of a design.

A moment's reflection, however, makes it clear that all work will have to be done under some model. In other words, we cannot altogether get out of asserting a model. However, this fact need not paralyze the further development of the subject. In any given scientific situation, we do not have to continue with the "narrow" model that we have been commonly asserting in that situation; we can assert a broader model that is free of the important shortcoming of the narrow model. Thus, instead of working blindly with the main-effect plans, claiming that all interactions are negligible, we can assert a more suitable model.

Suppose, from the previous experience in the field, we are almost certain that at most 5 interactions will be non-negligible in a particular situation; our broader model will simply be the model which will assert this fact. In other words, our broader model will assert that the possible non-negligible parameters include all the main effects, plus at most 5 interactions. Notice that this amounts to having a "family" of models. Similarly, consider a subject matter field where row-column designs (with two given cross-classified nuisance factors) have been previously used. Suppose past experience tells us that at most, 10% of the row-column pairs interact. Then, instead of asserting the narrower model, which altogether ignores these interactions, we can assert the proper model, which would include the fact that at most 10% of the cells may have interaction effects associated with them. From the above examples, it is also obvious that the "broad" model (i.e., the family of models) that we finally decided to assert should not be "unnecessarily broad" either. If we are quite certain that a particular group of parameters is going to be negligible, it would only be proper to assert the same. Thus, all the above arguments boil down to asking for the "narrowest broad model" (called "supermodel"), which we can claim to be almost certainly true.

We continue this discussion in the next section in a more general setting.

II. STAGES OF EXPERIMENTATION AND DESIGN OBJECTIVES

Although part of our discussion relates to more general situations, we shall confine our attention to multifactor, multiresponse experiments. (For a relatively elaborate discussion of these terms, see, for example, Roy, Gnanadesikan, and Srivastava, 1970.) We shall assume that some phenomenon is to be studied. The study is to be done by picking (randomly and/or otherwise) some "experimental material" and dividing it into "experimental units." (Presumably the units are "similar," though they do not have to be.) On each unit, one or more of the "responses" are measurable. Also, there are one or more "factors:" these are variables whose value or "level" we can vary ("discretely" or "continuously," "precisely" or to "within certain limits"). The object is to study the "effect" of the factors on the responses, and also the interrelationship among the various responses. An experiment is said to be "uniresponse" or "multiresponse" according as the number of responses is one, or more than one. Problems that arise in the uniresponse case have a (usually much more complex) counterpart in the multiresponse case. However, there are problems arising in the latter case which have no analog in the former.

It is assumed that the value of an observation on any response on a given experimental unit depends upon the "state" of the unit (at the time when the observation is taken), the "state" being characterized by the levels of a (usually) large number of factors. Some of the important (controllable) factors are presumably included in the experiment for study, others are kept controlled at certain fixed levels for all units. Among the uncontrolled factors, the experimenter hopefully would identify some of the major ones and would be able to record their level for every unit. Besides all these, there may be other major uncontrollable factors unknown to the experimenter. Furthermore, there are (almost always) a rather large number of minor controllable and uncontrollable factors, which taken individually may influence the response to a minute extent, but taken together may influence it considerably. Collectively, these are given the name "chance fluctuations," "random variation," etc.

By "a model" of the phenomenon, we shall mean any postulate which (i) specifies the functional relationship between each observed response on the one hand, and the levels of the various factors, and the random fluctuation on the other; (ii) partly or wholly specifies the "nature" of the random component, and (iii) partly or wholly specifies any relationship between the responses. However, all such specifications usually involve a number of mathematical constants called "parameters." For simplicity, we shall assume that (i) the value of any given parameter is either known, unknown, or partially known, and (ii) the parameters are "independent," i.e., the "parametric space" (or the set of all values of the parameters) is a (possibly multidimensional) rectangle.

By a "supermodel," we mean a family, or collection of models, such that the set of factors and responses is the same for each member of the family. Whatever is common to all models in the family will be called the "Common Part" (CP). Each model in the family would then consist of the CP and an "Individual Part" (IP). Finally, we may also consider a "hypermodel." This will mean a collection of supermodels, such that no two supermodels refer to the same set of factors and responses. Thus, different models inside a supermodel have the same frame of reference, while different supermodels inside a hypermodel have different frames of reference.

We shall carry out the discussion in terms of "models," statements, etc., which are "reasonably valid," a phrase which we proceed to elaborate.

A model will be said to be "reasonably valid" when the probability that the model is correct is so high that for all practical purposes, we can take the model to be correct. Notice that whether or not a probability will be considered high enough will itself depend upon the circumstances. For example, a model for predicting whether or not it will rain on a certain day in a certain area may be quite acceptable under normal circumstances if the probability that the model is true is at least 90%. On the other hand, the same model may not be acceptable if we are planning to hold a large open air concert, since it would then be desirable to be able to predict with more surety.

Similarly, a procedure will be called "reasonably correct" (or, a statement "reasonably true") if the probability that the procedure is correct (or, the statement is true) is so high that for all practical purposes, we can consider the procedure to be correct (or, the statement to be true).

We now consider the stages through which experimentation proceeds. A subject matter field may be said to be in "Stage Zero" with respect to a phenomenon, when the phenomenon is not yet "sufficiently identified," or its "relevance" is not yet "sufficiently appreciated" so that as a consequence, the desire to study the phenomenon has not yet developed and hence, not much attention has been focused on it. Next comes "Stage I." This is when the phenomenon is identified and is judged to be worthy of study. Theoretical guesses, educated or otherwise, based upon past experience, or perhaps intuition alone, are made concerning the "structure" of the phenomenon. Many possible approaches are considered, each with its own frame of reference. In due course, a reasonably valid hypermodel is developed.

Next, in Stage II different supermodels inside the hypermodel are examined more carefully. Out of the various possible approaches to the study of the phenomenon, one approach is selected. Finally, conditioned on this approach being itself "legitimate," experimentation (under this approach) is envisaged and a reasonably valid supermodel is decided upon. Next, experimentation begins and Stage III starts. The experiments, firstly, shed light on the legitimacy of the approach. If the approach is considered unfruitful, it is abandoned and a different approach (with its own supermodel) adopted. Otherwise, experiments are done to thrash out a reasonably valid model (out of the supermodel under consideration); when this is achieved, we are in Stage IV. More experiments are performed until the parameters (of interest) under the model are determined to a sufficiently high degree of accuracy. It may be remarked that in Stages III and IV, depending upon the complexity of the phenomenon, sometimes only a single (suitably large) experiment may suffice. On the other hand, in some cases, a sequence, or chain of experiments, may be necessary. In case the factors are discrete (or are continuous, but interest lies in only a discrete set of levels of each factor) experimentation stops here. In other cases, experimentation would usually proceed through many more stages until all important aspects of the true model have been successfully studied.

We now consider criteria for good statistical design of the experiments in Stages III and IV. Although most comments would be valid (if necessary, after minor modifications or generalization) in the continuous case, we shall, for simplicity, assume that all factors that we are dealing with are discrete. Firstly, consider Stage IV. We have a reasonably valid model. There are some controllable factors of interest, and also perhaps uncontrollable factors, some of interest (often called "concomitant variables") and some not of interest (the "nuisance factors"). This is an example of the classical situation where we seek an optimal design, optimal in the sense of minimizing some variance-based (or rather, mean square error-based) criterion. When there are many variables, this amounts to minimizing some suitable functional of the mean-square-error matrix. What we are emphasizing here is that the context in which optimal design theory arises in the situation is Stage IV.

Now, consider the situation in Stage III. We are dealing with a (reasonably valid) supermodel. We need to figure out the model (within the supermodel under consideration) which would be reasonably valid. Each model has, of course, the CP, as well as its own IP. Let IPI denote that (sub) part of the IP which is of interest to the experimenter, (i.e., involves factors which we wish to study) and let $IP\bar{I}$ denote the remaining (sub) part of the IP which is <u>not</u> of interest (i.e., involves only "nuisance factors"). Similarly, let CPI and $CP\bar{I}$, respectively, denote the subparts of the CP which are, and are not, of interest. Then, we would be in one of the following six substages of Stage III, denoted by (3.0)-(3.6) in the table below.

		$IP\bar{I}$ empty		$IP\bar{I}$ nonempty	
		IPI		IPI	
		Empty	Nonempty	Empty	Nonempty
CPI	Empty	3.0	3.2	3.0	3.5
	Nonempty	3.1	3.3	3.4	3.6

The table is self-explanatory, except for (3.0). Clearly, when both CPI and IPI are empty, there is nothing of interest in the supermodel and no experiment is needed. Thus, the two corresponding cells have the same number (3.0). Also, note that both the subcases where $CP\bar{I}$ is empty or nonempty go with each of the above 7 cases.

Consider now the design criteria for the stages (3.1)-(3.6), each of which correspond to the situation when there is something of interest. In stage (3.1), both the IPI and $IP\bar{I}$ are empty, for each model inside the supermodel. This in turn means that we are already in Stage IV, and in possession of a reasonably valid model. Any design which can help us determine the unknown parameters within this model will be called a "nonsingular design," and among these, one which in some sense does its task most efficiently, is an "optimal design." So stage (3.1) leads to optimal design theory.

In stage (3.2), $\mathrm{IP\bar{I}}$ is empty (for each model, within the supermodel) which means, for example, that there are no unknown nuisance factors (i.e., nuisance factors whose levels for any particular unit are unknown) to cope with. Also, CPI is empty, which means that in the common part (of <u>all</u> models within the supermodel) there is nothing of interest. What is nonempty is IPI, which refers to something of interest in the individual part of a reasonably valid model (within the supermodel). But, this reasonably valid model is itself not known and is to be "determined" or "searched." Any design which accomplishes this is called a "search design." A design which accomplishes this task with maximum efficiency (in some sense) will be called a "most revealing search design," or briefly, a "most revealing design," or a "design with maximum revealing power." Thus, if we are in stage (3.2), we shall seek designs of this last type.

In stage (3.3), both CPI and IPI are nonempty, though $\mathrm{IP\bar{I}}$ is still empty. This means, firstly, that we need a design which is both a nonsingular design and a search design. We can go two ways. Firstly, among the class of all search designs, we look for an optimal one, where for "optimality," we use the same criteria as in stage (3.1); such a design is called an "optimal search design." The other way is to first consider all nonsingular designs (as under stage (3.1)) and then to find among these the one which is most revealing. A design of this kind will be called a "most revealing nonsingular design."

Next, we come to the three situations in which the $\mathrm{IP\bar{I}}$ are nonempty, so that we are confronted with unknown nuisance factors. First, assume that we consider these factors (or parameters) to be "major" in their effect on the response, ignoring which would make the "signal/noise ratio" to be too small to be heard. Any design which can cope with these unknown trouble makers will be called a "sensitive design." More precisely, a design will be called "sensitive" if it allows us to determine the CPI, even though (some of) the $\mathrm{IP\bar{I}}$ are nonempty. One design may be termed more sensitive than another, if it allows the determination of the CPI in a better way in some sense. For example, in situations where we can not eliminate "the effect" or "the influence" of the IP$\bar{I}$ with certainty, a design may be called more sensitive if it minimizes some suitable probabilistic function related to the effect of the IP$\bar{I}$. It is clear that when IP$\bar{I}$ are nonempty, the first requirement from a design would have to be that it be "sensitive." Among sensitive designs, one would then look for nonsingular designs, search designs, etc. Thus, in an obvious terminology, we should be seeking "optimal sensitive design" in stage (3.4) and "most revealing sensitive designs" in stage (3.5). In stage (3.6), there are two ways open, as in stage (3.3). The first one is to look for an "optimal search design which is sensitive," briefly called "optimum revealing sensitive design." The other one is to seek a "most revealing nonsingular sensitive design."

In the last paragraph, we are requiring that when IP$\bar{I}$ are nonempty, we should look for "qualities" (or "attributes") like "nonsingular" and/or "revealing power" only in the class of "sensitive" designs. However, this does not mean that cases where the three qualities are to be sought in a different order would never arise and ought to be totally discouraged. The author's position is far from that. In case the role of the IP$\bar{I}$ is assessed to be relatively "minor," one could, for example, look for a "most sensitive optimal design" in stage (3.4). This will be particularly desirable when (as happens quite often) a large number of designs are simultaneously optimal (ignoring the IP$\bar{I}$), in which case we would try to select one which is expected to be most sensitive. Similarly in stage (3.5), a "most sensitive search design" would occasionally be quite suitable. Finally, for stage (3.6), there will obviously be six permutations of the order in which the three qualities are sought, where each permutation can be named in an obvious manner. Two of these are given earlier. Similarly, a "most revealing sensitive nonsingular design" is wanted when it is most important to be able to throw light on the CPI, coping with the IP$\bar{I}$ is considered next in importance, and furthermore, it is expected that the "unknown aspects of the signal" (i.e., the IPI) are quite strong and will be brought to light relatively easily by most designs. There are three other cases, namely, "most sensitive revealing nonsingular design," "optimal sensitive search design," and "most sensitive nonsingular search design," with obvious contexts in which they would arise.

To be precise, from the above discussion it is clear that the relative order in which the attributes of nonsingularity, revealing power, and sensitivity are sought in any given situation will depend upon the relative importance which is given (rightly or wrongly) to the CPI, IPI, and $\mathrm{IP\overline{I}}$ respectively. In case of doubts regarding $\mathrm{IP\overline{I}}$, a conservative approach would be to seek sensitivity first, since otherwise the signals in CPI and IPI may be drowned out.

It should be pointed out that there are other "intermediate" ways of seeking extrema with respect to combinations of the above attributes. For example, one may obtain a "most revealing optimal design" for stage (3.3), which means that one would first look for the class of all optimum designs (ignoring the IPI) and among these obtain the one whose revealing power is a maximum. Notice that the revealing power of such a design is less than, or equal to, the revealing power of a "most revealing nonsingular design" since the class of optimal designs is a subclass of the class of nonsingular designs. Also, notice that a "most revealing optimal design" need not also be an "optimal most revealing design." Similarly, we could have an "optimal most revealing, most sensitive design" for stage (3.6), and such a design may not necessarily be identical with one in which the adjectives occur in a different order (such as, for example, a "most sensitive, most revealing, optimal design") though one would hope so (at least "approximately").

We have not talked much about the $\mathrm{CP\overline{I}}$. If $\mathrm{CP\overline{I}}$ is empty, we are lucky. If not, it is understood that in each of the cases of stages (3.1)-(3.6), the designs that we are seeking will have to (first and foremost) satisfy the condition that they can cope with the $\mathrm{CP\overline{I}}$. Though important, this problem should generally not cause severe hurdles, in case "enough" experimental material is available (in an "economic" manner) for each level-combination of the known nuisance factors. The helpful feature of $CP\bar{I}$ (as opposed to the $IP\bar{I}$) is that for every experimental unit, the level of each nuisance factor included herein is known. This allows the grouping of the experimental material into relatively homogeneous subsets. It should also be pointed out that the factors in $CP\bar{I}$, though they and their levels be known, may "interact" in unknown ways, causing $IP\bar{I}$ to appear in the supermodel.

Before ending this section, it must be emphasized that the above is a rather oversimplified picture of the complexities through which the study of a phenomenon proceeds in most cases. However, though simplistic, it does introduce the concept of "sensitivity" and puts it in perspective with that of "search" (Srivastava, 1975) and "optimality." The last one among these is undoubtedly very important, since eventually one must pass through stage IV. Before "optimality" became fashionable, mere "nonsingularity" was sought, and it still is by some. On the other hand, now it sometimes appears as if "optimality" is regarded as the primary aim, the end-all of statistical experimental design. This implicitly presumes a far more simplistic picture than the one offered here. Indeed, our discussion establishes that in order to be realistic, "revealing power" and "sensitivity" also need to be reckoned with, particularly in the more exploratory stages of scientific experimental research.

Some mathematical theory already exists for dealing with "search," with or without "nonsingularity." "Sensitivity," with or without the other attributes, will be dealt with elsewhere. We end the chapter with a few examples in the next section to help clarify the above discussion.

III. EXAMPLES

It would make this paper too long if we presented examples for each type of design arising in each stage. Instead, we present three examples to illustrate the concepts of "sensitivity" and "revealing power," and the combination of these with optimality. It should be evident from these that anyone looking for challenging design problems, which combine realism with mathematical elegance, will find here a rich field to mine.

Example 1

We are planning a 2^7 experiment. We believe that quite a few main effects and two factor interactions may be nonnegligible and we have no idea as to which factors may be involved in these. Among the 99 higher-order factorial effects, past experience shows that it is almost certain that at most 6 effects are nonnegligible. Also, it is almost certain that in these 6 effects neither of the factors #6 and 7 will be involved (perhaps since they are of a "different type"). In this case, the CPI relates to the estimation of all the 29 effects involving 2 factors or less. Leaving factors #6 and 7 aside, we are left with 5 factors, among which there are $\binom{5}{3} + \binom{5}{4} + \binom{5}{5} =$ 16 interactions involving 3 or more factors. Out of these, any 6 effects could possibly be nonnegligible. Thus, there are $\binom{16}{6}$ possible models and their IPI's correspond to some set of 6 out of the above 16 effects being nonnegligible.

What is the best approach in this case? We notice that if we combine all the IPI's, we still have only 16 parameters, which is not much larger than the six parameters involved in any individual IPI. Moreover, search design theory tells us that we need at least $29 + 2 \times 6 = 41$ observations, a lower bound, which may not always be attained. Also, some degrees of freedom for error are needed. The appropriate way in this situation is to combine all the IPI's and the CPI into one model (involving $29 + 16 = 45$ parameters), which, of course, is reasonably valid. Under this model, we should obtain an optimal design for estimating the 45 parameters and use it to conduct this experiment. This illustrates a situation where sensitivity does not arise and the search aspect can be ignored.

Now consider a variant of the above in which it is known that there are two factors which are not involved in the higher order effects, but we do not know which. Notice that this is equivalent to saying that no 6-factor or 7-factor effect is significant and at most, 6 out of the remaining $\binom{7}{3} + \binom{7}{4} + \binom{7}{5} = 91$ are so. In this case, combining all the IPI's would require estimation of at least 91 extra parameters when we know that at most 6 of them may be non-negligible. In most situations, it would be unwise to do such a large experiment. Thus, both revealing power and optimality are called for. Notice also that in this case, our supermodel has $\binom{91}{6}$ models in it.

Suppose no particular information is available regarding the relative magnitude of the 29 + 6(=35) possible nonzero effects. Then we can go by the relative "sizes" of the CPI and IPI, which respectively involve 29 and 6 parameters. Now the CPI is much "bigger" than any IPI. We shall choose a number N (say, N = 50) considering the number of degrees of freedom for error that we may need, etc. Take the class of all optimal (or, say, optimal balanced) designs with N assemblies for estimating the 29 parameters of the CPI. Next in this class, find the most revealing (search) design for determining the (at most) 6 nonnegligible higher order effects. In other words, we should go for a "most revealing optimal (or optimal balanced) design."

It is important to caution the reader here concerning one matter. It is well known that "optimality" is not a single criterion; it is a class of criteria (though of the <u>same type</u>). For example, we have global-optimality, universal optimality, A-optimality, D-optimality, etc. Similarly, "revealing power" should also be looked upon as a class of attributes with its own variation and intricacies. The same holds for "sensitivity."

Example 2

Consider again a 5×5 Latin Square experiment. Clearly, there are two nuisance factors and one factor of interest. Under the assumption of additivity, we are in Stage IV, having only the CPI and $\overline{\text{CPI}}$.

For this situation, we already know that if we are willing to take 25 observations, one in each cell of the square, then the Latin Square design is optimal. What if we cannot afford more than 9 observations? It is easily seen that the class of nonsingular designs is nonempty, and hence, so is the class of optimal designs.

Now suppose it is known that levels 4 and 5 of the first nuisance factor do not necessarily obey the assumption of additivity with respect to the same two levels of the second nuisance factor. It will be seen that we can avoid these levels and obtain an optimal design with 5 treatments as before. This shows that even though $IP\bar{I}$ are nonempty, we can sometimes "avoid" them. On the other hand, suppose we want some d.f. for error. We can still avoid the 4 "odd" cells (involving levels 4 and 5 of each nuisance factor) and work with the remaining 21 cells, and obtain an optimal design with 8 d.f. for error. Notice that in the example in this paragraph, the $IP\bar{I}$ (though nonempty) are "small" and "localized," enabling us to "bypass" them.

Consider now the case where we know that some two levels of each of the two nuisance factors interact, causing 4 cells to be non-additive, but we do not know which pairs of levels these are. In other words, there is a 2×2 subsquare where interaction may be present in each cell, but we do not know which subsquare it is. Now the $IP\bar{I}$ are both non-empty and non-localized. We are in stage (3.4). Assume for a moment that all interactions are negligible (as if we are in stage (3.1)) and consider the class C_0 of nonsingular designs. The question arises whether, under stage (3.4), every design in C_0 is sensitive. Examples can be given to show that the answer to this question is in the negative. However, every Latin Square is sensitive since all treatment contrasts remain estimable if a 2×2 subsequence of the 5×5 square is missing. (This shows the connection between sensitive design problems of this type and the problem of robustness of designs when some observations are missing.) On the other hand, does <u>every</u> Latin Square constitute an optimal sensitive design? The answer is conjectured to be "No."

Obviously, the above example is rather simple. Problems of sensitivity, with varied complexity, arise only too often.

Example 3

Consider again a 2^7 experiment. Suppose we are told that on the basis of past experience, a reasonably valid supermodel (for some particular experimental situation) would have the following two features: (a) <u>At most</u> 10 out of the 128 factorial effects are nonnegligible and these do not include (i) any 5-factor or higher effects, and (ii) more than 5 main effects. (b) The set of nonnegligible effects has a "tree structure." (This structure actually occurs quite commonly in behavioral sciences, as noted by Srivastava (1983), and is defined as follows: A set of factorial effects is said to have a tree structure, if for every $k \geq 1$, whenever a particular $(k+1)$-factor interaction belongs to the set then there is a subset of k factors (out of these (k+1) factors) such that the corresponding k-factor interaction is also in the set.)

Suppose further, for simplicity, that the experimental units are homogeneous and we have no $IP\bar{I}$, or even $CP\bar{I}$ to worry about. However, the units are quite expensive and we do not have funds for more than 45 of them. Clearly, we need a design which has maximum revealing power in some sense, e.g., which maximizes the probability that we can accurately identify at least 7 out of 10 nonnegligible parameters.

Next, suppose units are not homogeneous and we have to lay them out in a 5×9 rectangle (the rows and columns representing two nuisance factors) knowing furthermore that some of the cells in some 5×2 subrectangle are quite non-additive, but we are not given the set of the 10 possible interactive cells. Clearly, here we need to develop sensitive designs with maximum revealing power.

Examples where all the three attributes of sensitivity, revealing power, and optimality come into play, would be generated if, for example, we were given that all the 7 main effects do need to be estimated.

REFERENCES

Addelman, S. and Kempthorne, O. (1961). Some main effect plans and orthogonal arrays of strength two. *Ann. Math. Stat. 32,* 1167-1176.

Bose, R.C. (1947). Mathematical theory of the symmetrical factorial design. *Sankhyā 8,* 107-166.

Bose, R.C. (1961). On some connections between the design of experiments and information theory. *Bull. Inst. Intern. Statist 38,* 257-271.

Bose, R.C. and Kishen, K. (1940). On the problem of confounding in the general symmetrical factorial designs. *Sankhyā 5,* 21-36.

Bose, R.C. and Nair, K.R. (1939). Partially balanced incomplete block designs. *Sankhyā 4,* 337-372.

Bose, R.C. and Shimamoto, T. (1953). Classification and analysis of partially balanced designs with two associate classes. *J. Am. Stat. Assoc. 24,* 151-184.

Box, G.E.P. and Hunter, J.S. (1961). The 2^{k-p} fractional factorial designs, Parts I and II. *Technometrics 3,* 311-351, 449-458.

Chakravarti, I.M. (1956). Fractional replication in asymmetrical factorial designs and partially balanced arrays. *Sankhyā 17,* 143-164.

Finney, D.J. (1945). The fractional replication of factorial experiments. *Ann. Eugenics 12,* 291-301.

Fisher, R.A. (1942). The theory of confounding in factorial experiments in relation to the theory of groups. *Ann. Eugenics 11,* 341-353.

Fisher, R.A. (1960). *The Design of Experiments* (7th ed.). New York: Hafner.

Kempthorne, O. (1952). *The Design and Analysis of Experiments.* New York: John Wiley.

Kiefer, J.C. (1959). Optimum experimental designs. *J. Roy. Stat. Soc. Ser. B 21,* 272-319.

Kiefer, J.C. (1975). On the construction and optimality of generalized Youden designs. In: *A Survey of Statistical Design and Linear Models*. J.N. Srivastava (ed.). Amsterdam: North-Holland, 333-353.

Kishen, K. and Srivastava, J.N. (1960). Mathematical theory of confounding in symmetrical and asymmetrical factorial designs. *J. Indian Soc. Agr. Stat. 11,* 73-110.

Plackett, R.L. and Burman, J.P. (1946). The design of optimum multifactorial experiments. *Biometrika 33,* 305-325.

Rao, C.R. (1946). Difference sets and combinatorial arrangements derivable from finite geometries. *Proc. Natl. Inst. Sci. 12,* 123-135.

Roy, S.N., Gnanadesikan, R., and Srivastava, J.N. (1971). *Analysis and Design of Certain Quantitative Multiresponse Experiments*. New York: Pergamon Press.

Srivastava, J.N. (1961). Contributions to the construction and analysis of designs. University of North Carolina, Inst. Statist. Mim. Series No. 301.

Srivastava, J.N. (1975). Designs for searching nonnegligible effects. In: *A Survey of Statistical Designs and Linear Models*. J.N. Srivastava (ed.). Amsterdam: North Holland, 507-519.

Srivastava, J.N. (1977). A review of some recent work on discrete optimal factorial designs for statisticians and experimenters. In: *Developments in Statistics, Vol. 1*. Paruchuri R. Krishnaiah (ed.). New York: Academic Press, 267-329.

Srivastava, J.N. (1981). Some problems in experiments with nested nuisance factors. International Statistical Institute 43rd Session, Buenos Aires, Argentina.

Srivastava, J.N. (1983). On certain structures arising in the set of nonnegligible effects in many factorial experiments in the field of psychology and related sciences. To be published.

Yamamoto, S., Shirakura, T., and Kuwada, M. (1975). Balanced arrays of strength 2ℓ and balanced fractional 2^m factorial designs. *Ann. Inst. Math. 27,* 143-157.

Yates, F. (1937). *The Design and Analysis of Factorial Experiments.* Imperial Bureau of Soil Science, Technical Communication No. 35, Harpenden, England.

Chapter 9

ANALYSIS OF RANDOMIZED BLOCK DESIGN WITH INORDINATE RIGHT CENSORSHIP

N.R. Bohidar
Merck, Sharp, and Dohme Research Laboratories West Point, Pennsylvania

I. INTRODUCTION

The statistical analysis of restricted randomized designs, such as randomized block designs, Latin square designs, etc., becomes complex when one encounters missing observations. This complexity arises because the expected value of the random variables δ_{ij}^{k} (see Kempthorne, 1952), which assigns a value of 1 if the treatment k falls on the experimental unit j in the ith block and a value of zero otherwise, is no longer equal to t^{-1}, where t is the number of treatments per block. Under the missing value situation, the analysis generally involves the estimation of the missing values, the augmented analysis of variances, and the "exact" analysis of variance for correcting the upward bias associated with the treatment sum of squares. This procedure is extremely cumbersome and time consuming and is ineffective if there is no missing value formula available. The introduction of the multiple covariance analysis as an algorithm for the analysis of restricted randomized designs with multiple missing values (Coons, 1957) alleviated the computational burden, especially with the availability of the high-speed computer statistical softwares.

In our experience with the real world situations, however, we find that the multiple covariance algorithm is appropriate only for the loss of a moderate number of observations (say 10-15%). For the situation in which there is an inordinate amount of missing values (say, more than 15%), the multiple covariance algorithm does not lend itself to a unique solution. The matrix of the error sum of squares and sum of products suffers from the malady of multicollinearity and singularity.

Let us consider a real-world experiment. In the pharmaceutical industry, a long-term toxicity study is conducted to assess the safety of each drug to be marketed. This is required by the regulatory agency. Two species of rodents are used involving both sexes. Each experiment runs for approximately two years (120 weeks in rats and 96 weeks in mice). The treatments consist, in general, of two controls and three to four graded doses of the drug. The drug is administered on a daily basis. Most often, littermates are used in these experiments to achieve homogeneity among the experimental units since each member of the litter shares the same maternal environment with its littermates. Each treatment is assigned at random to a member of the litter. This random assignment of treatment levels to experimental units within each litter forms the basis of a randomized block design. Generally, 50-100 litters are used in a study. This provides 50-100 animals per treatment.

The responses measured during the long-term toxicity study are as follows: (1) Bodyweight (determined once a week), (2) Food consumption (determined once a week), (3) Terminal bodyweight only for animals alive at the terminal stage of the study, and (4) Terminal organ weight, such as liver, kidney, spleen, etc., only for animals alive at the terminal stage of the study. The right censorship (since all the animals are available and alive at the beginning of the study) is caused by the competing risk which is in this case the mortality caused during the course of the study. The analysis of organ weights (expressed as the % of bodyweight) is going to be complex because of the missing values caused by mortality. Approximately 45-55% of the animals are lost during the study due to mortality.*

This paper presents an alternate procedure for the situation in which there is an inordinate amount of right censorship. The multiple covariance algorithm, however, should be used wherever it is appropriate and meaningful (see e.g., Anderson, 1946; Nelder, 1954; Norton, 1955; Smith, 1957), involving only a modest amount of right censorship.

II. CONCEPTS AND DEFINITIONS

The procedure entails the construction of pre-specified contrasts and the development of statistical inference about the contrasts. Certain specific contrasts might have been suggested by the scientists as well as the regulatory agency.

A. Contrast

Consider that there are t treatments in the study, denoted by $T_1, T_2, \ldots, T_t$. Then a contrast consisting of K treatments where K = 2,3,...,t is expressed by:

$$\lambda_1 T_1 + \lambda_2 T_2 + \ldots + \lambda_K T_K$$

with

$$\lambda_1 + \lambda_2 + \ldots + \lambda_K = 0$$

*All of this information was presented here for the reader to appreciate the importance of the study, the cost considerations, and the need for providing any inferential information to the regulatory agency since no study of such dimension can be repeated without an outlay of considerable expenditure. The same situation also arises when one intends to study the stability of a pharmaceutical product over many years involving several batches of the dosage forms and finds that not all batches were tested in all the years considered in the study.

B. Block Segments

The definition of "block segments" and their structural properties are fundamental to the development of the analysis. It is required that the data be presented in a specific configuration.

Structural Property I (SP-1). Let $\beta(1)$ be the set of blocks in which only the treatment T_1 occurs and none of the other treatments considered in the contrast occur. Let N_1 and $\bar{Y}_1$ be the number of blocks in the segment and the mean of T_1 over these blocks, $[\beta(1)]$, respectively. Let $\beta(1,2)$ be the set of blocks in which both T_1 and T_2 occur and none of the other treatments considered in the contrast occur. Let N_{12}, $\bar{Y}_{\underline{1}2}$ and $\bar{Y}_{\underline{2}1}$ be respectively the number of blocks in the segment and the mean of T_1 and T_2 over the set of $\beta(1,2)$. Thus we have the following:

Occurrence of Treatments	Block Segment	No. of Blocks in the Segment	Treatment Means in the Segment
T_1	$\beta(1)$	N_1	$\bar{Y}_1$
T_2	$\beta(2)$	N_2	$\bar{Y}_2$
T_1 and T_2	$\beta(1,2)$	N_{12}	$\bar{Y}_{\underline{1}2}$ or $\bar{Y}_{\underline{2}1}$
T_2 and T_3	$\beta(2,3)$	N_{23}	$\bar{Y}_{\underline{2}3}$ or $\bar{Y}_{\underline{3}2}$
T_1, T_2, and T_3	$\beta(1,2,3)$	N_{123}	$\bar{Y}_{\underline{1}23}$, $\bar{Y}_{\underline{2}13}$, $\bar{Y}_{\underline{3}12}$

Note that $E(\bar{Y}_1) = \mu_1$, $E(\bar{Y}_{\underline{1}2}) = \mu_1$, $E(\bar{Y}_{\underline{1}23}) = \mu_1$, $E(\bar{Y}_2) = \mu_2$, $E(\bar{Y}_{\underline{2}1}) = \mu_2$, and so on.

Structural Property II (SP-2). If there are K treatments in the contrast then there are (2^K-1) block segments associated with the contrast.

For example, consider T_1 and T_2 in the contrast; then $(2^2-1) = 3$ and a schematic diagram of SP-2 is given below.

	T_1	T_2	
	x		
	x		
(1)	x		$\beta(1)$
	x		
	x		
	x		
	x	x	
(2)	x	x	$\beta(1,2)$
	x	x	
	x	x	
		x	
		x	
(3)		x	$\beta(2)$
		x	
		x	
		x	
		x	

Schematics for K = 3 and K = 4 are given in Fig. 9.1 and 9.2.

Structural Property III (SP-3). The number of block-segments associated with each treatment considered in the contrast equals 2^{K-1}.

C. Convex Combination of Segment Means

For ease of notation, let the segment means associated with T_1 be denoted by $\overline{T}_{11}, \overline{T}_{12}, \overline{T}_{13}, \ldots, \overline{T}_{1\theta}$ where $\theta = 2^{K-1}$ (see SP-3) and let $a_1, a_2, a_3, \ldots, a_\theta$ be the coefficients associated with the linear combination of the segment means of T_1. Similarly, let the segment means associated with T_2 be denoted by: $\overline{T}_{21}, \overline{T}_{22}, \overline{T}_{23}, \ldots, \overline{T}_{2\theta}$ and let $b_1, b_2, b_3, \ldots, b_\theta$ be the coefficients associated with the linear combination of the segment means of T_2, etc.

Our interest, here, is to construct a treatment contrast of the weighted linear combination of the block-segments associated with each treatment. The coefficients associated with the linear combination of the block-segments of treatments would be such that the

	T_1	T_2	T_3	
	x			
	x			
(1)	x			β(1)
	x			
	x			
		x		
		x		
(2)		x		β(2)
		x		
		x		
		x		
			x	
			x	
(3)			x	β(3)
			x	
			x	
	x	x		
(4)	x	x		β(1,2)
	x	x		
	x	x		
	x		x	
	x		x	
(5)	x		x	β(1,3)
	x		x	
	x		x	
		x	x	
		x	x	
(6)		x	x	β(2,3)
		x	x	
		x	x	
		x	x	
	x	x	x	
	x	x	x	
	x	x	x	
(7)	x	x	x	β(1,2,3)
	x	x	x	
	x	x	x	
	x	x	x	

Figure 9.1 Schematic Diagram of SP-2 for K = 3

	T_1	T_2	T_3	T_4	
(1)	x				β(1)
	x				
	x				
	x				
(2)		x			β(2)
		x			
		x			
(3)			x		β(3)
			x		
			x		
			x		
(4)				x	β(4)
				x	
				x	
(5)	x	x			β(1,2)
	x	x			
	x	x			
(6)	x		x		β(1,3)
	x		x		
	x		x		
	x		x		
(7)	x			x	β(1,4)
	x			x	
	x			x	
(8)		x	x		β(2,3)
		x	x		
		x	x		
(9)		x		x	β(2,4)
		x		x	
		x		x	
(10)			x	x	β(3,4)
			x	x	
			x	x	
(11)	x	x	x		β(1,2,3)
	x	x	x		
	x	x	x		
(12)	x	x		x	β(1,2,4)
	x	x		x	
	x	x		x	
(13)	x		x	x	β(1,3,4)
	x		x	x	
	x		x	x	
	x		x	x	
(14)		x	x	x	β(2,3,4)
		x	x	x	
		x	x	x	
(15)	x	x	x	x	β(1,2,3,4)
	x	x	x	x	
	x	x	x	x	

Figure 9.2 Schematic Diagram of SP-2 for K = 4

estimated contrast is a minimum variance unbiased estimator of the population contrast.

Symbolically, we have the contrast based on the convex combinations of the block segments associated with the treatments under consideration.

Let C denote a contrast. Then

$$C = \lambda_1 \left(\sum_{i=1}^{\theta} a_i \overline{T}_{1i}\right) + \lambda_2 \left(\sum_{i=1}^{\theta} b_i \overline{T}_{2i}\right)$$

$$+ \lambda_3 \left(\sum_{i=1}^{\theta} c_i \overline{T}_{3i}\right) + \dots + \lambda_K \left(\sum_{i=1}^{\theta} m_i \overline{T}_{Ki}\right)$$

where

$$a_\theta = 1 - \sum_{i=1}^{\theta-1} a_i$$

$$b_\theta = 1 - \sum_{i=1}^{\theta-1} b_i$$

$$\vdots$$

$$m_\theta = 1 - \sum_{i=1}^{\theta-1} m_i$$

$$\sum_{j=1}^{K} \lambda_j = 0$$

III. CONSTRUCTION OF CONTRASTS, DERIVATION OF VARIANCE, AND ESTIMATION OF CONVEX COEFFICIENTS

Because of practical considerations, it is helpful to demonstrate the developments in this section by considering special cases. The symmetry associated with the algebraic expressions readily lends itself to the extension to the general cases.

Table 9.1 and Table 9.2 provide the contrast expressions, C, and its variance, var(C), for K = 4 and K = 5, respectively.

The variance expressions in the tables can be compactly written in matrix notation as:

$$\text{var}(C) = \Lambda' \, V(C)\Lambda$$

where Λ' is the $1\times K$ row vector $(\lambda_1,\lambda_2,\ldots,\lambda_K)$ and $V(C)$ is the $(K\times K)$ variance covariance matrix of the linear combinations of block segment means.

Now we generate a set of simultaneous equations for the solutions of the convex coefficients for K treatments in the contrast, with $K(2^{K-1}-1) = K(\theta-1)$ equations as follows:

$$1: \quad \frac{\partial[\Lambda'V(C)\Lambda]}{\partial a_1} = 0$$

$$2: \quad \frac{\partial[\Lambda'V(C)\Lambda]}{\partial a_2} = 0$$

$$\cdots$$

$$(\theta-1): \quad \frac{\partial[\Lambda'V(C)\Lambda]}{\partial a_{(\theta-1)}} = 0$$

$$\cdots$$

$$(K-1)(\theta-1): \quad \frac{\partial[\Lambda'V(C)\Lambda]}{\partial m_1} = 0$$

$$\cdots$$

$$K(\theta-1): \quad \frac{\partial[\Lambda'V(C)\Lambda]}{\partial m_{(\theta-1)}} = 0$$

After we obtain the derivatives of the expressions given, we set them to zero, and then we rearrange and simplify the equation-expressions in the following fashion:

$$\begin{bmatrix} \\ \text{LEFT} \\ \text{HAND} \\ \text{SIDE} \\ \\ \end{bmatrix} \begin{bmatrix} a_1 \\ a_2 \\ \vdots \\ a_{\theta-1} \\ \vdots \\ m_1 \\ \vdots \\ m_{\theta-1} \end{bmatrix} = \begin{bmatrix} \\ \text{RIGHT} \\ \text{HAND} \\ \text{SIDE} \\ \\ \end{bmatrix}$$

TABLE 9.1

The Convex Combination of Block Segments (C) and Its Variance Var(C) for K = 4

$$C = \lambda_1[a_1\bar{T}_1 + a_2\bar{T}_{\underline{1}2} + a_3\bar{T}_{\underline{1}3} + a_4\bar{T}_{\underline{1}4} + a_5\bar{T}_{\underline{1}23} + a_6\bar{T}_{\underline{1}24} + a_7\bar{T}_{\underline{1}34} + (1 - \sum_{i=1}^{7} a_i)\bar{T}_{\underline{1}234}]$$

$$+ \lambda_2[b_1\bar{T}_2 + b_2\bar{T}_{1\underline{2}} + b_3\bar{T}_{\underline{2}3} + b_4\bar{T}_{\underline{2}4} + b_5\bar{T}_{1\underline{2}3} + b_6\bar{T}_{1\underline{2}4} + b_7\bar{T}_{\underline{2}34} + (1 - \sum_{i=1}^{7} b_i)\bar{T}_{1\underline{2}34}]$$

$$+ \lambda_3[c_1\bar{T}_3 + c_2\bar{T}_{1\underline{3}} + c_3\bar{T}_{2\underline{3}} + c_4\bar{T}_{\underline{3}4} + c_5\bar{T}_{12\underline{3}} + c_6\bar{T}_{1\underline{3}4} + c_7\bar{T}_{2\underline{3}4} + (1 - \sum_{i=1}^{7} c_i)\bar{T}_{12\underline{3}4}]$$

$$+ \lambda_4[d_1\bar{T}_4 + d_2\bar{T}_{1\underline{4}} + d_3\bar{T}_{2\underline{4}} + d_4\bar{T}_{3\underline{4}} + d_5\bar{T}_{12\underline{4}} + d_6\bar{T}_{13\underline{4}} + d_7\bar{T}_{23\underline{4}} + (1 - \sum_{i=1}^{7} d_i)\bar{T}_{123\underline{4}}]$$

$$\mathrm{Var}(C) = \lambda_1^2\sigma_1^2\left[\frac{a_1^2}{N_1} + \frac{a_2^2}{N_{12}} + \frac{a_3^2}{N_{13}} + \frac{a_4^2}{N_{14}} + \frac{a_5^2}{N_{123}} + \frac{a_6^2}{N_{124}} + \frac{a_7^2}{N_{134}} + \frac{(1 - \sum_{i=1}^{7} a_i)^2}{N_{1234}}\right]$$

$$+ \lambda_2^2\sigma_2^2\left[\frac{b_1^2}{N_2} + \frac{b_2^2}{N_{12}} + \frac{b_3^2}{N_{23}} + \frac{b_4^2}{N_{24}} + \frac{b_5^2}{N_{123}} + \frac{b_6^2}{N_{124}} + \frac{b_7^2}{N_{234}} + \frac{(1 - \sum_{i=1}^{7} b_i)^2}{N_{1234}}\right]$$

$$+ \lambda_3^2\sigma_3^2 \left[\frac{c_1^2}{N_3} + \frac{c_2^2}{N_{13}} + \frac{c_3^2}{N_{23}} + \frac{c_4^2}{N_{34}} + \frac{c_5^2}{N_{123}} + \frac{c_6^2}{N_{134}} + \frac{c_7^2}{N_{234}} + \frac{(1 - \sum_{i=1}^{7} c_i)^2}{N_{1234}} \right]$$

$$+ \lambda_4^2\sigma_4^2 \left[\frac{d_1^2}{N_4} + \frac{d_2^2}{N_{14}} + \frac{d_3^2}{N_{24}} + \frac{d_4^2}{N_{34}} + \frac{d_5^2}{N_{124}} + \frac{d_6^2}{N_{134}} + \frac{d_7^2}{N_{234}} + \frac{(1 - \sum_{i=1}^{7} d_i)^2}{N_{1234}} \right]$$

$$+ 2\lambda_1\lambda_2\sigma_{12} \left[\frac{a_2b_2}{N_{12}} + \frac{a_5b_5}{N_{123}} + \frac{a_6b_6}{N_{124}} + \frac{(1 - \Sigma a_i)(1 - \Sigma b_i)}{N_{1234}} \right] + 2\lambda_1\lambda_3\sigma_{13} \left[\frac{a_3c_2}{N_{13}} + \frac{a_5c_5}{N_{123}} + \frac{a_7c_6}{N_{134}} + \frac{(1 - \Sigma a_i)(1 - \Sigma c_i)}{N_{1234}} \right]$$

$$+ 2\lambda_1\lambda_4\sigma_{14} \left[\frac{a_4d_2}{N_{14}} + \frac{a_7d_6}{N_{124}} + \frac{a_7d_6}{N_{134}} + \frac{(1 - \Sigma a_i)(1 - \Sigma d_i)}{N_{1234}} \right] + 2\lambda_2\lambda_3\sigma_{23} \left[\frac{b_3c_3}{N_{23}} + \frac{b_5c_5}{N_{123}} + \frac{b_7c_7}{N_{234}} + \frac{(1 - \Sigma b_i)(1 - \Sigma c_i)}{N_{1234}} \right]$$

$$+ 2\lambda_2\lambda_4\sigma_{24} \left[\frac{b_4d_3}{N_{24}} + \frac{b_6d_5}{N_{124}} + \frac{b_7d_7}{N_{234}} + \frac{(1 - \Sigma b_i)(1 - \Sigma d_i)}{N_{1234}} \right] + 2\lambda_3\lambda_4\sigma_{34} \left[\frac{c_4d_4}{N_{34}} + \frac{c_6d_6}{N_{134}} + \frac{c_7d_7}{N_{234}} + \frac{(1 - \Sigma c_i)(1 - \Sigma d_i)}{N_{1234}} \right]$$

TABLE 9.2

The Convex Combination of Block Segments for K = 5 (in Matrix Notation)

$$C = \lambda_1[A'T_1^*] + \lambda_2[B'T_2^*] + \lambda_3[C'T_3^*] + \lambda_4[D'T_4^*] + \lambda_5[E'T_5^*]$$

Where

$$A' = [a_1\ a_2\ a_3\ a_4\ a_5\ a_6\ a_7\ a_8\ a_9\ a_{10}\ a_{11}\ a_{12}\ a_{13}\ a_{14}\ a_{15}\ a_{16}]$$

$$B' = [b_1\ b_2\ b_3\ b_4\ b_5\ b_6\ b_7\ b_8\ b_9\ b_{10}\ b_{11}\ b_{12}\ b_{13}\ b_{14}\ b_{15}\ b_{16}]$$

$$C' = [c_1\ c_2\ c_3\ c_4\ c_5\ c_6\ c_7\ c_8\ c_9\ c_{10}\ c_{11}\ c_{12}\ c_{13}\ c_{14}\ c_{15}\ c_{16}]$$

$$D' = [d_1\ d_2\ d_3\ d_4\ d_5\ d_6\ d_7\ d_8\ d_9\ d_{10}\ d_{11}\ d_{12}\ d_{13}\ d_{14}\ d_{15}\ d_{16}]$$

$$E' = [e_1\ e_2\ e_3\ e_4\ e_5\ e_6\ e_7\ e_8\ e_9\ e_{10}\ e_{11}\ e_{12}\ e_{13}\ e_{14}\ e_{15}\ e_{16}]$$

$$T'^*_1 = [\overline{T}_1\ \overline{T}_{\underline{1}2}\ \overline{T}_{\underline{1}3}\ \overline{T}_{\underline{1}4}\ \overline{T}_{\underline{1}5}\ \overline{T}_{\underline{1}23}\ \overline{T}_{\underline{1}24}\ \overline{T}_{\underline{1}25}\ \overline{T}_{\underline{1}34}\ \overline{T}_{\underline{1}35}\ \overline{T}_{\underline{1}45}\ \overline{T}_{\underline{1}234}\ \overline{T}_{\underline{1}245}\ \overline{T}_{\underline{1}235}\ \overline{T}_{\underline{1}345}\ \overline{T}_{\underline{1}2345}]$$

$$T'^*_2 = [\overline{T}_2\ \overline{T}_{1\underline{2}}\ \overline{T}_{\underline{2}3}\ \overline{T}_{\underline{2}4}\ \overline{T}_{\underline{2}5}\ \overline{T}_{1\underline{2}3}\ \overline{T}_{1\underline{2}4}\ \overline{T}_{1\underline{2}5}\ \overline{T}_{\underline{2}34}\ \overline{T}_{\underline{2}35}\ \overline{T}_{\underline{2}45}\ \overline{T}_{1\underline{2}34}\ \overline{T}_{1\underline{2}45}\ \overline{T}_{1\underline{2}35}\ \overline{T}_{\underline{2}345}\ \overline{T}_{1\underline{2}345}]$$

$$T'^*_3 = [\overline{T}_3\ \overline{T}_{1\underline{3}}\ \overline{T}_{2\underline{3}}\ \overline{T}_{\underline{3}4}\ \overline{T}_{\underline{3}5}\ \overline{T}_{12\underline{3}}\ \overline{T}_{1\underline{3}4}\ \overline{T}_{1\underline{3}5}\ \overline{T}_{2\underline{3}4}\ \overline{T}_{2\underline{3}5}\ \overline{T}_{\underline{3}45}\ \overline{T}_{12\underline{3}4}\ \overline{T}_{12\underline{3}5}\ \overline{T}_{1\underline{3}45}\ \overline{T}_{2\underline{3}45}\ \overline{T}_{12\underline{3}45}]$$

$$T'^*_4 = [\overline{T}_4\ \overline{T}_{1\underline{4}}\ \overline{T}_{2\underline{4}}\ \overline{T}_{3\underline{4}}\ \overline{T}_{\underline{4}5}\ \overline{T}_{12\underline{4}}\ \overline{T}_{13\underline{4}}\ \overline{T}_{1\underline{4}5}\ \overline{T}_{13\underline{4}}\ \overline{T}_{2\underline{4}5}\ \overline{T}_{3\underline{4}5}\ \overline{T}_{123\underline{4}}\ \overline{T}_{12\underline{4}5}\ \overline{T}_{13\underline{4}5}\ \overline{T}_{23\underline{4}5}\ \overline{T}_{123\underline{4}5}]$$

$$T'^*_5 = [\overline{T}_5\ \overline{T}_{1\underline{5}}\ \overline{T}_{2\underline{5}}\ \overline{T}_{3\underline{5}}\ \overline{T}_{4\underline{5}}\ \overline{T}_{12\underline{5}}\ \overline{T}_{13\underline{5}}\ \overline{T}_{14\underline{5}}\ \overline{T}_{23\underline{5}}\ \overline{T}_{24\underline{5}}\ \overline{T}_{34\underline{5}}\ \overline{T}_{124\underline{5}}\ \overline{T}_{123\underline{5}}\ \overline{T}_{134\underline{5}}\ \overline{T}_{234\underline{5}}\ \overline{T}_{1234\underline{5}}]$$

The left-hand side (LHS) matrix has a dimension of $[K(\theta-1) \times K(\theta-1)]$. The right-hand side (RHS) expression is a column vector with $[K(\theta-1)]$ components.

For example,

Value of K	Dimension of LHS Matrix	Dimension of RHS Matrix
2	2×2	2×1
3	9×9	9×1
4	28×28	28×1
5	75×75	75×1

The explicit expressions of the LHS matrix (28×28), and the RHS vector (28×1) have been presented in Tables 9.3 and 9.4, respectively. In Table 9.3 we define $N_1^* = N_1 + N_{1234}$, $N_{12}^* = N_{12} + N_{1234}$, etc. Also note: there are 16 submatrices each of dimension (7×7). Inside each submatrix there is a rectangular inset containing a sigma quantity (σ_1^2, σ_2^2, or σ_{12} ... etc.). Multiply each of the 49 elements of a submatrix by the sigma quantity in the rectangular inset inside that submatrix.

After simplification the simultaneous equations for the solution of the convex coefficients, such as $a_1, a_2, \ldots, m_1, m_2, \ldots$ can be written compactly in matrix notation as follows:

$$[\Lambda\Lambda' \otimes N^{**}]A = [RHS]$$

For K = 4, $[\Lambda\Lambda']$ is a (28×28) matrix and

$$\Lambda' = [\lambda_1, \lambda_1, \ldots, \lambda_1, \lambda_2, \ldots, \lambda_2, \lambda_3, \ldots, \lambda_3, \lambda_4, \ldots, \lambda_4]$$

with each contrast coefficient repeated seven times, consecutively. The symbol $\otimes$ represents the element-by-element product of two matrices of the same dimension. The structure of N^{**} (28×28) matrix has been presented in Table 9.3 for K = 4. A is a (28×1) vector whose components are the unknown convex coefficients. The [RHS] vector which is of (28×1) dimension has been presented in Table 9.4.

TABLE

Explicit Expression of LHS Matrix

	a_1	a_2	a_3	a_4	a_5	a_6	a_7	b_1	b_2	b_3	b_4	b_5	b_6	b_7
a_1	N_1^*	N_1	N_1	N_1	N_1	N_1	N_1	N_1	N_1	N_1	N_1	N_1	N_1	N_1
a_2	N_{12}	N_{12}^*	N_{12}	N_{12}	N_{12}	N_{12}	N_{12}	N_{12}	N_{12}^*	N_{12}	N_{12}	N_{12}	N_{12}	N_{12}
a_3	N_{13}	N_{13}	N_{13}^*	N_{13}	N_{13}	N_{13}	N_{13}	N_{13}	N_{13}	N_{13}	N_{13}	N_{13}	N_{13}	N_{13}
a_4	N_{14}	N_{14}	N_{14}	N_{14}^*	N_{14}	N_{14}	N_{14}	N_{14}	N_{14}	N_{14}	N_{14}	N_{14}	N_{14}	N_{14}
a_5	N_{123}	N_{123}	N_{123}	N_{123}	N_{123}^*	N_{123}	N_{123}	N_{123}	N_{123}	N_{123}	N_{123}	N_{123}^*	N_{123}	N_{123}
a_6	N_{124}	N_{124}	N_{124}	N_{124}	N_{124}	N_{124}^*	N_{124}	N_{124}	N_{124}	N_{124}	N_{124}	N_{124}	N_{124}^*	N_{124}
a_7	N_{134}	N_{134}	N_{134}	N_{134}	N_{134}	N_{134}	N_{134}^*	N_{134}	N_{134}	N_{134}	N_{134}	N_{134}	N_{134}	N_{134}
	σ_1^2							σ_{12}						
b_1	N_2	N_2	N_2	N_2	N_2	N_2	N_2	N_2^*	N_2	N_2	N_2	N_2	N_2	N_2
b_2	N_{12}	N_{12}^*	N_{12}	N_{12}	N_{12}	N_{12}	N_{12}	N_{12}	N_{12}^*	N_{12}	N_{12}	N_{12}	N_{12}	N_{12}
b_3	N_{23}	N_{23}	N_{23}	N_{23}	N_{23}	N_{23}	N_{23}	N_{23}	N_{23}	N_{23}^*	N_{23}	N_{23}	N_{23}	N_{23}
b_4	N_{24}	N_{24}	N_{24}	N_{24}	N_{24}	N_{24}	N_{24}	N_{24}	N_{24}	N_{24}	N_{24}^*	N_{24}	N_{24}	N_{24}
b_5	N_{123}	N_{123}	N_{123}	N_{123}	N_{123}^*	N_{123}	N_{123}	N_{123}	N_{123}	N_{123}	N_{123}	N_{123}^*	N_{123}	N_{123}
b_6	N_{124}	N_{124}	N_{124}	N_{124}	N_{124}	N_{124}^*	N_{124}	N_{124}	N_{124}	N_{124}	N_{124}	N_{124}	N_{124}^*	N_{124}
b_7	N_{234}	N_{234}	N_{234}	N_{234}	N_{234}	N_{234}	N_{234}	N_{234}	N_{234}	N_{234}	N_{234}	N_{234}	N_{234}	N_{234}^*
	σ_{21}							σ_2^2						
c_1	N_3	N_3	N_3	N_3	N_3	N_3	N_3	N_3	N_3	N_3	N_3	N_3	N_3	N_3
c_2	N_{13}	N_{13}	N_{13}^*	N_{13}	N_{13}	N_{13}	N_{13}	N_{13}	N_{13}	N_{13}	N_{13}	N_{13}	N_{13}	N_{13}
c_3	N_{23}	N_{23}	N_{23}	N_{23}	N_{23}	N_{23}	N_{23}	N_{23}	N_{23}	N_{23}^*	N_{23}	N_{23}	N_{23}	N_{23}
c_4	N_{34}	N_{34}	N_{34}	N_{34}	N_{34}	N_{34}	N_{34}	N_{34}	N_{34}	N_{34}	N_{34}	N_{34}	N_{34}	N_{34}
c_5	N_{123}	N_{123}	N_{123}	N_{123}	N_{123}^*	N_{123}	N_{123}	N_{123}	N_{123}	N_{123}	N_{123}	N_{123}^*	N_{123}	N_{123}
c_6	N_{134}	N_{134}	N_{134}	N_{134}	N_{134}	N_{134}	N_{134}^*	N_{134}	N_{134}	N_{134}	N_{134}	N_{134}	N_{134}	N_{134}
c_7	N_{234}	N_{234}	N_{234}	N_{234}	N_{234}	N_{234}	N_{234}	N_{234}	N_{234}	N_{234}	N_{234}	N_{234}	N_{234}	N_{234}^*
	σ_{31}							σ_{32}						
d_1	N_4	N_4	N_4	N_4	N_4	N_4	N_4	N_4	N_4	N_4	N_4	N_4	N_4	N_4
d_2	N_{14}	N_{14}	N_{14}	N_{14}^*	N_{14}	N_{14}	N_{14}	N_{14}	N_{14}	N_{14}	N_{14}	N_{14}	N_{14}	N_{14}
d_3	N_{24}	N_{24}	N_{24}	N_{24}	N_{24}	N_{24}	N_{24}	N_{24}	N_{24}	N_{24}	N_{24}^*	N_{24}	N_{24}	N_{24}
d_4	N_{34}	N_{34}	N_{34}	N_{34}	N_{34}	N_{34}	N_{34}	N_{34}	N_{34}	N_{34}	N_{34}	N_{34}	N_{34}	N_{34}
d_5	N_{124}	N_{124}	N_{124}	N_{124}	N_{124}	N_{124}^*	N_{124}	N_{124}	N_{124}	N_{124}	N_{124}	N_{124}	N_{124}^*	N_{124}
d_6	N_{134}	N_{134}	N_{134}	N_{134}	N_{134}	N_{134}	N_{134}^*	N_{134}	N_{134}	N_{134}	N_{134}	N_{134}	N_{134}	N_{134}
d_7	N_{234}	N_{234}	N_{234}	N_{234}	N_{234}	N_{234}	N_{234}	N_{234}	N_{234}	N_{234}	N_{234}	N_{234}	N_{234}	N_{234}^*
	σ_{41}							σ_{42}						

9.3

(N^{**}) of Dimension (28×28)

c_1	c_2	c_3	c_4	c_5	c_6	c_7	d_1	d_2	d_3	d_4	d_5	d_6	d_7
N_1	N_1	N_1	N_1	N_1	N_1	N_1	N_1	N_1	N_1	N_1	N_1	N_1	N_1
N_{12}	N_{12}	N_{12}	N_{12}	N_{12}	N_{12}	N_{12}	N_{12}	N_{12}	N_{12}	N_{12}	N_{12}	N_{12}	N_{12}
N_{13}	N_{13}	N^*_{13}	N_{13}	N_{13}	N_{13}	N_{13}	N_{13}	N_{13}	N_{13}	N_{13}	N_{13}	N_{13}	N_{13}
N_{14}	N_{14}	N_{14}	N_{14}	N_{14}	N_{14}	N_{14}	N_{14}	N^*_{14}	N_{14}	N_{14}	N_{14}	N_{14}	N_{14}
N_{123}	N_{123}	N_{123}	N_{123}	N^*_{123}	N_{123}	N_{123}	N_{123}	N_{123}	N_{123}	N_{123}	N_{123}	N_{123}	N_{123}
N_{124}	N_{124}	N_{124}	N_{124}	N_{124}	N_{124}	N_{124}	N_{124}	N_{124}	N_{124}	N_{124}	N^*_{124}	N_{124}	N_{124}
N_{134}	N_{134}	N_{134}	N_{134}	N_{134}	N^*_{134}	N_{134}	N_{134}	N_{134}	N_{134}	N_{134}	N_{134}	N^*_{134}	N_{134}
σ_{13}							σ_{14}						
N_2	N_2	N_2	N_2	N_2	N_2	N_2	N_2	N_2	N_2	N_2	N_2	N_2	N_2
N_{12}	N_{12}	N_{12}	N_{12}	N_{12}	N_{12}	N_{12}	N_{12}	N_{12}	N_{12}	N_{12}	N_{12}	N_{12}	N_{12}
N_{23}	N_{23}	N^*_{23}	N_{23}	N_{23}	N_{23}	N_{23}	N_{23}	N_{23}	N_{23}	N_{23}	N_{23}	N_{23}	N_{23}
N_{24}	N_{24}	N_{24}	N_{24}	N_{24}	N_{24}	N_{24}	N_{24}	N_{24}	N^*_{24}	N_{24}	N_{24}	N_{24}	N_{24}
N_{123}	N_{123}	N_{123}	N_{123}	N^*_{123}	N_{123}	N_{123}	N_{123}	N_{123}	N_{123}	N_{123}	N_{123}	N_{123}	N_{123}
N_{124}	N_{124}	N_{124}	N_{124}	N_{124}	N_{124}	N_{124}	N_{124}	N_{124}	N_{124}	N_{124}	N^*_{124}	N_{124}	N_{124}
N_{234}	N_{234}	N_{234}	N_{234}	N_{234}	N_{234}	N^*_{234}	N_{234}	N_{234}	N_{234}	N_{234}	N_{234}	N_{234}	N^*_{234}
σ_{23}							σ_{24}						
N^*_3	N_3	N_3	N_3	N_3	N_3	N_3	N_3	N_3	N_3	N_3	N_3	N_3	N_3
N_{13}	N^*_{13}	N_{13}	N_{13}	N_{13}	N_{13}	N_{13}	N_{13}	N_{13}	N_{13}	N_{13}	N_{13}	N_{13}	N_{13}
N_{23}	N_{23}	N^*_{23}	N_{23}	N_{23}	N_{23}	N_{23}	N_{23}	N_{23}	N_{23}	N_{23}	N_{23}	N_{23}	N_{23}
N_{34}	N_{34}	N_{34}	N^*_{34}	N_{34}	N_{34}	N_{34}	N_{34}	N_{34}	N_{34}	N^*_{34}	N_{34}	N_{34}	N_{34}
N_{123}	N_{123}	N_{123}	N_{123}	N^*_{123}	N_{123}	N_{123}	N_{123}	N_{123}	N_{123}	N_{123}	N_{123}	N_{123}	N_{123}
N_{134}	N_{134}	N_{134}	N_{134}	N_{134}	N^*_{134}	N_{134}	N_{134}	N_{134}	N_{134}	N_{134}	N_{134}	N^*_{134}	N_{134}
N_{234}	N_{234}	N_{234}	N_{234}	N_{234}	N_{234}	N^*_{234}	N_{234}	N_{234}	N_{234}	N_{234}	N_{234}	N_{234}	N^*_{234}
σ^2_3							σ_{34}						
N_4	N_4	N_4	N_4	N_4	N_4	N_4	N^*_4	N_4	N_4	N_4	N_4	N_4	N_4
N_{14}	N_{14}	N_{14}	N_{14}	N_{14}	N_{14}	N_{14}	N_{14}	N^*_{14}	N_{14}	N_{14}	N_{14}	N_{14}	N_{14}
N_{24}	N_{24}	N_{24}	N_{24}	N_{24}	N_{24}	N_{24}	N_{24}	N_{24}	N^*_{24}	N_{24}	N_{24}	N_{24}	N_{24}
N_{34}	N_{34}	N_{34}	N^*_{34}	N_{34}	N_{34}	N_{34}	N_{34}	N_{34}	N_{34}	N^*_{34}	N_{34}	N_{34}	N_{34}
N_{124}	N_{124}	N_{124}	N_{124}	N_{124}	N_{124}	N_{124}	N_{124}	N_{124}	N_{124}	N_{124}	N^*_{124}	N_{124}	N_{124}
N_{134}	N_{134}	N_{134}	N_{134}	N_{134}	N^*_{134}	N_{134}	N_{134}	N_{134}	N_{134}	N_{134}	N_{134}	N^*_{134}	N_{134}
N_{234}	N_{234}	N_{234}	N_{234}	N_{234}	N_{234}	N^*_{234}	N_{234}	N_{234}	N_{234}	N_{234}	N_{234}	N_{234}	N^*_{234}
σ_{43}							σ^2_4						

TABLE 9.4

Explicit Expression of [RHS] Vector (28×1) for K = 4

$$
\begin{bmatrix}
\begin{bmatrix}
N_1\sigma_1^2 & N_1\sigma_{12} & N_1\sigma_{13} & N_1\sigma_{14} \\
N_{12}\sigma_1^2 & N_{12}\sigma_{12} & N_{12}\sigma_{13} & N_{12}\sigma_{14} \\
N_{13}\sigma_1^2 & N_{13}\sigma_{12} & N_{13}\sigma_{13} & N_{13}\sigma_{14} \\
N_{14}\sigma_1^2 & N_{14}\sigma_{12} & N_{14}\sigma_{13} & N_{14}\sigma_{14} \\
N_{123}\sigma_1^2 & N_{123}\sigma_{12} & N_{123}\sigma_{13} & N_{123}\sigma_{14} \\
N_{124}\sigma_1^2 & N_{124}\sigma_{12} & N_{124}\sigma_{13} & N_{124}\sigma_{14} \\
N_{134}\sigma_1^2 & N_{134}\sigma_{12} & N_{134}\sigma_{13} & N_{134}\sigma_{14}
\end{bmatrix}
\begin{bmatrix}
\lambda_1^2 \\ \lambda_1\lambda_2 \\ \lambda_1\lambda_3 \\ \lambda_1\lambda_4
\end{bmatrix} \\
\begin{bmatrix}
N_2\sigma_{12} & N_2\sigma_2^2 & N_2\sigma_{23} & N_2\sigma_{24} \\
N_{12}\sigma_{12} & N_{12}\sigma_2^2 & N_{12}\sigma_{23} & N_{12}\sigma_{24} \\
N_{23}\sigma_{12} & N_{23}\sigma_2^2 & N_{23}\sigma_{23} & N_{23}\sigma_{24} \\
N_{24}\sigma_{12} & N_{24}\sigma_2^2 & N_{24}\sigma_{23} & N_{24}\sigma_{24} \\
N_{123}\sigma_{12} & N_{123}\sigma_2^2 & N_{123}\sigma_{23} & N_{123}\sigma_{24} \\
N_{124}\sigma_{12} & N_{124}\sigma_2^2 & N_{124}\sigma_{23} & N_{124}\sigma_{24} \\
N_{234}\sigma_{12} & N_{234}\sigma_2^2 & N_{234}\sigma_{23} & N_{234}\sigma_{24}
\end{bmatrix}
\begin{bmatrix}
\lambda_2\lambda_1 \\ \lambda_2^2 \\ \lambda_2\lambda_3 \\ \lambda_2\lambda_4
\end{bmatrix} \\
\begin{bmatrix}
N_3\sigma_{13} & N_3\sigma_{23} & N_3\sigma_3^2 & N_3\sigma_{34} \\
N_{13}\sigma_{13} & N_{13}\sigma_{23} & N_{13}\sigma_3^2 & N_{13}\sigma_{34} \\
N_{23}\sigma_{13} & N_{23}\sigma_{23} & N_{23}\sigma_3^2 & N_{23}\sigma_{34} \\
N_{34}\sigma_{13} & N_{34}\sigma_{23} & N_{34}\sigma_3^2 & N_{34}\sigma_{34} \\
N_{123}\sigma_{13} & N_{123}\sigma_{23} & N_{123}\sigma_3^2 & N_{123}\sigma_{34} \\
N_{134}\sigma_{13} & N_{134}\sigma_{23} & N_{134}\sigma_3^2 & N_{134}\sigma_{34} \\
N_{234}\sigma_{13} & N_{234}\sigma_{23} & N_{234}\sigma_3^2 & N_{234}\sigma_{34}
\end{bmatrix}
\begin{bmatrix}
\lambda_3\lambda_1 \\ \lambda_3\lambda_2 \\ \lambda_3^2 \\ \lambda_3\lambda_4
\end{bmatrix} \\
\begin{bmatrix}
N_4\sigma_{14} & N_4\sigma_{24} & N_4\sigma_{34} & N_4\sigma_4^2 \\
N_{14}\sigma_{14} & N_{14}\sigma_{24} & N_{14}\sigma_{34} & N_{14}\sigma_4^2 \\
N_{24}\sigma_{14} & N_{24}\sigma_{24} & N_{24}\sigma_{34} & N_{24}\sigma_4^2 \\
N_{34}\sigma_{14} & N_{34}\sigma_{24} & N_{34}\sigma_{34} & N_{34}\sigma_4^2 \\
N_{124}\sigma_{14} & N_{124}\sigma_{24} & N_{124}\sigma_{34} & N_{124}\sigma_4^2 \\
N_{134}\sigma_{14} & N_{134}\sigma_{24} & N_{134}\sigma_{34} & N_{134}\sigma_4^2 \\
N_{234}\sigma_{14} & N_{234}\sigma_{24} & N_{234}\sigma_{34} & N_{234}\sigma_4^2
\end{bmatrix}
\begin{bmatrix}
\lambda_4\lambda_1 \\ \lambda_4\lambda_2 \\ \lambda_4\lambda_3 \\ \lambda_4^2
\end{bmatrix}
\end{bmatrix}
$$

Solving the equation above, we have:

$$A = [\Lambda\Lambda' \otimes N^{**}]^{-1} \cdot [RHS] \quad (1)$$

It may be noted that N** is not a symmetric matrix. But the structural layout of the elements of N** and [RHS] are so symmetric that one will be able to construct the matrix and the vector by following a set of rules for the construction. No such rules will be provided at this time.

IV. STATISTICAL TESTS FOR CONTRASTS

It is observed from the expression in (1) that the quantities on the right-hand side are all computable from the data. And thus we have the numerical values of the convex coefficients. The element of $\Lambda\Lambda'$ matrix are the functions of the contrast coefficients. The elements of the N** matrix and the components of RHS vector are functions of S_i^2 (replacing σ_i^2) (i=1,2,...) where S_i^2 is the estimated variance associated with the ith treatment in the contrast and S_{ij} (replacing σ_{ij}) is the estimated covariance associated with the observations of the ith treatment and the corresponding observations of the jth treatment occurring in the same block segment, and N_i, N_{ij}, N_{ijk}, etc., are the numbers of blocks in a given block segment, as defined earlier.

Since we are dealing with contrasts in this development, the null hypotheses to be tested are:

H_o: $\mu_1 - \mu_2 = 0$ (for comparing T_1 with T_2)

H_o: $2\mu_1 - \mu_2 - \mu_3 = 0$ (for comparing T_1 with T_2 and T_3)

and so on.

The usual assumptions implicit in the analysis of variance for restricted randomization designs have to be considered for the

TABLE

Numerical Values of N-Quantities and T-Quantities

	T_1	T_2	T_3	T_4		
(1)	x x x x				$\beta(1)$	$N_1 = 0$
(2)		x x x			$\beta(2)$	$N_2 = 0$
(3)			x x x x		$\beta(3)$	$N_3 = 1$
(4)				x x x	$\beta(4)$	$N_4 = 18$
(5)	x x x	x x x			$\beta(1,2)$	$N_{12} = 3$
(6)	x x x x		x x x x		$\beta(1,3)$	$N_{13} = 1$
(7)	x x x			x x x	$\beta(1,4)$	$N_{14} = 0$
(8)		x x x	x x x		$\beta(2,3)$	$N_{23} = 1$
(9)		x x x		x x x	$\beta(2,4)$	$N_{24} = 0$
(10)			x x x	x x x	$\beta(3,4)$	$N_{34} = 0$
(11)	x x x	x x x	x x x		$\beta(1,2,3)$	$N_{123} = 16$
(12)	x x x	x x x		x x x	$\beta(1,2,4)$	$N_{124} = 0$
(13)	x x x x		x x x x	x x x x	$\beta(1,3,4)$	$N_{134} = 1$
(14)		x x x	x x x	x x x	$\beta(2,3,4)$	$N_{234} = 1$
(15)	x x x	x x x	x x x	x x x	$\beta(1,2,3,4)$	$N_{1234} = 19$

9.5

Associated with Comparison of T_1, T_2, and T_3, with T_4

$\bar{T}_1 = 0$			
$\bar{T}_2 = 0$			
$\bar{T}_3 = 12.6$			
$\bar{T}_4 = 9.867$			
$\bar{T}_{\underline{1}2} = 5.2$	$\bar{T}_{1\underline{2}} = 6.367$		
$\bar{T}_{\underline{1}3} = 5.3$	$\bar{T}_{1\underline{3}} = 3.3$		
$\bar{T}_{\underline{1}4} = 0$	$\bar{T}_{1\underline{4}} = 0$		
$\bar{T}_{\underline{1}3} = 5.3$	$\bar{T}_{2\underline{3}} = 6.3$		
$\bar{T}_{\underline{2}4} = 0$	$\bar{T}_{2\underline{4}} = 0$		
$\bar{T}_{\underline{3}4} = 0$	$\bar{T}_{3\underline{4}} = 0$		
$\bar{T}_{\underline{1}23} = 7.3687$	$\bar{T}_{1\underline{2}3} = 7.85$	$T_{12\underline{3}} = 7.7125$	
$\bar{T}_{\underline{1}24} = 0$	$\bar{T}_{1\underline{2}4} = 0$	$T_{12\underline{4}} = 0$	
$\bar{T}_{\underline{1}34} = 6.2$	$\bar{T}_{1\underline{3}4} = 6.7$	$\bar{T}_{13\underline{4}} = 10.6$	
$\bar{T}_{\underline{2}34} = 4.9$	$\bar{T}_{2\underline{3}4} = 10.3$	$\bar{T}_{23\underline{4}} = 16.8$	
$\bar{T}_{\underline{1}234} = 7.7474$	$\bar{T}_{1\underline{2}34} = 7.7947$	$\bar{T}_{12\underline{3}4} = 8.0947$	$\bar{T}_{123\underline{4}} = 11.0053$

TABLE 9.6

Variance - Covariance-Matrix

For Comparison of T_1, T_2, and T_3 with T_4

$$\begin{bmatrix} \sigma_1^2 & \sigma_{12} & \sigma_{13} & \sigma_{14} \\ \sigma_{21} & \sigma_2^2 & \sigma_{23} & \sigma_{24} \\ \sigma_{31} & \sigma_{32} & \sigma_3^2 & \sigma_{34} \\ \sigma_{41} & \sigma_{42} & \sigma_{43} & \sigma_4^2 \end{bmatrix} = \begin{bmatrix} 4.7546 & 0.8207 & 0.6882 & 1.9839 \\ 0.8207 & 4.2665 & -0.2244 & -1.7478 \\ 0.6882 & -0.2244 & 4.6476 & 1.5239 \\ 1.9839 & -1.7478 & 1.5239 & 6.8141 \end{bmatrix}$$

development of an appropriate test statistic. The tests of validity, such as the Wilk-Shapiro test for normality, the Levene test for equality of variances, and Tukey's test for additivity, should be based on the set of blocks with no missing observations.

The test statistics for the null hypotheses associated with the contrasts is:

$$F_{\nu_1,\nu_2} = \frac{[C]^2}{\text{var}(C)}$$

where $\nu_1 = 1$ and ν_2 is estimated by the Welch (1947) type approximation, as follows:

$$\nu_2 \approx \frac{(W_1S_1^2 + W_2S_2^2 + \cdots + W_KS_K^2)^2}{\dfrac{S_1^4}{DF_1} + \dfrac{S_2^4}{DF_2} + \cdots + \dfrac{S_K^4}{DF_K}}$$

where DF_1, DF_2, etc. are the degrees of freedom associated with S_1^2, $S_2^2, \ldots, S_K^2$, respectively. $W_1, W_2, \ldots, W_K$ are the respective weights expressed as a function of the number of blocks in the block segments considered in the contrast. If there are sufficient degrees

TABLE 9.7

Numerical Values of Coefficients $a_1, a_2, \ldots, a_7$, $b_1, b_2, \ldots, b_7$, $c_1, \ldots, c_7$, $d_1, d_2, \ldots, d_7$ Derived From $A = [\Lambda\Lambda' \otimes N^{**}]^{-1} \cdot [RHS]$ Associated with Comparison of T_1, T_2, and T_3 With T_4

a_1 = 7.5287 E-19	b_1 = -1.2746 E-51	c_1 = 1.2819 E-2	d_1 = 3.9966 E-1
a_2 = 2.0005 E-2	b_2 = 1.3366 E-1	c_2 = 1.0977 E-2	d_2 = 9.6366 E-18
a_3 = 1.2777 E-2	b_3 = 4.6627 E-2	c_3 = 1.5070 E-2	d_3 = 7.0614 E-18
a_4 = 1.5974 E-17	b_4 = 1.7633 E-17	c_4 = 1.9787 E-18	d_4 = 9.3117 E-18
a_5 = 7.0089 E-2	b_5 = 7.3197 E-1	c_5 = 2.3006 E-1	d_5 = 1.0558 E-17
a_6 = 1.4876 E-17	b_6 = -4.2688 E-18	c_6 = 3.4794 E-2	d_6 = 2.9256 E-2
a_7 = 4.5944 E-2	b_7 = 1.9127 E-2	c_7 = 3.6663 E-2	d_7 = 2.3301 E-2

NOTE: The numerical values are in scientific floating point notation.

of freedom associated with $S_1^2, S_2^2, \ldots, S_K^2$, a reasonable approximation will be to express ν_2 as $[(DF_1) + (DF_2) + \ldots + (DF_K)]$.

A numerical example to illustrate the estimation procedure for the convex coefficients is presented in Tables 9.5, 9.6, and 9.7. In this example, $K = 4$ and $\lambda_1 = \lambda_2 = \lambda_3 = -1$, $\lambda_4 = 3$.

This procedure can be extended easily to other restricted randomized designs such as incomplete randomized block designs, Latin square designs, etc.

REFERENCES

Anderson, R.L. (1946). Missing plot techniques. *Biometrics 2:* 41-47.

Coons, I. (1957). The analysis of covariance as a missing plot technique. *Biometrics 13:* 387-405.

Kempthorne, O. (1952). *The Design and Analysis of Experiments.* New York: John Wiley.

Nelder, J.A. (1954). A note on missing plot values. *Biometrics 10:* 400-401.

Norton, H.W. (1955). A further note on missing data. *Biometrics 11:* 110.

Smith, H.F. (1957). Missing plot estimates. *Biometrics 13:* 115-118.

Welch, B.L. (1947). Generalization of Student's problem when several different population variances are involved. *Biometrika 34:* 28-35.

Chapter 10

TREATMENT × UNIT INTERACTIONS IN THE COMPLETELY RANDOMIZED AND RANDOMIZED BLOCK DESIGNS

C.Z. Roux
Animal and Dairy Science Research Institute, Irene
Republic of South Africa

INTRODUCTION

The possibility of interaction between treatment and experimental unit arises naturally in the derivation of linear models by the use of an identity in the (conceptual) treatment responses in the randomization theory of experimental inference (vide Kempthorne, 1952, and Wilk and Kempthorne, 1955, 1956). Perhaps the most graphic description of such a situation would be in medicine where one patient could die from a drug while another benefits from it.

THE COMPLETELY RANDOMIZED DESIGN

Consider the situation with p = rt experimental units, in which there are t treatments to compare. It is, then, advantageous to imagine a conceptual population of responses, Y_{uk}, say, where the subscript u indicates the u-th experimental unit and the subscript k indicates the k-th treatment. A simple randomized design is then realized by randomization, by which certain unit/treatment combinations are drawn for actual experimentation.

The identity in the conceptual population of responses, Y_{uk}, can be used to derive a population model,

$$Y_{uk} = Y_{\cdot\cdot} + (Y_{u\cdot} - Y_{\cdot\cdot}) + (Y_{\cdot k} - Y_{\cdot\cdot}) + (Y_{uk} - Y_{u\cdot} - Y_{\cdot k} + Y_{\cdot\cdot}) \tag{1}$$

where a dot indicates averaging over a subscript. The population model is then related to the experimental model by the use of design random variables, which are Kronecker deltas taking on the value unity if a treatment occurs on a certain unit, and zero otherwise.

Consider now an analysis of covariance type of situation with relationships

$$E(Y_{uk} - Y_{\cdot k}) = \beta_k(X_{uk} - X_{\cdot k}) \tag{2}$$

or

$$E(Y_{uk} - Y_{\cdot k}) = \beta(X_{uk} - X_{\cdot k}) \tag{3}$$

where E indicates expectation of errors of measurement or technique over their distribution.

Theorem. Assume no treatment × experimental unit interaction in X, i.e., assume that

$$X_{uk} - X_{u\cdot} - X_{\cdot k} + X_{\cdot\cdot} \equiv 0$$

Then there will be treatment effects on slopes [see (2)] if and only if there is treatment × experimental unit interaction in E(Y).

Proof. First, assume no treatment effects on slopes. Averaging in (3) then results in

$$E(Y_{u\cdot} - Y_{\cdot\cdot}) = \beta(X_{u\cdot} - X_{\cdot\cdot}) \tag{4}$$

Hence (3) minus (4) gives

$$E(Y_{uk} - Y_{u\cdot} - Y_{\cdot k} + Y_{\cdot\cdot}) = \beta(X_{uk} - X_{u\cdot} - X_{\cdot k} + X_{\cdot\cdot}) \tag{5}$$

and the assumption of no interaction in X implies no interaction in E(Y).

Second, assume no interaction in E(Y). Then

$$E(Y_{uk} - Y_{\cdot k}) = E(Y_{u\cdot} - Y_{\cdot\cdot}) \tag{6}$$

which also holds in X. Substitute (6) and its dual in X in (2). Hence

$$E(Y_{u\cdot} - Y_{\cdot\cdot}) = \beta_k(X_{u\cdot} - X_{\cdot\cdot}) \tag{7}$$

must hold for all k, so that there can be no treatment effects on slopes.

NOTES ON THE APPLICATION OF THE THEOREM

1. Covariates not influenced by treatment, and hence not showing treatment × experimental unit interaction, are readily obtainable from measurements on experimental units before treatment application. In animal experiments on the influence of treatment on, say, growth rate, pre-experimental body mass of animals is perhaps the most obvious example.
2. Denote the residual sums of squares and products by R_{xx} and R_{xy}, with the subscripts indicating the variables. From Roux (1982) it follows that the variance of $\hat{\beta} = R_{xy}/R_{xx}$ is equal to $2(t-1)\sigma^2/t(r-1)E(R_{xx})$, with the expectation taken over randomizations. By symmetry a similar type of expression should hold for the variance of $\hat{\beta}_k$. It follows that the usual test for homogeneity of regression in a one-way analysis of variance

would be biassed, with the heterogeneity mean square coefficient of σ^2 being of $O\{2(t-1)/t(r-1)\}$ instead of unity. As the higher moments of this mean square are unknown, it seems that, at present, a permutation test on some function indicative of the degree of heterogeneity would be the only reliable procedure. The biassedness of the F-test is in line with experience in animal experiments where the heterogeneity F-ratio seems to be smaller than one in a disproportionately large number of cases.

3. From the usual ideas on regression and covariance analysis one would expect reasonable sensitivity in the detection of interaction only in the situation of a considerable range in X-values, i.e., only in experimental material with a considerable degree of heterogeneity. The strength of the relationship between X and Y is, of course, also important for sensitivity of the test.

4. The theorem gives a general explanation for regression lines differing in slopes in the situation of the completely randomized design. The standard textbooks offer no explanation.

RANDOMIZED BLOCKS

There are situations where one might have groups with little experimental unit variation between members of such a group, and relatively large differences between groups. A situation in which this may be true might be a test for treatment differences on different strains of inbred mice. Here one would use a randomized blocks design, with blocks identified with strains of inbred mice. In such a situation interest in treatment × experimental unit interaction might center on treatment × block interaction, rather than treatment × plot (within block) interaction. A relatively mild assumption is here all that is necessary to provide a unbiassed test for interaction.

This simplest model capable of handling interaction is, of course, a multiplicative model. Here there are two approaches.

The first dates back to Fisher and Mackenzie (1923), culminated in Mandel (1971), and is, essentially, singular value decomposition of the matrix of residuals. The distribution theory associated with this approach is, however, difficult and Mandel (1971) had to resort to Monte Carlo methods to obtain degrees of freedom for the different sums of squares. The second approach was discovered by Yates and Cochran (1938), with the mathematical ideas and distribution theory explicated by Mandel (1961). In comparison with the first method the second treats the rows and columns of a two-way table asymmetrically in that it is, essentially, a regression analysis of the row observations on the column means, or vice versa.

Inspection of the expected mean squares in Kempthorne (1952) shows that the finite model analysis of variance of randomized blocks shows asymmetry between treatments and blocks in that blocks do not contain a contribution from between plot error variance. It is this asymmetry between treatments and blocks which is exploited in the present development.

To eliminate unnecessary description, the notation of Kempthorne (1952, Chapter 8) is followed. The population model is

$$Y_{ijk} = Y_{\cdot\cdot\cdot} + (Y_{i\cdot\cdot} - Y_{\cdot\cdot\cdot}) + (Y_{ij\cdot} - Y_{i\cdot\cdot}) + (Y_{\cdot\cdot k} - Y_{\cdot\cdot\cdot}) + (Y_{i\cdot k} - Y_{i\cdot\cdot} - Y_{\cdot\cdot k} + Y_{\cdot\cdot\cdot}) + (Y_{ijk} - Y_{ij\cdot} - Y_{i\cdot k} + Y_{i\cdot\cdot}) \quad (8)$$

where Y_{ijk} is the (conceptual) yield of treatment k on plot j of block i. The last term represents plot × treatment interaction and is assumed to be identically zero. In light of Mandel's (1961) results it seems reasonable to describe the interaction between block and treatment by writing

$$Y_{i\cdot k} - Y_{\cdot\cdot k} = \beta_k(Y_{i\cdot\cdot} - Y_{\cdot\cdot\cdot}) + d_{ik} \quad (9)$$

or

$$Y_{i\cdot k} - Y_{\cdot\cdot k} - Y_{i\cdot\cdot} + Y_{\cdot\cdot\cdot} = (\beta_k - 1)(Y_{i\cdot\cdot} - Y_{\cdot\cdot\cdot}) + d_{ik}$$

where β_k is defined by the relationship

$$\beta_k = \sum_i Y_{i \cdot k}(Y_{i..} - Y_{...})/\sum_i(Y_{i..} - Y_{...})^2$$

Denote the experimental observables by y_{ik}, so that the experimental (or sample) model becomes

$$y_{ik} = \sum_j \delta_{ij}^k Y_{ijk}$$

where

$$\delta_{ij}^k = 1 \text{ if treatment k occurs on plot (ij)}$$

$$= 0 \text{ otherwise,}$$

with $i = 1,2,\ldots,r$ and $j,k = 1,2,\ldots,t$

Hence

$$y_{ik} = \mu + b_i + t_k + (bt)_{ik} + \sum_j \delta_{ij}^k e_{ij} \tag{10}$$

where the relationship of the symbols to those of (8) is indicated by the correspondences between subscripts and $(bt)_{ik}$ is accommodated by (9). Since $E(y_{ik}) = Y_{i \cdot k}$ the unbiassed estimate of β_k is

$$\hat{\beta}_k = \sum_i y_{ik}(y_{i.} - y_{..})/\sum_i(y_{i.} - y_{..})^2 \tag{11}$$

as $\sum_k \delta_{ij}^k = 1$, so that $y_{i.} - y_{..} = b_i$, without any error terms. Note furthermore that $\hat{\beta}_. = 1$, so that $\sum_k(\hat{\beta}_k - \hat{\beta}_.)^2 = \sum_k(\hat{\beta}_k - 1)^2$. Hence consider

$$E(R) = E\sum_k(\hat{\beta}_k - 1)^2 \sum_i b_i^2$$

$$= E\sum_k\{\sum_{ij} \delta_{ij}^k e_{ij} b_i + \sum_i (bt)_{ik} b_i\}^2/\sum_i b_i^2$$

$$= E[\sum_k(\sum_{ij} \delta_{ij}^k e_{ij} b_i)^2 + \sum_k\{\sum_i (bt)_{ik} b_i\}^2]/\sum_i b_i^2$$

$$= \sum_{ij} e_{ij}^2/r + \sum_k(\beta_k - 1)^2 \sum_i b_i^2 \tag{12}$$

from Kempthorne (1952, p. 138) on the assumption that the sums of squares of errors within blocks are homogeneous, and from (9). Also

$$E(I) = E\sum_{ik}(y_{ik} - y_{i\cdot} - y_{\cdot k} + y_{\cdot\cdot})^2$$

$$= (r-1)\sum_{ij} e_{ij}^2/r + \sum_k(\beta_k - 1)^2\sum_i b_i^2 + \sum_{ik} d_{ik}^2 \tag{13}$$

which follows from Kempthorne (1952) and Mandel (1961).

Define

$$\sigma_d^2 = \sum_{ik} d_{ik}^2/(r-2)(t-1) \text{ and } \sigma^2 = \sum_{ij} e_{ij}^2/r(t-1)$$

Then

$$E(R)/(t-1) = \sigma^2 + \sum_k(\beta_k - 1)^2 \sum_i b_i^2/(t-1)$$

and (14)

$$E(I-R)/(t-1)(r-2) = \sigma^2 + \sigma_d^2$$

It follows from (9) that, under the hypothesis of $(bt)_{ik} \equiv 0$, $\beta_k \equiv 1$ and $d_{ik} \equiv 0$. Hence (14) provides a unbiassed test for interaction. The sensitivity of the test, however, is oDviously connected to the relative sizes of $\sum_k(\beta_k - 1)^2 \sum_i b_i^2/(t-1)$ and σ_d^2.

Since the corresponding beta ratio has a variable instead of a constant in the divisor, the permutation test variance of the F-ratio of (14) is probably less well approximated by the variance of the F-test, than the usual permutation F-ratio for treatments. However, asymptotically the moments of the permutation and F-tests would be equal, as can be obtained by the approach of Robinson (1975). The unbiassed nature, in the Yatesian sense, of the F-test from (14) in the Mandel (1961) approach probably makes it of greater value in the finite (or permutation) situation than Tukey's one degree of freedom for interaction, which is dealt with by Robinson (1975). Tukey's procedure provides a biased test, since the treatment values and means in the divisor of the single degree of freedom sum of squares contain contributions from the error variance.

In an interactive situation attention is usually focused on the y_{ik} values rather than the treatment means. Hence a further point of interest is to estimate $Y_{i \cdot k}$ by the equation

$$\hat{Y}_{i \cdot k} = y_{\cdot k} + \hat{\beta}_k (y_{i \cdot} - y_{\cdot \cdot}) \tag{15}$$

with variance

$$\text{var}(\hat{Y}_{i \cdot k}) = \sigma^2/r + b_i^2 \, \sigma^2 / \sum_i b_i^2 \tag{16}$$

from the derivation of (12) and Kempthorne (1952), under the assumption of homogeneous intrablock error sums of squares. It follows that the average variance of the treatments in a block is equal to

$$\text{ave var}(\hat{Y}_{i \cdot k}) = 2 \, \sigma^2/r \tag{17}$$

The variance of y_{ik} is equal to σ^2, so that with $r > 2$, it will be advantageous to estimate $Y_{i \cdot k}$ by $\hat{Y}_{i \cdot k}$ from (15). From (9) it is clear that the bias resulting from the use of $\hat{Y}_{i \cdot k}$ is equal to d_{ik}. The average magnitude for the upper bound of the bias is apparent from the estimates of the right hand sides of (14) and from the corresponding expressions for the individual regression line of the treatment involved. It is also clear from (14) that (I-R)/(t-1)(r-2) will be an overestimate of σ^2, although the overestimation can possibly be far more serious from I/(t-1)(r-1). The overestimation of σ^2 leads, as usual, to loss of sensitivity in inference. The possible existence of σ_d^2 was ignored in Mandel's (1961) formulation of the breakdown of the interaction sum of squares. If an unbiased estimate of σ^2 is required, replication of treatments within a block is called for.

NOTES ON APPLICATION

The fitting of multiplicative effects is the simplest approximation to an interactive situation. It is shown here that a multiplicative model is consonant with the randomization or finite model approach to the analysis of variance. Possible applications follow from the discussion by Mandel (1961). It seems also that multiplicative models may be used advantageously for the description of genotype-environment interaction (Freeman, 1973). In such an application of a multiplicative model, environments take the place of blocks, and are therefore considered to be fixed effects, rather than random effects, as is often assumed for the purpose of description by variance components.

Considering environments fixed is probably often in line with practice, as plant breeders tend to test new varieties on stations representative of various rainfall/fertility combinations. The large gains in efficiency that may be possible from the use of a multiplicative model in the prediction of yield for a certain locality is evident from (17), where in such a situation σ^2 denotes the variance of a variety mean in a certain environment and r is equal to the number of environments.

Freeman (1973) suggests that it would be better to test the β_k for a particular genotype by use of the appropriate sum of squares of deviations from regression of that genotype, instead of the pooled formulation developed here. However, heterogeneity of error variances associated with different genotypes would imply interaction between plots and treatments in the finite model formulation. From Kempthorne (1952, Section 8.3) it is clear that such interaction will complicate matters considerably. On the simplest level the problem will be that the divisor in $\hat{\beta}_k$ becomes a random variable. It is possible that an asymptotic approach, like that of Robinson (1975) may be of value here.

As regards the assumption of homogeneity of intrablock sums of squares, it is clear that in the derivation of (12) and (16), it can be replaced by lack of correlation between d^2_{ij} and b^2_i. With negative correlation the right hand side of (12) will be underestimated and with positive correlation it will be overestimated. The same is likely to happen in (16). It is, however, of interest to note that the assumption of homogeneous intrablock error sums of squares is necessary for a close approximation of the permutation test by the F-test.

Mandel (1961) noted that instead of fitting a straight line for each treatment, orthogonal polynomials of higher order could be used. It is readily apparent that this is also true for the finite situation. It is of some interest to note that (17) generalizes to

$$\text{ave var } (\hat{Y}_{i \cdot k}) = (1 + s)\ \sigma^2/r \tag{18}$$

where s is the degree of the constructed orthogonal polynomial used in the estimation of $Y_{i \cdot k}$. Note that (18) becomes equal to σ^2 if $s = r - 1$, which implies, as it should, that there is no gain in efficiency of estimation from a perfect polynomial fit.

PURELY MULTIPLICATIVE EFFECTS

In the situation where significant interaction is indicated by a test based on (14) attention should also be given to fitting a model with purely multiplicative effects

$$y_{ik} = \mu\rho_i\tau_k + \sum_j \delta^k_{ij}\ e_{ij} \tag{19}$$

with

$$\sum_i \rho_i/r = 1 = \sum_k \tau_k/t$$

since

$$\mu\rho_i\tau_k \equiv \mu\{1 + (\rho_i - 1) + (\tau_k - 1) + (\rho_i - 1)(\tau_k - 1)\}$$

is in an obvious relationship to (10).

Equation (19) can also be related to (8) by assuming the last term on the right hand side of (8) equal to zero, and assuming

$$Y_{i\cdot k} \equiv \mu\rho_i\tau_k \tag{20}$$

Hence $y_{i\cdot} \equiv \mu\rho_i$, so that

$$\hat{\tau}_k = \sum_i y_{ik}y_i/\sum_i y_{i\cdot}^2 \tag{21}$$

is the unbiassed simple least squares estimator of τ_k. Consider

$$E(R) = E\{\sum_k(\hat{\tau}_k - 1)^2\Sigma y_{i\cdot}^2\}$$

$$= \sum_{ij} e_{ij}^2/r + \mu^2 \sum_i \rho_i^2 \sum_k(\tau_k - 1)^2 \tag{22}$$

like (12) under the assumption of homogeneous errors within blocks. Likewise

$$E(T) = E\{\sum_{ik} (y_{ik} - y_{i\cdot})^2\} \tag{23}$$

$$= \sum_{ij} e_{ij}^2 + \mu^2 \sum_i \rho_i^2 \sum_k(\tau_k - 1)^2$$

From (21) the error attached to $\hat{\tau}_k$ is

$$\sum_i \sum_j \delta_{ij}^k e_{ij} y_{i\cdot}/\sum_i y_{i\cdot}^2 \tag{24}$$

Hence, precisely analogous to Kempthorne's (1952) derivation of the variance of treatment mean:

$$\mathrm{var}(\hat{\tau}_k - \hat{\tau}_{k'}) = 2\sigma^2/\sum_i y_{i\cdot}^2 \tag{25}$$

under the assumption of homogeneous intrablock errors. Likewise it follows from (24) that R/T (defined in 22,23) provides a beta test ratio with precisely the same properties under H_0 and homogeneous intrablock errors as in the usual additive randomized block analysis of variance. From (22) and (23) the corresponding F test is R/(t-1) ÷ (T-R)/(r-1)(t-1).

DISCUSSION

From Zyskind's (1967) Theorem 3 it is easy to verify that (21) and (11) are also best linear unbiased estimators with the error covariance structure induced by randomization in the randomized block situation. Describe the relevant linear model by $y = X\beta + \varepsilon$, $E(\varepsilon\varepsilon') = \sigma^2 V$, then the relevant part of Zyskind's (1967) theorem is: A linear function w'y is a best linear unbiassed estimator of its expectation E(w'y) if and only if the vector Vw belongs to the column space of X. Take $w' = (y_{1\cdot}, y_{2\cdot}, \ldots, y_{r\cdot})$ and the conclusion that the simple least squares estimator is also the best linear unbiased estimator follows immediately from the form of V given by Zyskind (1967). In contrast the infinite analog of (19) allows further development in both the least squares and maximum likelihood senses, as follows from e.g., Mandel (1971).

It is of some importance to distinguish between (19) and

$$y_{ik} = \mu\rho_i\tau_k\varepsilon_{ik} \tag{26}$$

which can be linearized by the taking of logarithms, while (19) should be analyzed as it stands. The need to emphasize homogeneous within block errors in the derivation of (25) suggests ploting of within block standard deviations, calculated from estimates of residual effects, against block means, as a way of distinguishing between (19) and (26). A linear relationship through the origin would indicate (26) as the appropriate model, with a conventional randomized blocks analysis on the logtransformed data.

REFERENCES

Fisher, R.A. and Mackenzie, W.A. (1923). Studies in crop variation. II. The manurial response of different potato varieties. *J. Agric. Sci. 13*, 311-320.

Freeman, G.H. (1973). Statistical methods for the analysis of genotype-environment interactions. *Heredity 31*, 339-354.

Kempthorne, O. (1952). Design and analysis of experiments. New York: Wiley.

Mandel, J. (1961). Non-additivity in two-way analysis of variance. *J. Am. Stat. Assoc. 56*, 878-888.

Mandel, J. (1971). A new analysis of variance model for non-additive data. *Technometrics 13*, 1-18.

Robinson, J. (1975). On the test for additivity in a randomized block design. *J. Am. Stat. Assoc. 70*, 184-185.

Roux, C.Z. (1982). The analysis of covariance in randomization theory. *South African Stat. J. 16*, 1-23.

Wilk, M.B. and Kempthorne, O. (1955). Fixed, mixed and random models. *J. Am. Stat. Assoc. 50*, 1144-1167.

Wilk, M.B. and Kempthorne, O. (1956). Some aspects of the analysis of factorial experiments in a completely randomized design. *Ann. Math. Stat. 27*, 950-985.

Yates, F. and Cochran, G.W. (1938). The analysis of groups of experiments. *J. Agric. Sci. 28*, 556-580.

Zyskind, G. (1967). On canonical forms, non-negative covariance matrices and best and simple least squares linear estimators in linear models. *Ann. Math. Stat. 38*, 1109.

Chapter 11

THE USE OF MULTIPLE COPIES OF DATA IN FORMING AND INTERPRETING ANALYSIS OF VARIANCE

Robin Thompson
University of Edinburgh
Edinburgh, Scotland

I. INTRODUCTION

In this paper the use of multiple copies of data is illustrated as an aid in both forming and interpreting analyses of variance. My interest in this technique arose from trying to analyse diallel crosses using the computer program GENSTAT (Alvey, Galwey and Lane, 1982). This program includes a general algorithm for the analyses of data with a high degree of balance and orthogonality (Wilkinson, 1970; Payne and Wilkinson, 1977). Whilst Wilkinson's algorithm is very powerful and is said to be able to analyse all the designs in Cochran and Cox (1957) (Heiberger, 1981) one cannot even specify the models appropriate for diallel crosses. Whilst one could use regression techniques to analyse such designs, these ignore the symmetry of the data and so can be computationally inefficient and less easy to interpret.

It was discovered that if the data was extended by taking two copies then this general algorithm can still be used. One associates different fixed effects with each copy and introduces another random factor. Then the terms in the diallel cross analysis of variance correspond to terms in the analysis of variance of this extended set. In Section II this example is explained.

In field experiments the use of blocking to reduce the effect of soil heterogeneity has been used for many years. An alternative approach (P) originally suggested by Papadakis (1937) is based on adjusting yields by covariance on the residuals of neighbouring plots. Bartlett (1938) stressed the theoretical complications and the method was little used for some time, partly because of the tedious calculation (Yates, 1970).

Interest in the method has been stimulated by two recent discussion papers read to the Royal Statistical Society. In the first Bartlett (1978) extended Atkinson's (1969) results to show that the Papadakis treatment estimates are close to the maximum likelihood estimates under an autoregressive model. Usually solutions of the maximum likelihood equations for the variance parameters have to be found iteratively and this suggests estimating the Papadakis covariate regression coefficient iteratively using the hopefully improved estimates of residuals in each iteration.

In the second discussion paper Wilkinson, Hancock, Eckert and Mayo (1983) found that iterating the Papadakis method led to biased tests of treatment effects. The average adjusted treatment mean square was not equal to the expectation of the residual mean square for simulations using several sets of uniformity data. This led Wilkinson et al. to introduce an alternative procedure (NN) based on using 'sliding' blocks rather than fixed blocks. They used for each plot a block centered on each plot and including neighbours of that plot. This idea is appealing because with fixed blocks weights given to treatment comparisons from adjacent plots depend on whether or not the adjacent plots are deemed to be in the same block. In some circumstances one would want to give the same weight to both comparisons.

Wilkinson et al. (1983) derived estimators using intuitive, if not altogether consistent, arguments based on the classical analogue of recovery of inter-block information from incomplete blocks (Yates,

1936a). They produced evidence, again from uniformity trials, that suggested that tests of their treatment estimates were less biased. However, it was pointed out in the discussion that the NN treatment estimating equations were in fact equivalent to the P estimating equations, and the difference in the methods lay in the method of estimating the variance parameters.

This leads to the question of whether there are any other links between neighbour analysis and incomplete block analysis. There are at least two reasons why this is important. Firstly, there is instinctive prejudice against adjustment by covariates based on residuals (Kempton and Howes, 1981) suggesting that this and incomplete block methods are completely different, yet the link uncovered by the discussants of the Wilkinson et al. paper suggests common ground underlying the methods. Secondly, there are presumably circumstances when neighbour analysis, giving equal weight to treatment comparisons from adjacent plots is more efficient than incomplete block analysis. Then one would like to allocate treatments to plots in an optimum manner. If one can uncover more links between neighbour analyses and incomplete block analyses then perhaps optimum incomplete block designs, of which there is a huge literature, can easily be modified to give efficient designs for neighbour analysis.

These ideas are developed in Section III and again the technique of multiple copies of data is found useful using blocks of two superimposed on the data.

It is perhaps appropriate to explain why this paper should be in a volume dedicated to Professor Kempthorne. Firstly, all the classical designs mentioned in this paper are in his books on experimental design and genetic statistics. Secondly, several years ago after I had attended a talk advocating iteratively reweighted least squares as the solution to every statistical problem Kemp pointed out to me that many sets of genetic data, perhaps the simplest being ABO blood group data, did not fit into this framework. Later Bob Baker and I (1981) introduced the idea of composite link functions that allowed these genetic models to be embedded into the iteratively reweighted least squares framework. Composite link functions link an expanded expectation of observation vector to the observation vector. The technique in this paper is closely related for here both observations and their expectation are expanded.

II. APPLICATION TO DIALLEL CROSS MODELS AND OTHER SIMILAR MODELS

In this section Wilkinson's (1970) algorithm for analysis of experimental designs is briefly discussed. Diallel cross models are introduced and it is shown how these can be fitted into the algorithm. The same procedure can be used on other designs and examples are given.

Wilkinson's Algorithm

Wilkinson's algorithm is a recursive procedure for the analysis of experimental designs. It involves a finite sequence of sweeps, in each of which a set of effects for a factor are calculated and an updated residual vector formed. The calculated effects are usually means calculated from the current residual vector, or 'effective' means which are means divided by an efficiency factor (Yates, 1936a). It is obvious that the procedure can be used for orthogonal experiments, with only two sweeps, one for blocks and one for treatments required for randomized blocks. Wilkinson uses a preliminary analysis of special dummy variates to calculate efficiency factors and eliminate redundant sweeps. The algorithm can deal with stratified designs having more than one error component by modifying the input observations for each stratum.

Wilkinson in his description of the algorithm allows a factor to have k associated efficiency factors and then k sweeps are carried out each with a different efficiency. Because of difficulties in calculating degrees of freedom associated with these k efficiencies the GENSTAT implementation of the algorithm only allows each factor to have one efficiency factor. Pseudo-factors can often be introduced, to allow GENSTAT analyses. Sometimes the necessary pseudo-factors have been used in generating the design, as in lattices (Yates, 1936b).

To use the algorithm in GENSTAT the linear model to be fitted is specified in terms of qualitative fixed factors (called "Treatments" in GENSTAT), qualitative random factors ("Blocks") and quantitative variates ("Covariates").

Diallel Crosses

In plant breeding work on p different parental lines often all reciprocal crosses can be made and the p^2 progeny used for comparison of parental lines. Various models, analyses and interpretations have been given for data with this structure (Kempthorne, 1956; Griffing, 1956; Jinks and Mather, 1971). My limited interest here is in showing how analyses can be constructed using Wilkinson's algorithm and start by using a simple model that exhibits the difficulty in fitting these classes of models. Suppose that when a female of line i is crossed with a male of line j there is a progeny with observation y_{ij}. A factorial model $y_{ij} = \mu + f_i + m_j + e_{ij}$, where μ, f_i, m_j and e_{ij} are respectively mean, female, male and error terms. Sometimes, on genetic or other arguments, it can be argued that the male and female effects are equivalent and then a model

$$y_{ij} = \mu + \ell_i + \ell_j + e_{ij} \tag{1}$$

where ℓ_i represents a line effect, is appropriate. Now each cross $(i \neq j)$ observation includes two line terms and pure $(i=j)$ plants includes the same term twice. Hence the model (1) cannot be specified directly using qualitative factors in GENSTAT terms.

Diallel crosses are not the only models that have more than one level of a factor related to an "observation". Yates (1936a) in his discussion of incomplete block experiments suggested an inter-block analysis as a first stage in making use of information from blocks. This analysis can be regarded as fitting a linear model to block totals and of course the expectation of the totals will include sums of individual treatment effects. As Wilkinson's algorithm can do analyses in several strata this suggests extending the diallel cross data and constructing a model so that an inter-block analysis corresponds to model (1). This can be done by taking two copies of the data with $m_{ijI} = m_{ijII} = y_{ij}$ and using the models $m_{ijI} = \mu + \ell_i + u_{ij} + e_{ijI}$, $m_{ijII} = \mu + \ell_j + u_{ij} + e_{ijII}$, where u_{ij} is a random unit effect and e_{ijI} and e_{ijII} are random unit × copy effects. Female line effects are associated with the first copy and male effects with the second copy. Then an inter-unit analysis of m_{ij} gives multiples of the required line effects and sums of squares. The line effects have efficiency (p-2)/(p-1) as noted by Wilkinson (1970).

Other models and designs can be fitted using the same technique. Hayman (1954) introduced a model that includes a comparison of mean (and line) pure performance with mean (and line) cross performance and also line effects on reciprocal differences $(y_{ij} - y_{ji})$. The extended model for the two copies of the data are

$$m_{ijI} = \mu + \ell_i + p_k + \ell p_{ik} + c_I + \ell.c_{iI} + r_s + r.h_{st} + e_{ijI}$$

$$m_{ijII} = \mu + \ell_j + p_k + \ell p_{jk} + c_{II} + \ell.c_{jII} + r_s + r.h_{st} + e_{ikII}$$

where p_k is a fixed effect with two levels representing pure and cross (k=1 if i=j, k=2 if k≠j) and h_t is a fixed effect with two levels representing halves of the table (t=1 if i<j, t=2 if i≥j), c_i a fixed effect representing copies of the data (i=I, II), r_s is a random factor with p × (p+1)/2 levels representing reciprocals with crosses i,j and j,i at level i × (i-1)/2+j. Some interactions of line, copy, reciprocal and half are also in the model. This gives the following skeleton analysis of variance:

Source	Degrees of freedom
Reciprocal Stratum	
Lines	p-1
Pure versus Cross	1
Lines × Pure versus Cross	p-1
Residual	p(p-3)/2
Total	p(p+1)/2-1

Half within Reciprocal Stratum	
Lines × Copy	p-1
Residual	(p-1)(p-2)/2
Total	p(p-1)/2
Within Half within Reciprocal Stratum	
Lines	p-1
Copy	1
Lines × Copy	p
Residual	$(p-1)^2$
Total	p^2

The sums of squares in the lowest stratum are zero because they are comparisons between the two copies and the other sums of squares are four times (because two copies of the data are used) those in Hayman's analysis.

Other Designs

Elsewhere (1983) the author has applied the same idea to other designs. I make use of the fact that a pseudo-factorial model similar to (1) can be imposed on partially balanced incomplete blocks with triangular structure (Clatworthy, 1973). With a little more difficulty rectangular lattices (Harshbarger, 1947) can be dealt with in a similar manner.

III. INCOMPLETE BLOCK AND NEIGHBOUR ANALYSIS

In this section links between Incomplete Block and Neighbour analysis are explored. A multiple copy model is introduced to interpret terms arising in an autoregressive model. This assists in the explanation of terms in the NN and P methods, and allows the consideration of efficiency factors and helps in constructing designs. The contributions of various strata to variance estimates and extensions to more appropriate models are discussed.

One Dimensional Auto-Regressive Model

Suppose there are n=rt plots, comprising r for each of t treatments, laid out in one long strip and that the ith plot receives treatment j and a model

$$y_i = \tau_j + e_i \quad (i=1,\ldots,n;\ j=1,\ldots,t) \tag{2}$$

is appropriate, where e_i is a plot error with mean zero, variance σ^2 and covariance with e_k of $\rho^{|i-k|}\sigma^2$. This autoregressive variance structure has been used to explicate the P method by Atkinson (1969) and Bartlett (1978). This model (2) cannot be defended in all circumstances. For example, the plots normally should be arranged in replicates and separate replicate effects often need to be included. This layout in replicates or other experimental constraints in the field will sometimes imply that the autoregressive structure applies to groups of less than rt plots. However, this model (2) will be used because it illustrates several important points that carry over to other, more appropriate and complicated, models.

It is convenient to write (2) in matrix form

$$E(y) = X\tau,\ \mathrm{var}(e) = \sigma^2 H \tag{3}$$

where X is a n×t matrix with elements zero and one and H is a n×n matrix with elements $H_{ij} = \rho^{|i-j|}$. Then if ρ and σ^2 were known the least squares estimates of the treatment effects, τ_A, satisfy

$$(X'H^{-1}X)\tau = X'H^{-1}y \text{ with } \mathrm{var}(\tau_A) = \sigma^2(X'H^{-1}X)^{-1} \tag{4}$$

If matrices D and L, of size n×n, are defined with all elements of D and L zero except $D_{11} = D_{nn} = 1$ and $L_{i,i+1} = L_{i+1,i} = 1$ then H^{-1} can be written as

$$(1-\rho^2)H^{-1} = [(1+\rho^2)(I-D) + D-\rho L] \tag{5}$$

Alternatively if $2B = 2I-D + L$ and $2W = 2I-D - L$ then

$$2(1-\rho^2)H^{-1} = (1-\rho)^2 B + (1+\rho)^2 W + (1-\rho^2)D \tag{6}$$

$$(1-\rho^2)H^{-1} = (1+\rho)^2[I - 1/2\ D + (1-\rho)/2(1+\rho)D - 2\rho/(1+\rho)^2 B] \tag{7}$$

By comparison a mixed model often used with incomplete block experiments is $y = X\tau + Zb + e$ where Z represents allocation of plots to blocks and b represents block effects with variance Γ and e are independent plot errors with variance σ_e^2. Using both intra- and inter-block information this leads to estimating equations similar to (5) with H^{-1} now given by

$$(I - Z(Z'Z + I\Gamma^{-1})Z') \tag{8}$$

and σ_e^2 replacing σ^2. If Γ^{-1} is put equal to zero in (8) then the intra-block estimating equations result.

Comparison of (7) and (8) suggests considering a conceptual model. Suppose two copies of the data are taken, with $m_I = m_{II} = y$, and two sets of blocks of two are superimposed. In the first copy (I) plots 2i and 2i+1 are in the ith I block and in the second copy (II) plots 2i-1 and 2i are in the ith II block (i=1,...,n). Diagramatically

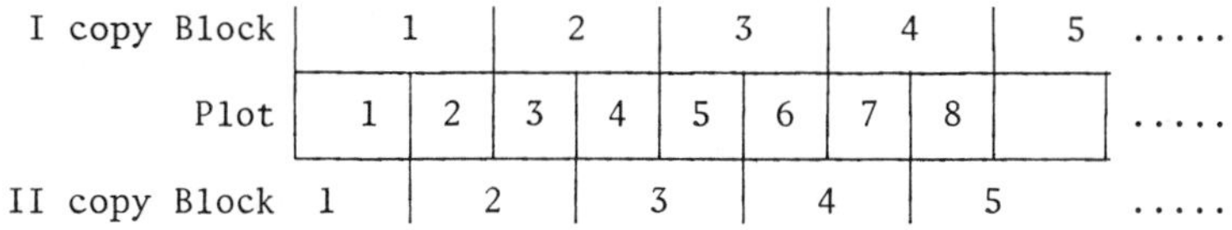

I copy Block	1		2		3		4		5	.....
Plot	1	2	3	4	5	6	7	8		.....
II copy Block	1	2		3		4		5		.....

Then consider the incomplete block model similar to the one above with

$$m_{Ii} = \tau_j + b_{Ir} + e_{Ii},\; m_{IIi} = \tau_j + b_{IIs} + e_{IIi}$$

or (9)

$$m_I = X\tau + Z_I b_I + e_I,\; m_{II} = X\tau + Z_{II} b_{II} + e_{II}$$

where r and s are integer parts of (i+1)/2 and (i+2)/2 respectively. Plots 1 and n will each be in blocks of size one once and without loss of generality suppose treatment 1(2) is applied to plot 1(n).

If the block effects are independent and size two (one) block effects have variance $\gamma_2\sigma_e^2(\gamma_1\sigma_e^2)$ then $Z(Z'Z + \Gamma^{-1})Z'$ for this model is $1/(1+\gamma_1^{-1})D + 2/(2+\gamma_2^{-1})B$. So if $\gamma_1 = 2\rho/(1-\rho)$, $\gamma_2 = 2\rho/(1-\rho)^2$ and $\sigma_e^2 = 2\sigma^2(1-\rho^2)/(1+\rho^2)$ the parameters of the incomplete block model can be chosen to mimic the autoregressive model. Equations (6) and (9) suggests that τ_A combines information from between blocks of two (B), within blocks of two (W) and from the end plots (D), the weighting depending on ρ.

Model (9) can also be used to interpret terms arising in the estimation of σ^2 and ρ by maximizing the likelihood of error contrasts (REML) (Patterson and Thompson, 1971) if e in (5) is normally distributed. Terms arise in sums of squares of residuals $r_c' r_c$ and $\hat{b}_c' D_c \hat{b}_c$ where $\hat{b}_c = (Z_c'Z_c + \Gamma_c^{-1})^{-1} Z_c'm_c$, $r_c = m_c - X\tau - Z_c\hat{b}_c$ are residuals and $D_c = \Gamma_c^{-1}(d\Gamma_c/d\rho)\Gamma_c^{-1}$ $(c = I, II)$. These terms also arise if (9) were the appropriate model. In both models these sums of squares of residuals are equated to their expectation. The expectations under (5) and (9) differ but there are similar terms in both models associated with corrections for τ being estimated.

Connection with P and NN Methods

An approximation, suggested by (5), is to use V^{-1} proportional to $(\Delta-\beta L)$ where $\Delta = I - 1/2\, D$ and $\beta = \rho/(1+\rho^2)$ helps to explain the P and NN methods. Then (4) becomes

$$X'(\Delta-\beta L)X\tau = X'(\Delta-\beta L)y \qquad (10)$$

or

$$X'\Delta X\tau = X'(\Delta y-\beta L(y-X\tau)) \qquad (11)$$

As $L(y-X\tau)$ has ith element the sum of residuals from the (i-1) and (i+1) plots then (10) can be thought of as using this vector as a covariate and this is one justification for the P method.

Wilkinson et al. (1983) consider using 'sliding' blocks that is for plot i considering a block of plots (i-1), i and (i+1), which also leads, apart from differential weighting to border plots, to (11). In particular they discuss intra-(sliding) block treatment estimates with $2\beta=1$ which corresponds to replacing V^{-1} by W in (5) and is equivalent to the intra-block treatment estimates from the multiple copy model.

The methods differ on variance estimation. The P method uses analogues with the analysis of covariance to estimate β although no correction is made for the estimation of τ in the covariate $L(y-X\tau)$. Both P and NN use sums of squares of residuals of the form $R = (y-X\tau)'(I-\beta L)^2(y-X\tau)$. The use of R arises naturally and the residuals can be related to the sum of residuals in the multiple copy model, $(r_I + r_{II})$. R will be an efficient statistic under some variance models, for example $V^{-1} \propto (I-\beta L)^2$, but it is not an efficient statistic when the treatment estimates are efficient (i.e., $V^{-1} \propto (I-\beta L)$). The implicit use of these two variance models in the P method partly explains the problems in estimating β iteratively.

Efficiency Factors

For incomplete block models it is often useful to consider contrasts that are independently estimated. Their efficiency factors play a fundamental role in the structure of the design (James and Wilkinson, 1971; Pearce, Caliński and Marshall, 1974). The intrablock estimate of τ from (9) satisfies $(2R-X'BX)\tau_w = X'(2I-D-B)$ where $R = X'(I - 1/2\, D)X$ is a diagonal matrix. Because $I - (2R)^{-1/2}X'BX(2R)^{-1/2} = A$ is a symmetric matrix there exists an orthogonal matrix such that $A = P\varepsilon P'$, where ε is a diagonal matrix containing the eigenvalues of A. The columns of P are eigenvectors of A and the columns can be ordered according to the eigenvalues. If the design is connected there will be only one zero eigenvalue corresponding to an eigenvector $(2R)^{1/2}/(2n)^{1/2}$. The matrix $C = (2R)^{1/2}P$ can be constructed and it can be shown (Pearce et al., 1974) that the first t-1 elements of $C'\tau_w$ represent contrasts of τ_w. The variance of $C'\tau_w$ is $\sigma_e^2\varepsilon^-$, ε^- is a diagonal matrix with elements ε_i^{-1} (or zero if $\varepsilon_i = 0$). So $C'\tau_w$ represent independent contrasts. The variance of

$C'(2R)^{-1}X'(2I-D)y$ is $I\sigma_e^2$ if $\Gamma = 0$ so that ε_i can be thought of as an efficiency factor for the i-th contrast.

Similarly if a combined estimate τ_R is found using the intra- and inter-block (of size two) information then for model (9) the variance of $C'\tau_c$ is $\sigma_e^2(1+2\gamma_2)[I+2\gamma_2\varepsilon]^{-1}$ so that $\varepsilon_{Ri} = (1+2\gamma_2\varepsilon_i)/(1+2\gamma_2)$ is an efficiency factor for $C'\tau_i$.

The variance matrix for $C'\tau_A$ using the autoregressive model is slightly more complicated due to the contribution of end plots. Then using (7) we find that

$$(1-\rho^2)X'H^{-1}X = (1+\rho^2)R^{1/2}[(1-\rho^2)/(1+\rho^2)I + 4\rho/(1+\rho)^2A$$

$$+ (1-\rho)/2(1+\rho)R^{-1/2}X'DXR^{-1/2}]R^{1/2}$$

Then it can be shown that the i,j element of the variance matrix of τ_A is

$$2(1-\rho^2)\sigma^2/(1+\rho)^2[\varepsilon_{Ri}^{-1}\delta_{ij} - \varepsilon_{Ri}^{-1}\varepsilon_{Rj}^{-1}\sum_{m,n=1}^{2} P_{mi}P_{mj}F_{mn}] \tag{12}$$

where F and G are 2×2 matrices with $F = G^{-1}$ and elements of G given by $G_{mn} = \sum_i P_{mi}\varepsilon_{Ri}^{-1}P_{ni} + \delta_{mn}(2r-1)(1+\rho)/(1-\rho)$, $\varepsilon_{Ri} = (1+2\gamma_2\varepsilon_i)/(1+2\gamma_2) = [(1-\rho)^2 + 4\rho\varepsilon_i]/(1+\rho)^2$ and δ_{ij} is the Kronecker delta. The first term in G_{mn} relates to the covariance of τ_{Rm} and τ_{Rn}. The second term in (12) represents contributions to the covariance from the end plots, this term should be negligible for most contrasts, and reduce as ρ increases, especially if the end treatments, 1 and 2, do not contribute much to the contrast. The second term can contribute substantially to the variance of $2R/(2n)^{1/2}\tau_A$, especially as ρ increases, because $\varepsilon_t = 0$. When $\rho = 0$ then (12) reduces to $2\sigma^2\{1-(P_{1i}P_{1j} + P_{2i}P_{2j})/2r\}$. This suggests the contrasts are approximately uncorrelated and that $\varepsilon_{Ai} = [(1-\rho)^2 + 4\rho\varepsilon_i]/(1-\rho^2)$ is an efficiency factor.

To quantify the gain from using two copies of the data consider the case when the auto-regressive model is appropriate but an incomplete block model is used with blocks of two. Then $\sigma_e^2 = (1-\rho)\sigma^2$ and σ_b^2 is approximately $\rho\sigma^2$. This ignores the average covariance between plots not in the same block, of the order of $1/[2t(1-\rho)]$ if ρ is not too large. For a contrast with intra-block efficiency ε_i then using the block information increases the efficiency (relative to σ_e^2) to $[(1-\rho)+2\rho\varepsilon_i]/(1+\rho)$ and relative to σ^2 is $[(1-\rho)+2\rho\varepsilon_i]/(1-\rho^2)$ so the gain from using all the information is of the order of $\rho(2\varepsilon_i-1+\varepsilon)/(1-\rho^2)$. Note that the average efficiency from a 2r replicate design can be higher than from an r replicate design because there is more scope for balancing comparisons within blocks.

Variances for NN Method

Wilkinson et al. (1983) used a variance matrix given approximately, neglecting some border information, by

$$(I-\beta L)V(I-\beta L) = \sigma^2[(I-\beta L)(I-\beta L) + \Omega(I+L+L^2-2I)]$$

where $\Omega = (1-2\beta)^2 w$ and w and β are variance parameters, to give variances of their treatment estimates. Then

$$[(I-\beta L)(I-\beta L)+\Omega(I+L+L^2-2I)]$$

$$= (1+2\beta^2+\Omega)I + (\Omega-2\beta)(2B-2I) + (\Omega+\beta^2)(2B_1-2I)$$

where $2B_1 = 2I+(L^2-2I)$ and can be thought of as block information when 2 copies of the data and blocks of two are formed from plots one apart i.e., plots j and j+2. Suppose ε_1 a diagonal matrix exists such that $X(2R-B_1)X = P\varepsilon_1 P'$ that is the estimated contrasts $C'\tau$ would be independent with both blocking schemes. Then

$$X'(I-\beta L)X = rP[(1-2\beta)I + 4\beta\varepsilon]P'$$

and $X'(I-\beta L)V(I-\beta L) = \sigma^2 rP[((1-2\beta)^2 + \Omega)I + 8\beta\varepsilon - 4\beta\varepsilon_1 - 4\Omega(I-\varepsilon-\varepsilon_1)]P'$ so that an efficiency factor for the ith contrast is approximately $\varepsilon_{iNN} = [(1-2\beta)+4\beta\varepsilon_i]^2/[(1-2\beta)^2 + \Omega + 8\beta\varepsilon_i - 4\beta\varepsilon_{1i} - 4\Omega(1-\varepsilon_i-\varepsilon_{1i})]$.

Wilkinson et al. (1983) give a numerical example of a Rothamsted experiment (Jenkyn et al., 1979). This is an experiment balanced for first and second neighbours with efficiencies $e_i = 2/3$, $e_{1i} = 4/9$. Wilkinson et al. give efficiencies of 1.22 for an intra-block analysis which is approximately $(3/2)\ \varepsilon_{iNN}$ using $2\beta=1$, $\Omega=0$. Wilkinson et al. use the factor 3/2 to scale their variances.

These efficiencies also arise in expressions for the bias in the P covariate.

Variance Parameter Estimation

The subdivision of the treatments into independent contrasts helps in subdividing the residual sums of squares $R = (y-X\tau_A)'H^{-1}(y-X\tau_A)$. By relating τ_A to τ_R and then relating τ_R to the estimates from $Wy(\tau_W)$ and $By(\tau_B)$ then it can be shown that

$$R = [(1-\rho)^2/(1+\rho)^2](R_W+T_{RD}) + R_B + T_{WB}$$

where R_W and R_B are residual sums of squares $y'Wy - \tau_W'C'\varepsilon C\tau_W$ and $y'By - \tau_B'C'(I-\varepsilon)C\tau_B$ and T_{WB} is a sum of squares for treatment comparisons between strata W and B, i.e.,

$$T_{WB} = (\tau_W-\tau_B)'C\ \varepsilon\ \varepsilon_R^-(I-\varepsilon)C'(\tau_W-\tau_B)$$

and similarly

$$T_{RD} = (2r-1)(y_1-\tau_{R1},\ y_n-\tau_{R2})F \begin{pmatrix} y_1 - \tau_{R1} \\ y_n - \tau_{R2} \end{pmatrix}$$

is a comparison between τ_R and estimates from the end plots. Yates (1940) when recovering inter-block information interpreted his analysis of variance in a similar way.

In REML estimation of σ^2 and ρ can be thought of as equating R and $dR/d\rho$ to their expectation. The contributions of R_W, R_B and T_{BW} to $\hat{\sigma}^2$ and $\hat{\rho}$ can be fairly easily calculated showing, for instance that the weight given to T_{BW}, relative to R_B, increases as ρ increases. The contribution of T_{RD} is harder to quantify as it depends on $dF/d\rho$ but in some numerical examples T_{RD} seems to become more influential than T_{BW} as ρ increases. I am a little uneasy about this, given that T_{RD} is a comparison of border plots with others and border plots might behave slightly differently just because they are on the boundary.

Implications for Design

Efficient balanced designs are available for this variance model and for a limited number of treatment and replicate combinations (Williams, 1952; Freeman, 1979). The efficiencies of a derived 2r replicate incomplete block model play a key role in quantifying the efficiencies of an r replicate design with an autoregressive model. Much work has been done on constructing efficient incomplete block designs. These facts suggest that these 2r incomplete blocks designs might be converted into good r replicate designs.

For example, John, Wolock and David (1972) suggest a cyclic design with t = 18, k = 2 and r = 6 constructed cyclically from initial blocks of 1 2, 1 4 and 1 9. This suggests a 3 replicates autoregressive design.

1	2	5	6	9	10	13	14	17	18	3	4	5	8	11	12	15	16
3	6	16	13	5	8	18	15	7	10	2	17	9	12	4	1	11	14
4	5	15	14	6	7	17	16	8	9	1	18	10	11	3	2	12	13

where the rows represent replicates and where there is a constant difference between the ith and (i+4)th elements (i=1,...,14) in each row. Taking two copies of this design and imposing two sets of blocks of two as in (10) gives blocks included in the 6 replicate design.

Extension to Other Variance Structures

The autoregressive variance model was introduced to illustrate the use of a multiple copy model and relate it to other results but it will not necessarily be appropriate. Patterson and Hunter (1983) from an analysis of 166 cereal trials in the United Kingdom suggested that an error model with $V_{ii} = (1+\theta)\sigma^2$, $V_{ij} = \rho^{|i-j|}\sigma^2$ might be more appropriate. It is interesting to show how this model fits into the multiple copy framework.

For this alternative model

$$V = \sigma^2\theta I + \sigma^2(1-\rho^2)[(1+\rho^2)(I-D) + D-\rho L]^{-1}$$

$$= \sigma^2\theta I + \sigma^2[(1-\rho)/(1+\rho)]V_A^{-1}$$

so that

$$V^{-1} \propto V_A - V_A[V_A + [(1-\rho)/(1+\rho)\theta]I]^{-1} V_A$$

The least squares equations for τ then satisfy

$$\begin{bmatrix} X'V_A X & X'V_A \\ V_A X & V_A + [(1-\rho)/(1+\rho)\theta]I \end{bmatrix} \begin{bmatrix} \tau \\ u \end{bmatrix} = \begin{bmatrix} X'V_A y \\ V_A y \end{bmatrix} \qquad (13)$$

This is rather like (4) and suggests taking 2 copies of the data and imposing blocks of two on the data as in model (9). Then (13) can be manipulated to give

$$\begin{bmatrix} 2X'X & X'Z_I & X'Z_{II} & 2X' \\ Z_I'X & Z_I'Z_I + \Gamma^{-1} & 0 & Z_I' \\ Z_{II}'X & 0 & Z_{II}'Z_{II} + \Gamma^{-1} & Z_{II}' \\ 2X & Z_I & Z_{II} & 2[1+(1-\rho)/(1+\rho)\theta]I \end{bmatrix} \begin{bmatrix} \tau \\ b_I \\ b_{II} \\ u \end{bmatrix} = \begin{bmatrix} 2X'y \\ Z_I' y \\ Z_{II}' y \\ 2y \end{bmatrix}$$

This suggests that a multiple copy model to mimic this error model is

$$m_I = X\tau + Z_I b_I + u + e_{Ii}$$
$$m_{II} = X\tau + Z_{II} b_{II} + u + e_{IIi} \qquad (14)$$

where u has variance $2\{\theta(1+\rho)\sigma^2/(1-\rho)\}I$. Hence there are two copies of the data, blocks of two superimposed and a unit effect u_i associated with each copy of plot i. The parameter θ in a sense measures the closeness of the two copies and if $\theta = 0$ or $-2(1-\rho)/(1+\rho)$ then auto-regressive or moving average models result. Although (14) is a conceptual model one can speculate whether the parameters have any physical interpretation. The layout suggests that if m_I and m_{II} were observations on plots harvested in halves then perhaps θ, ρ and σ^2 could be interpreted in terms of these observations.

Often designs are laid out in two dimensions, and again one can introduce more copies and blocks of two to mimic the structure.

Randomization

There remains the question of whether randomization of the multiple copies gives added justification for the use of specific variance models. Certainly randomizations are possible that mimic the structure of V^{-1}.

IV. CONCLUSION

It has been shown that multiple copies give added insight into two specific examples. The idea can be applied in other areas. Space only allows me to say that problems of symmetry in x and y can sometimes be fruitfully expressed in terms of x, y and y, x and that the use of unit, copy and unit × copy information can be used to investigate extra variation in exponential family models (Hinde, 1982).

ACKNOWLEDGMENTS

I am grateful for encouragement from Dr. H.D. Patterson and other members of the ARC Unit.

REFERENCES

Alvey, N., Galwey, N., and Lane, P. (1982). *An Introduction to Genstat*. London: Academic Press.

Atkinson, A.C. (1969). The use of residuals as a concomitant variable. *Biometrika 56,* 33-41.

Bartlett, M.S. (1938). The approximate recovery of information from field experiments with large blocks. *J. Agric. Sci. 28,* 418-427.

Bartlett, M.S. (1978). Nearest neighbour models in the analysis of field experiments. *J. Roy. Stat. Soc. B 40,* 147-174.

Clatworthy, W.H. (1973). *Tables of two-associate-class partially balanced designs. Nat. Bur. Standards Appl. Math. Ser.* 63.

Cochran, W.G. and Cox, G.M. (1957). *Experimental Designs*. New York: John Wiley.

Freeman, G.H. (1979). Some two-dimensional designs, balanced for nearest neighbours. *J. Roy. Stat. Soc. B 41,* 88-95.

Griffing, B. (1956). A generalized treatment of the use of diallel crosses in quantitative inheritance. *Heredity 10,* 31-50.

Harshbarger, B. (1947). Rectangular Lattices. Virginia Agricultural Experimental Station Memoir 1.

Hayman, B.I. (1954). The analysis of variance of diallel tables. *Biometrics 10,* 235-244.

Heiberger, R.M. (1981). The specification of experimental designs to ANOVA programs. *American Statistician 35,* 98-104.

Hinde, J. (1982). *Compound Poisson Regression Models. GLIM82,* R. Gilchrist (ed.). New York: Springer.

James, A.T. and Wilkinson, G.N. (1971). Factorisation of the residual operator and canonical decomposition of nonorthogonal factors in the analysis of variance. *Biometrika 58,* 279-294.

Jenkyn, J.F., Bainbridge, A., Dyke, G.V. and Todd, A.D. (1979). An investigation into inter-plot interactions, in experiments with mildew on experiments with mildew on barley, using balanced designs. *Ann. Appl. Biol. 92,* 11-28.

John, J.A., Wolock, F.W. and David, H.A. (1972). *Cyclic designs. Nat. Bur. Standards Appl. Math.* Ser., 62.

Kempthorne, O. (1956). The theory of the diallel cross. *Genetics 41,* 451-459.

Kempton, R.A. and Howes, C.W. (1981). The use of neighbouring plot values in the analysis of variety trials. *Applied Statistics 30,* 59-70.

Mather, K. and Jinks, J.L. (1971). *Biometrical Genetics.* Ithaca: Cornell Univ. Press.

Papadakis, J.S. (1937). Méthode statistique pour des expériences sur champ. *Bull Inst. Amél. Plantes á Salonique 23.*

Patterson, H.D. and Hunter, E.A. (1983). The efficiency of incomplete block designs in national list and recommended list cereal variety trials. *J. Agric. Sci.* (to be published).

Patterson, H.D. and Thompson, R. (1971). Recovery of inter-block information when block sizes are unequal. *Biometrika 58,* 545-554.

Payne, R.W. and Wilkinson, G.N. (1977). A general algorithm for analysis of variance. *J. Roy. Stat. Soc.* Ser. C *26,* 251-260.

Pearce, S.C., Calinski, T. and Marshall, T.F. de C. (1974). The basic contrasts of an experimental design with special reference to the analysis of data. *Biometrika 61,* 449-460.

Thompson, R. (1983). Diallel Cross, Incomplete Block designs and rectangular Lattices. *Genstat Newsletter 9* (in press).

Thompson, R. and Baker, R.J. (1981). Composite link functions in generalized linear models. *Appl. Statistics 30,* 125-131.

Wilkinson, G.N. (1970). A general recursive procedure for analysis of variance. *Biometrika 57,* 19-46.

Wilkinson, G.N., Eckert, S.R., Hancock, T.N. and Mayo, O. (1983). Nearest neighbour (NN) analysis of field experiments. *J. Roy. Stat. Soc. B* (in press).

Williams, R.M. (1952). Experimental designs for serially correlated observations. *Biometrika 39,* 151-167.

Yates, F. (1936a). Incomplete randomised blocks. *Ann. Eugen.* *7*, 121-140.

Yates, F. (1936b). A new method of arranging variety trials involving a large number of varieties. *J. Agric. Sci* *26*, 424-455.

Yates, F. (1940). The recovery of inter-block information in balanced incomplete block designs. *Ann. Eugen.* *10*, 223-230.

Yates, F. (1970). *Experimental Design*. London: Griffin.

PART III

LINEAR AND NONLINEAR MODELS

Chapter 12

ESTIMATING VARIANCE COMPONENTS USING ALTERNATIVE MINQE'S IN SELECTED UNBALANCED DESIGNS

John Brocklebank
SAS Institute
Cary, North Carolina

F.G. Giesbrecht
North Carolina State University
Raleigh, North Carolina

I. INTRODUCTION

The problem of estimating variance components in random and mixed linear models has received much attention in the statistics literature during the past 30 or 40 years. In the case of "balanced" data sets, estimates that are optimal in some broad sense were quickly obtained (Graybill, 1954; and Graybill and Wortham, 1956). Unbalanced data sets proved to be much more difficult. A major thrust in the research work was to obtain estimates that were unbiased, especially in the presence of fixed effects. Pioneer work in this area includes papers by S.L. Crump (1946) and C.R. Henderson (1953). Excellent reviews of the large body of research that has been developed are found in Harville (1969, 1977), Kleffe (1977, 1978), Rao (1979) and Searle (1979). Also Anderson (1975, 1978) reviews a series of papers that deal with the inter-related problems of selecting a design and selecting a method for estimating variance components. In particular, he evaluates alternative nested and cross classification random models when several different methods of estimation are used.

It is not surprising that the simplest (except for the simple random sample) design, the among and within classification has received the most attention. Optimal designs for the balanced and unbalanced case have been studied by Hammersley (1949), Herrendörfer (1974, 1979), Ginsburg (1973), Crump (1954), Anderson and Crump (1967) and Prairie (1962). Optimal designs for both the balanced and unbalanced case using estimators based on analysis of variance type arithmetic have been studied by many authors, including Hammersley (1949), Prairie (1962), Crump (1954), Ginsburg (1973) and Herrendörfer (1974, 1979). Extensions and comparisons with other methods of estimation include papers by Thompson and Anderson (1975), Anderson (1975, 1978), Ahrens (1978), Townsend and Searle (1971), LaMotte (1973a, 1973b), Bush and Anderson (1963), Swallow and Searle (1978), Hess (1979) and Swallow (1981). Extensions to more complex designs can be found in Prairie (1962), Gaylor (1960), Goldsmith (1969), Goldsmith and Gaylor (1970), Muse (1974), Muse and Anderson (1978), and Thitakomal (1977).

The object in this paper is to compare the utility of two-level nested designs for estimating variance components when using invariant quadratic estimators.

II. QUADRATIC ESTIMATION OF VARIANCE COMPONENTS

Let Y denote an n-variate normally distributed random variable with mean vector

$$E[Y] = X\beta$$

where β is a p-vector of fixed parameters and X an $n \times p$ matrix of known constants and covariance matrix

$$E[(Y - X\beta)(Y - X\beta)'] = V$$

where V can be expressed as $V = \theta_1 V_1 + \dots + \theta_k V_k$, $\{\theta_i\}$ are unknown variance components and $\{V_i\}$ are known matrices. The general class of quadratic estimators, $Y'AY$ of the components $\{\theta_i\}$ is obtained by considering all possible $n \times n$ symmetric A matrices. The restriction that estimators be unbiased for $\Sigma p_i \theta_i$ and translation invariant, i.e., remain unchanged when Y is replaced by $Y - X\beta^*$ requires A be such that $AX = 0$ and $tr(AV_i) = p_i$ for $i=1,\ldots,k$. Rao (1971b) has shown that the minimum variance quadratic unbiased invariant estimator of $\Sigma p_i \theta_i$ is obtained by selecting A of the form $\Sigma \lambda_i R(\theta) V_i R(\theta)$ where $R(\theta) = V^{-1} - V^{-1}X(X'V^{-1}X)^{-}X'V^{-1}$. The $\{\lambda_i\}$ are constants selected to insure unbiasedness. Unfortunately, this result is of limited utility since $R(\theta)$ depends on the true but unknown variance components. Replacing $R(\theta)$ by $R(\alpha)$, i.e., replacing V by $\Sigma \alpha_j V_j$ leads to an infinite class of quadratic estimators. The minimum variance property is lost, though invariance is preserved and unbiasedness maintained by the definition of $\{\lambda_i\}$. Following the terminology suggested by the work of Rao (1971a, 1971b) and Rao and Mitra (1979) we denote these estimators by MINQE(U, I, α). Note that with the definitions used in this paper, estimates are unchanged if the scale on the $\{\alpha_i\}$ is changed by an arbitrary constant. Common practice is to replace α_i by α_i/α_k for $i=1,\ldots,k-1$ and then set $\alpha_k = 1$.

The variance of an invariant quadratic unbiased estimate $Y'AY$, equal to 2 tr(AVAV) depends on the $\{\alpha_i\}$ as well as the exact nature of the unbalance in the design. For balanced data sets all will give the same estimates. Ahrens (1978), Hess (1979) and Swallow (1981) study the variances of the estimated variance components for among and within designs (one-stage nested) with selected patterns of unbalance, prior information and true variance components.

III. THE TWO-STAGE NESTED DESIGN

In order to develop some feel for the inter-relationship of methods of estimation, i.e., selection of $\{\alpha_i\}$ and the sample design we now specialize our development to the two-stage nested design with model

$$y_{ijk} = \mu + a_i + b_{ij} + e_{ijk}$$

for $k=1,\ldots,n_{ij}$, $j=1,\ldots,m_i$, $i=1,\ldots,h$ and where $\{y_{ijk}\}$ are the observations, μ is an unknown constant, $\{a_i\}$ are independent $N(0,\theta_a)$, $\{b_{ij}\}$ are independent $N(0,\theta_b)$, $\{e_{ijk}\}$ are independent $N(0,\theta_e)$ and the latter three sets are also mutually independent. The convention of denoting a summation by replacing the corresponding subscript with a "+" symbol will be followed. Occasionally it will also be convenient to write the model as

$$Y = X\mu + U_a a + U_b b + e$$

The connection between the two versions of the model is established by letting Y correspond to the elements $\{y_{ijk}\}$ in dictionary order, X is a column of 1's, and U_a and U_b are matrices of 0's and 1's. Occasionally it is also convenient to let U_e denote the $n_{++} \times n_{++}$ identity matrix. It follows that Y is multivariate $N(X\mu, V)$ where

$$V = \theta_a U_a U_a' + \theta_b U_b U_b' + \theta_e U_e U_e'$$

Now the MINQE(U, I, α) of θ_a, θ_b and θ_e are obtained as solutions to the system of equations:

$$\begin{bmatrix} Z_{11} & Z_{12} & Z_{13} \\ Z_{21} & Z_{22} & Z_{23} \\ Z_{31} & Z_{32} & Z_{33} \end{bmatrix} \begin{bmatrix} \hat{\theta}_a \\ \hat{\theta}_b \\ \hat{\theta}_e \end{bmatrix} = \begin{bmatrix} Y'R(\alpha)U_a U_a' R(\alpha)Y \\ Y'R(\alpha)U_b U_b' R(\alpha)Y \\ Y'R(\alpha)U_e U_e' R(\alpha)Y \end{bmatrix}$$

Explicit formulae for the $\{Z_{ij}\}$ are derived by computing expected values of the quadratic forms in the column on the right. We note in passing that the system is always consistent, though it may be singular. Since the $\{Z_{ij}\}$ depend only on the design and the $\{a_i\}$, the variances and covariances of $\hat{\theta}_a$, $\hat{\theta}_b$ and $\hat{\theta}_e$ can be computed as linear functions of the variances and covariances of the three quadratic forms in Y.

The first obvious competitor for the MINQE(U, I, α) is the set of estimates produced by applying the analysis of variance as given by Steel and Torrie (1980). These will be referred to as the ANOVA estimators. The second class of competitors, which we denote by MINQH(U, I, α) is obtained by replacing $Y'R(\alpha)U_eU_e'R(\alpha)Y$ in the above system by the error of sum of squares obtained from the analysis of variance. An advantage of the MINQH strategy is that the estimate of θ_e is unbiased, guaranteed positive and translation invariant.

The designs to be considered in this study are made up of five fundamental structures similar to those described in Goldsmith and Gaylor (1970). These are formed by taking all possible combinations of structures formed when splits of either one or two are allowed when passing from one stage to the next lower stage in the design. The five fundamental structures are pictorially represented by the stick diagrams shown in Table 12.1.

TABLE 12.1

Fundamental Structures

	Structure Number				
Stage	1	2	3	4	5
1 2	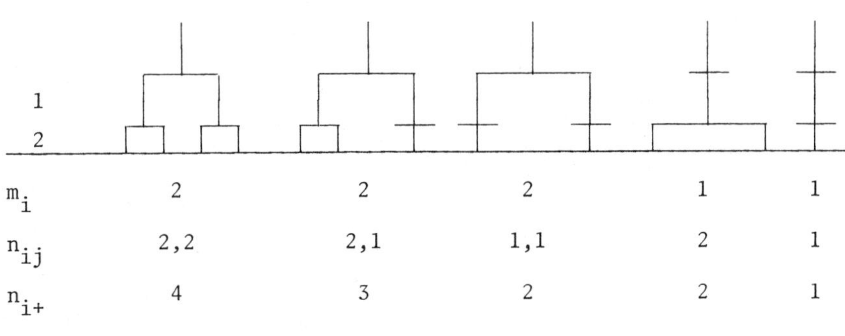				
m_i	2	2	2	1	1
n_{ij}	2,2	2,1	1,1	2	1
n_{i+}	4	3	2	2	1

The designs were restricted to contain 12r observations, r = 1 or 5, and as many as three types of the fundamental structures found in Table 12.1 so that each of the three variance components could be estimated by the ANOVA method. The number 12 was chosen as a multiplier of r since it is the lowest common multiple of the n_{i+} found in Table 12.1. Originally letting r go as high as 10 was contemplated; however, preliminary investigations showed very little change in the results between r = 5 and r = 10. Letting r = 1 and 5 gave sample sizes 12 and 60. These restrictions led to 61 different designs. Of these 61, 16 "optimal" design structures (see Goldsmith and Gay-

TABLE 12.2

Selected Two-Stage Nested Designs

	Structure No.						Coefficient Nos.			
Design No.	1	2	3	4	5	Code	c	d	e	g
1	1	0	0	0	8	(9,1,2)	60	8	12	2
2	1	0	1	0	6	(8,2,2)	54	8	13	3/2
4	1	0	3	0	2	(6,4,2)	42	8	15	5/4
5	1	0	4	0	0	(5,5,2)	36	8	16	2
6	2	0	0	0	4	(6,2,4)	48	10	18	2
8	2	0	2	0	0	(4,4,4)	36	10	20	3/2
9	3	0	0	0	0	(3,3,6)	36	12	24	2
10	0	1	0	0	9	(10,1,1)	64	7	9	4/3
13	0	1	3	0	3	(7,4,1)	46	7	12	13/12
17	0	2	2	0	2	(6,4,2)	44	8	14	7/6
18	0	2	3	0	0	(5,5,2)	38	8	15	17/15
21	0	4	0	0	0	(4,4,4)	40	10	18	4/3
53	0	0	2	2	4	(8,2,2)	60	8	10	1
55	0	0	2	4	0	(6,2,4)	60	10	12	1
60	0	0	4	2	0	(6,4,2)	48	8	12	1
61	0	0	5	1	0	(6,5,1)	42	7	12	1

lor, 1970) were selected for further study. They are identified in Table 12.2.

The designs in Table 12.2 are systematically numbered according to the identification code used in Goldsmith (1969). The numbers in the body of the table below the structure identification indicate (when multiplied by r) the number of times that structure is to be used in the design. The code that accompanies each design is a triplet (u, v, w) which serves as a design code as well as a way of identifying the degrees of freedom structure in the ANOVA table. The degrees of freedom and codes are illustrated in Table 12.3.

The coefficient numbers c, d, e, and g in Table 12.3 are used to determine coefficients for the expected mean squares in the ANOVA table of the two-stage nested design, as $k_1 = (cr-d)/6(h-1)$, $k_2 = (72r-e)/6(h-1)$, $k_3 = g$, where k_1, k_2, and k_3 are identified in Table 12.3.

As an example, suppose an experimenter decided to run an experiment with design 13. The integers 01303 beneath the structure numbers in Table 12.2 mean that the experimenter would select r replicates of structures 2 and 3r replicates of structures 3 and 5. Pictorially, the experimental plan for r=1 would be the appropriate randomization of the following pattern:

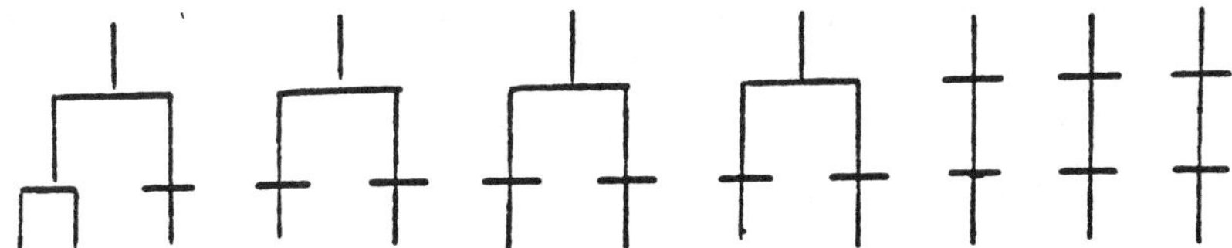

The design code (7, 4, 1) indicates that there are 7r-1 df for the A-classes, 4r df for the B in A-classes and r df for the error source in the ANOVA table. From the coefficient numbers, the expected mean square coefficients are:

$$k_1 = (46r-7)/6(7r-1),\ k_2 = (72r-12)/6(7r-1),\ k_3 = 13/12$$

TABLE 12.3

ANOVA Table

Source	df	df code	MS	E(MS)
A-classes	$h - 1$	$= ur-1$	MSA	$\theta_e^2 + k_1\theta_b^2 + k_2\theta_a^2$
B in A-classes	$\sum_{i=1}^{h} m_i - h$	$= vr$	MSB	$\theta_e^2 + k_3\theta_b^2$
Error	$\sum_{i=1}^{h} \sum_{j=1}^{m_i} n_{ij} - \sum_{i=1}^{h} m_i$	$= wr$	MSE	θ_e^2

IV. RISK STUDIES FOR THE MINQE(U, I, α), ANOVA, AND THE MINQH(U, I, α) ESTIMATORS

Expressions for the variances of these different estimators are too complicated to allow analytic comparison and consequently must be compared numerically. Unfortunately this involves evaluation of the variances under a variety of n-patterns for different design structures, a variety of values of unknown components $(\theta_a, \theta_b, \theta_e)$, and for the MINQE(U, I, α) and MINQH(U, I, α) estimators, a variety of prior weights denoted $(\alpha_a, \alpha_b, \alpha_e)$. For convenience we will scale the unknown components and prior weights by θ_e and α_e, respectively, to give the scaled configurations (Rho(A), Rho(B), 1) for the unknown components and (ρ(A), ρ(B), 1) for the prior weights. We note in passing that when ρ(A) = Rho(A) and ρ(B) = Rho(B) and a, b and e are multivariate normal then MINQE(U, I, α) are minimum variance unbiased, translation invariant, quadratic estimators.

In order to obtain numerical values for variances, it was necessary to specify a sample size, 12r, for r = 1 or 5, the scaled unknown variance component configuration and scaled prior value configuration. Sixty-four different scaled prior value configurations were used, formed by setting ρ(A), ρ(B) = 0 or 2^k, where k=-3,-2,..., 3. Also 4225 unknown component configurations were considered, formed by setting Rho(A), Rho(B) = .125k where k=0,1,...,64.

V. DESIGN EVALUATION FOR ANOVA, MINQE(U, I, α) AND MINQH(U, I, α) ESTIMATORS

Tables 12.4 and 12.5 summarize the results of a study to compare the 16 selected designs. Following Goldsmith and Gaylor (1970) we use the trace of the covariance matrix of the variance component estimators as the criterion to evaluate designs and estimators. This criterion has appeal since it tends to concentrate the sampling at the stage for which the variance component is large, relative to others, and is sensitive to changes in sample size. Consequently, given a design structure we would calculate

$$tr = var(\hat{\theta}_e) + var(\hat{\theta}_b) + var(\hat{\theta}_a)$$

as a function of θ_a, θ_b, θ_e, $\rho(A)$ and $\rho(B)$. A design is said to be optimal for a given estimation procedure for a group of designs if its trace value does not exceed that of any other design in the group.

Table 12.4 utilizes MINQE(U, I, α) under the idealized conditions, prior weights equal to true components and represents some sort of optimum. Table 12.5 utilizes the ANOVA estimators which do not require any prior values and are always available.

These tables can be interpreted in the following fashion. For example when Rho(B) = $\rho(B)$ = 4 and Rho(A) = $\rho(A)$ = 4 and the basic structure of the designs is replicated once, the design 5 with code (5,5,2) was selected for the ANOVA estimators.

With only one exception, both estimators suggest a balanced design when Rho(A) < 1 and Rho(B) $\leq$ 2 for both 1 and 5 replications. For Rho(B) = 4, 8 and Rho(A) $\leq$ 4, i.e., for relatively large θ_b the design favored by MINQE appears to be slightly more balanced than the design favored by ANOVA. However, both have the (5,5,2) code, the same distribution of degrees of freedom. Although the optimal designs selected are structurally different for the two estimation procedures, the design codes or degrees of freedom in the analysis of variance are always very similar if not identical. Optimal designs 5, 4, 1 and 2 in the MINQE(U, I, α) estimation are repeatedly

TABLE 12.4

Optimal Designs (Codes) for MIVQE(U,I) When All 16 Designs Are Compared

Rho(B) = ρ(B)	Rep 1	Rep 5	Rep 1	Rep 5	Rep 1	Rep 5	Rep 1	Rep 5	Rep 1	Rep 5	Rep 1	Rep 5	Rep 1	Rep 5	Rep 1	Rep 5
8	5 (5,5,2)	5	5 (5,5,2)	5	5 (5,5,2)	5	5 (5,5,2)	5	5 (5,5,2)	5	5 (5,5,2)	5	5 (5,5,2)	5	61 (6,5,1)	61
4	5 (5,5,2)	5	5 (5,5,2)	5	5 (5,5,2)	5	5 (5,5,2)	5	5 (5,5,2)	5	5 (5,5,2)	5	5 (5,5,2)	5	61 (6,5,1)	61
2	8 (4,4,4)	8	8 (4,4,4)	8	8 (4,4,4)	8	8 (4,4,4)	8	5 (5,5,2)	8 (4,4,4)	5 (5,5,2)	5	5 (5,5,2)	5	13 (7,4,1)	13
1	9 (3,3,6)	9	9 (3,3,6)	9	9 (3,3,6)	9	8 (4,4,4)	9 (3,3,6)	8 (4,4,4)	8	5 (5,5,2)	8 (4,4,4)	4 (6,4,2)	4	2 (8,2,2)	2
.5	9 (3,3,6)	9	9 (3,3,6)	9	9 (3,3,6)	9	9 (3,3,6)	9	8 (4,4,4)	9 (3,3,6)	6 (6,2,4)	8 (4,4,4)	2 (8,2,2)	2	1 (9,1,2)	1
.25	9 (3,3,6)	9	9 (3,3,6)	9	9 (3,3,6)	9	9 (3,3,6)	9	8 (4,4,4)	9 (3,3,6)	6 (6,2,4)	6	2 (8,2,2)	2	1 (9,1,2)	1
.125	9 (3,3,6)	9	9 (3,3,6)	9	9 (3,3,6)	9	9 (3,3,6)	9	8 (4,4,4)	9 (3,3,6)	6 (6,2,4)	6	1 (9,1,2)	1	1 (9,1,2)	1
0	9 (3,3,6)	9	9 (3,3,6)	9	9 (3,3,6)	9	9 (3,3,6)	9	8 (4,4,4)	9 (3,3,6)	6 (6,2,4)	6	1 (9,1,2)	1	1 (9,1,2)	1
Rho(A) = ρ(A)	0		.125		.25		.5		1		2		4		8	

TABLE 12.5

Optimal Designs (Codes) for ANOVA Estimators When All 16 Designs Are Compared

Rho(B) \ Rho(A)	0		.125		.25		.5		1		2		4		8	
	Rep 1	Rep 5	Rep 1	Rep 5	Rep 1	Rep 5	Rep 1	Rep 5	Rep 1	Rep 5	Rep 1	Rep 5	Rep 1	Rep 5	Rep 1	Rep 5
8	18 (5,5,2)	18	18 (5,5,2)	18	18 (5,5,2)	18	18 (5,5,2)	18	18 (5,5,2)	18	18 (5,5,2)	18	18 (5,5,2)	18	61 (6,5,1)	61
4	18 (5,5,2)	18	18 (5,5,2)	18	18 (5,5,2)	18	18 (5,5,2)	18	18 (5,5,2)	18	18 (5,5,2)	18	18 (5,5,2)	18	61 (6,5,1)	61
2	8 (4,4,4)	8	8 (4,4,4)	8	8 (4,4,4)	8	8 (4,4,4)	8	5 (5,5,2)	5	5 (5,5,2)	5	18 (5,5,2)	18	13 (7,4,1)	16 (6,5,1)
1	9 (3,3,6)	9	9 (3,3,6)	9	9 (3,3,6)	9	8 (4,4,4)	9 (3,3,6)	8 (4,4,4)	8	5 (5,5,2)	8 (4,4,4)	60 (6,4,2)	18 (5,5,2)	53 (8,2,2)	60 (6,4,2)
.5	9 (3,3,6)	9	9 (3,3,6)	9	9 (3,3,6)	9	9 (3,3,6)	9	8 (4,4,4)	9 (3,3,6)	6 (6,2,4)	21 (4,4,4)	2 (8,2,2)	60 (6,4,2)	10 (10,1,1)	53 (8,2,2)
.25	9 (3,3,6)	9	9 (3,3,6)	9	9 (3,3,6)	9	9 (3,3,6)	9	8 (4,4,4)	9 (3,3,6)	6 (6,2,4)	21 (4,4,4)	1 (9,1,2)	1	10 (10,1,1)	10
.125	9 (3,3,6)	9	9 (3,3,6)	9	9 (3,3,6)	9	9 (3,3,6)	9	9 (3,3,6)	9	6 (6,2,4)	6	1 (9,1,2)	1	10 (10,1,1)	10
0	9 (3,3,6)	9	9 (3,3,6)	9	9 (3,3,6)	9	9 (3,3,6)	9	6 (6,2,4)	9 (3,3,6)	6 (6,2,4)	6	1 (9,1,2)	1	10 (10,1,1)	10

replaced by designs 18, 60, 10, and 53, respectively, in the ANOVA estimation. MINQE tends to favor either the balanced design or a design that utilizes fundamental structures #1.

For MINQE(U, I, α) estimation, replicating the designs five times increased the region of optimality for the balanced design configuration and design 8 with code (4, 4, 4) occasionally replaced designs 5 (5,5,2) and 6 (6,4,2). Similarly when using the ANOVA estimators, increasing the number of replications to 5 increases the region of optimality for design 9 and introduced three new designs namely designs 16 (6,5,1), 60 (6,4,2), and 21 (4,4,4).

A table describing the optimal designs for the combined estimator MINQH(U, I, α) was not included since it was identical to Table 12.4.

When estimating variance components from balanced data such as design 9 (3,3,6), all estimators and all variances are identical for all three methods. When design 9 was excluded the selection process consistently replaced it with design 8 with code (4,4,4).

VI. VARIANCE COMPARISON OF MINQE(U, I, α), ANOVA, AND MINQH(U, I, α) ESTIMATION

Probably the most common method for estimating variance components is through equating observed and expected mean squares in the analysis (ANOVA) and solving the resulting equations for the estimators. Desirable properties of these estimators include simple computations, unbiasedness, positiveness of $\hat{\theta}_e$ and under normality and balanced data minimum variance in the class of unbiased estimators. However, for unbalanced data, optimal properties beyond unbiasedness and a positive $\hat{\theta}_e$ are largely unknown or lacking.

Because of the dependence on prior weights, MINQE(U, I, α) are only locally minimum variance, i.e., "best" in Townsend and Searle (1971) and La Motte (1973b) and "locally best" in Harville (1969). The fact that in applications one cannot correctly specify $\rho(A)$ = Rho(A) and $\rho(B)$ = Rho(B) and so cannot (given unbalanced data) use estimators which are truly minimum variance begs the question "Why use MINQE(U, I, α)?" First, no matter what values are used for $\rho(A)$ and $\rho(B)$, so long as the values are chosen independently of the data, the estimators based on them are unbiased. This is the only claim for the ANOVA estimates with unbalanced data. Second, if the $\rho(A)$ and $\rho(B)$ are not specified "correctly" it would be of interest to see for which unknowns Rho(A) and Rho(B), the resulting estimators, have smaller variances than ANOVA estimators.

Similarly, by construction, the MINQH(U, I, α) estimators are unbiased; but unlike the MINQE(U, I, α), the variance of the θ_e estimate is identical to the variance of θ_e using the ANOVA estimator. We would also like to see if $\rho(A)$ and $\rho(B)$ are incorrectly specified, for which unknowns Rho(A) and Rho(B), the resulting MINQH estimators, have smaller variances than ANOVA and MINQE(U, I).

Since 4,225 combinations of Rho(A) and Rho(B) are considered, enumerating those points is unreasonable and consequently we summarize by reporting the proportion of cases where specific ratios of the variances of specific estimators are less than 1.

Table 12.6 can be interpreted as follows. For design 5, with $\rho(B) = 8$ and $\rho(A) =$ either 2 or 8 we found that in 98% of the 4,225 cases examined the ratio of the variance of the MINQE(U, I, α) of θ_a divided by the variance of the ANOVA estimate of θ_a denoted by VA(M)/VA(N) was less than or equal to one. Similarly, for the estimate of θ_b, the corresponding ratio of the variances denoted by VB(M)/VB(N) $\leq$ 1 for 71 percent of the cases.

Finally, for the indicated priors $\rho(A)$ and $\rho(B)$ we find the variance of the MINQE(U, I, α) of θ_e is less than the variance of the ANOVA estimate of θ_e for all unknown components, i.e., VE(M)/VE(N) always $\leq$ 1.

The average of these three percentages is 90 and is reported under the column V(M)/V(N) $\leq$ 1. This value gives some indication of how well the MINQE(U, I, α) does in relation to the ANOVA estimates based on a risk function comparison over all unknowns. Those priors maximizing this average are then adopted and included in Table 12.6 for the 15 different unbalanced design structures.

TABLE

Priors That Maximize the Percentage of Points

(ANOVA Estimates) $\leq$ 1 over

DESIGN	INDICATED PRIOR VALUES ρ(B)	ρ(A)	$\frac{V(M)}{V(N)} \leq 1$ over 3 grids	$\frac{VA(M)}{VA(N)} \leq 1$	$\frac{VB(M)}{VB(N)} \leq 1$	$\frac{VE(M)}{VE(N)} \leq 1$	$\frac{V(MH)}{V(N)} \leq 1$ over 3 grids
1	8	8	78	60	73	100	78
2	8	8	89	94	74	100	89
4	8	8	89	96	73	100	89
5	8	2	90	98	71	100	90
	8	8	89	98	71	100	89
6	8	8	78	61	73	100	78
8	8	8	91	98	74	100	91
10	4	8	89	78	89	99	88
	8	8	82	71	76	100	82
13	8	8	91	99	74	100	91
17	8	8	91	99	74	100	91
18	8	8	89	99	68	100	89
21	8	8	98	94	100	100	98
53	4	8	97	100	92	100	97
	8	8	87	82	80	100	87
55	0	8	99	100	100	97	100
	8	8	87	81	81	100	87
60	1	8	98	97	100	96	99
	8	8	89	82	85	100	89
61	8	1	98	97	100	96	99
	8	8	89	82	85	100	89

Table 12.6 also includes the percentages V(MH)/V(N) $\leq$ 1 and V(MH)/V(M) $\leq$ 1 where (MH) indicates the combined MINQH(U, I, α) estimator. Note VE(MH) = VE(N) for all cases and is not recorded. There were few differences between r = 1, and r = 5, and hence no separate breakdown is recorded.

In terms of applications, perhaps an experimenter has unbalanced data like structure 21 and is without prior knowledge of estimates

12.6

Where the Variance (MINQE(U,I))/Variance

All Unknown Components

$\frac{VA(MH)}{VA(N)} \leq 1$	$\frac{VB(MH)}{VB(N)} \leq 1$	$\frac{V(MH)}{V(M)} \leq 1$ over 3 grids	$\frac{VA(MH)}{VA(M)} \leq 1$	$\frac{VB(MH)}{VB(M)} \leq 1$	$\frac{VE(MH)}{VE(M)} \leq 1$
60	73	32	45	51	0
94	74	30	57	33	0
96	72	25	65	10	0
98	71	11	32	0	0
98	71	31	93	0	0
61	73	31	43	51	0
98	74	19	58	0	0
76	88	30	21	68	<1
71	76	31	46	46	0
99	73	24	63	9	0
99	74	18	43	12	0
99	66	33	100	0	0
94	100	16	48	0	0
100	92	14	26	15	0
82	80	29	50	38	0
100	100	<1	0	0	3
81	81	30	46	42	0
98	100	3	6	<1	4
82	86	29	48	38	0
98	100	3	6	<1	4
81	86	29	48	38	0

of components determined independently of the data other than the strong feeling they are probably smaller than 8.0. Selecting priors $\rho(A) = \rho(B) = 8.0$ and using MINQE(U, I, α) would guarantee him a smaller risk function than using ANOVA estimators on the average 98 percent of the time over the three ratios of variances.

This procedure is somewhat restricted by the fact that only 64 combinations of priors are considered. It does appear, however, that of these 64 priors, several can be found whereupon implementing MINQE(U, I, α), the variance of the estimates will be smaller than those of ANOVA estimates for a majority of those 4,225 scaled unknown components.

Table 12.6 reveals that for priors $\rho(A) = \rho(B) = 8.0$ the proportion where $V(M)/V(N) \leq 1$ ranges from 78 percent for design 1 to 98 percent for design 21 and VE(M) is uniformly smaller than VE(N) for all unknown components and all designs. Although the above priors are not selected as those maximizing the proportion where $V(M)/V(N) \leq 1$ for designs 5, 10, 53, 55, 60, and 61, these priors appear to do well in a risk comparison as indicated in Table 12.6.

The proportion where $V(MH)/V(N) \leq 1$ is almost identical to that of $V(M)/V(N) \leq 1$ for all 15 designs. A closer examination of the proportion of $V(MH)/V(M) \leq 1$, however, suggests that the variance of the MINQE(U, I, α) is somewhat smaller than the variance of the MINQH(U, I, α) estimates for a large proportion of unknowns. In fact only in the estimation of θ_a does the variance of the combined estimator MINQE(U, I, α) exceed the variance of the MINQE(U, I, α) for approximately 50 percent of the cases over all 15 designs.

Figures 12.1-12.5 display the ratio of the variances of the different estimators over the scaled unknowns for design 17 with priors $\rho(A) = \rho(B) = 8.0$. Figure 12.4 indicates $VB(M) \leq VB(MH)$ for 88 percent of those scaled unknown components considered. By examining Figure 12.5 one concludes that the variance of the MINQE(U, I, α) is only slightly but universally smaller than the variance of the ANOVA estimate of θ_e over the unknowns considered.

Table 12.7 depicts those priors maximizing the average of the three percentages $VA(MH)/VA(N) \leq 1$, $VB(MH)/VB(N) \leq 1$ and 100 since $VE(MH) = VE(N)$ for all scaled unknowns and is denoted by $V(MH)/V(N) \leq 1$. For designs 2, 4, 5, 8, 13, and 17, the recommended priors are $\rho(B) = 1$ and $\rho(A) = 2$ when using the combined estimator MINQH(U, I, α). Designs 1, 6, 10, 18, 21, 53, 55, 60, and 61 suggest alternative priors; however, with the exception of design 1, use of priors 1 and 2, re-

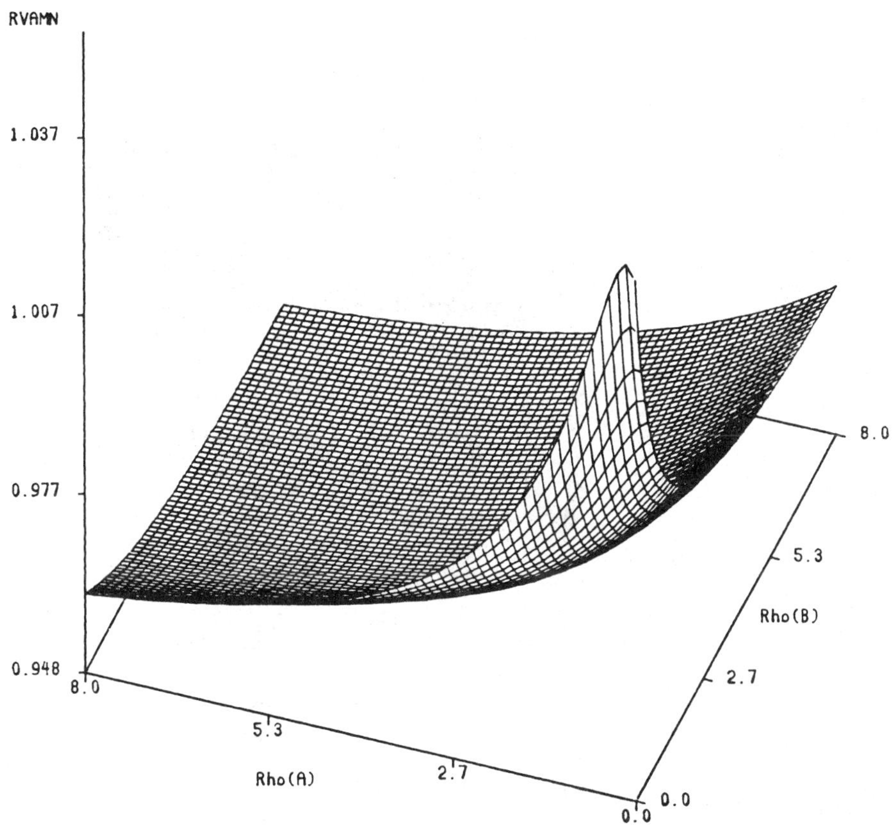

Figure 12.1 Ratio of the Variance (MINQE(U,I))/Variance (ANOVA) for Estimating the A Component (Using Design 17 the a priori Values Are $\rho(A) = 8$ and $\rho(B) = 8$ (99 Percent of Region ≤ 1.0))

spectively, appears to do an adequate job in a risk comparison with the ANOVA estimators.

Although the proportion of cases $V(M)/V(N) \leq 1$ is universally less than cases where $V(MH)/V(N) \leq 1$ for all designs considered, a close examination of the proportions $VA(M)/VA(N) \leq 1$, $VA(MH)/VA(N) \leq 1$,

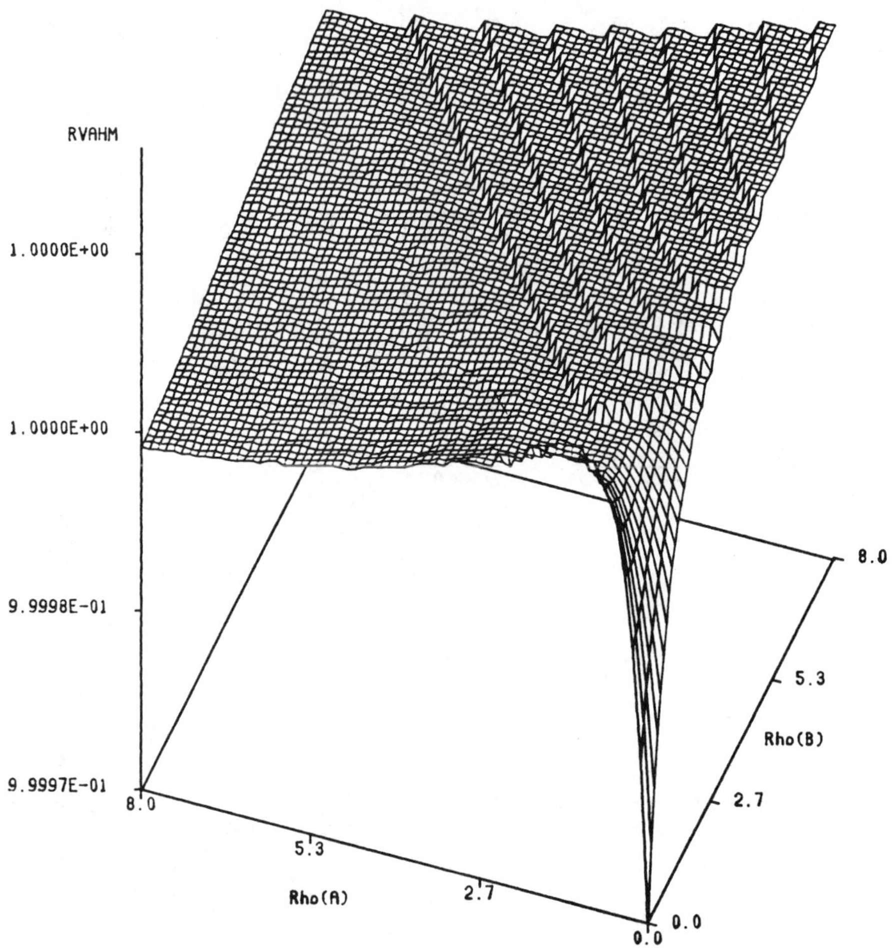

Figure 12.2 Ratio of the Variance (MINQH)/Variance (MINQE(U,I)) for Estimating the A Component (Using Design 17 the a priori Values Are $\rho(A) = 8$ and $\rho(B) = 8$ (43 Percent of Region ≤ 1.0))

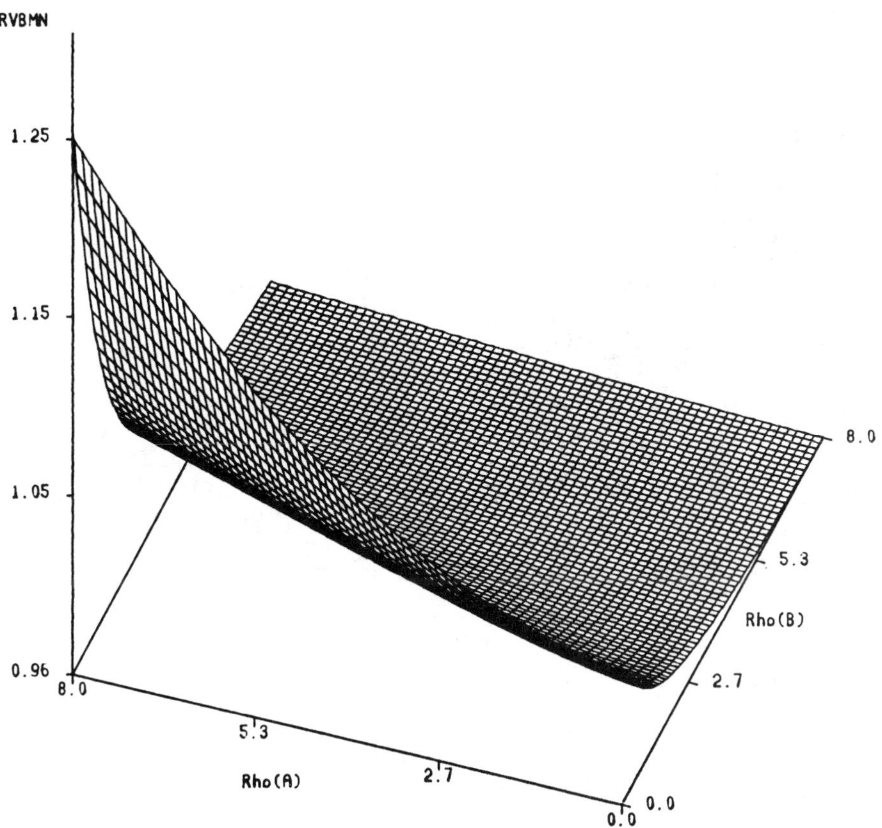

Figure 12.3 Ratio of the Variance (MINQE(U,I))/Variance (ANOVA) for Estimating the B Component (Using Design 17 the a priori Values Are $\rho(A) = 8$ and $\rho(B) = 8$ (74 Percent of Region ≤ 1.0))

and VA(MH)/VA(M) ≤ 1 indicates the variance of the MINQE estimate of θ_a is less than the variance of the combined MINQH estimator for a rather large proportion of cases. On the other hand, for those priors considered in Table 12.7 the proportion of unknown components

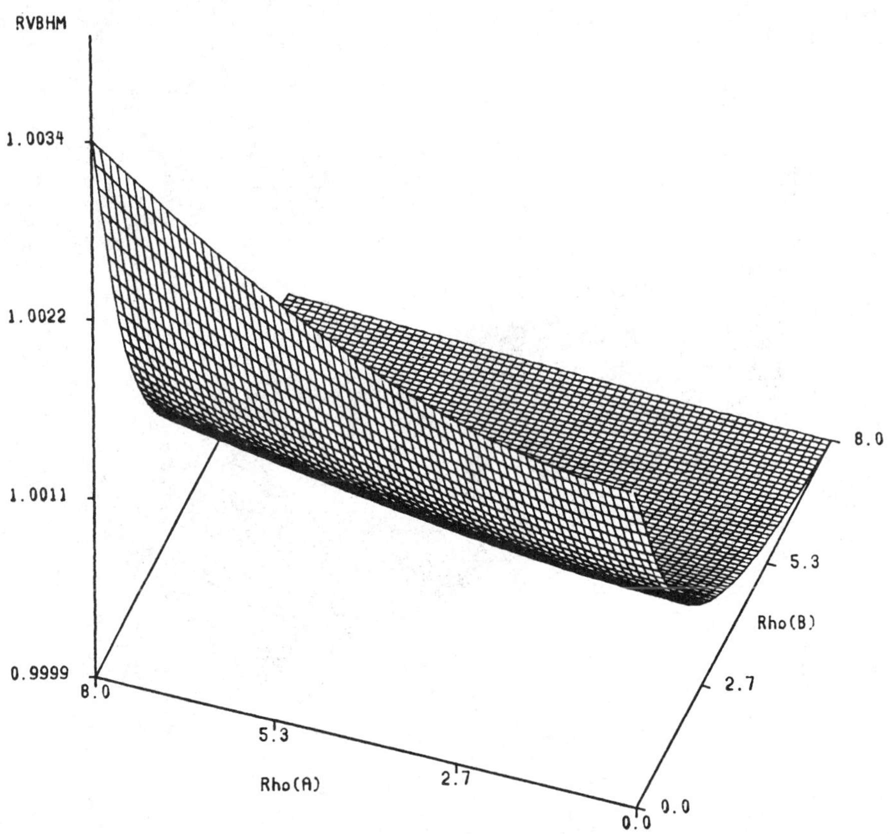

Figure 12.4 Ratio of the Variance (MINQH)/Variance (MINQE(U,I)) for Estimating the B Component (Using Design 17 the a priori Values Are $\rho(A) = 8$ and $\rho(B) = 8$ (12 Percent of Region ≤ 1.0))

where VE(M)/VE(N) ≤ 1 is significantly less than the proportion of unknown components where VE(MH)/VE(N) ≤ 1. This fact is further illustrated by examining the percentages in the VE(MN)/VE(M) ≤ 1 column of Table 12.7.

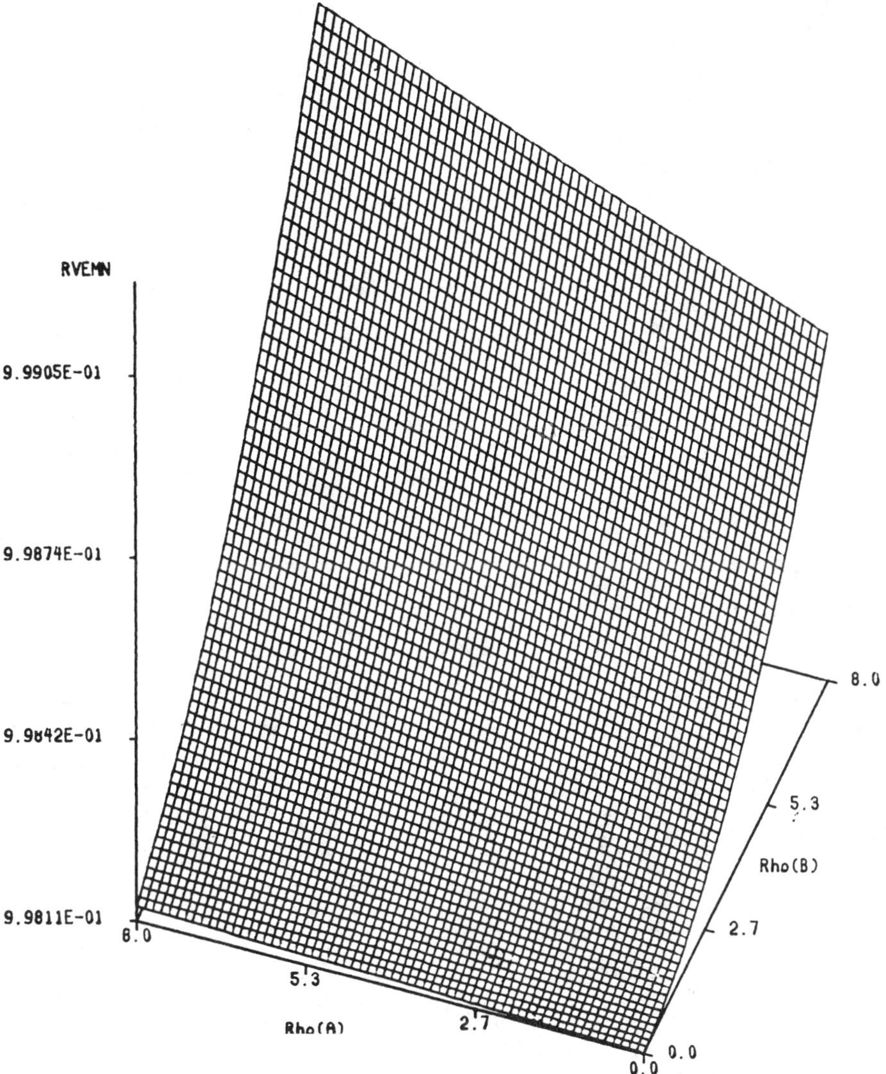

Figure 12.5 Ratio of the Variance (MINQE(U,I))/Variance (ANOVA) for Estimating the E Component (Using Design 17 the a priori Values Are $\rho(A) = 8$ and $\rho(B) = 8$ (100 Percent of Region ≤ 1.0))

TABLE

Priors That Maximize the Percentage of (ANOVA Estimates) $\leq$ 1 over

DESIGN	INDICATED PRIOR VALUES ρ(B)	ρ(A)	$\frac{V(MH)}{V(N)} \leq 1$ over 3 grids	$\frac{VA(MH)}{VA(N)} \leq 1$	$\frac{VB(MH)}{VB(N)} \leq 1$	$\frac{V(M)}{V(N)} \leq 1$ over 3 grids	$\frac{VA(M)}{VA(N)} \leq 1$
1	.125	1	83	64	86	52	60
	1	2	72	35	82	55	64
2	1	2	93	96	82	66	97
4	1	2	91	91	81	42	93
5	1	2	93	99	82	45	99
6	0	1	84	63	89	56	65
	1	2	81	60	82	54	60
8	1	2	94	99	85	53	99
10	.5	2	94	93	89	62	89
	1	2	86	76	82	58	75
13	1	2	94	99	83	47	99
17	1	2	95	99	85	54	99
18	1	4	99	99	100	49	99
	1	2	96	99	88	46	99
21	0	8	100	100	100	70	100
	1	2	85	96	57	50	96
53	1	4	98	100	95	79	100
	1	2	90	88	82	61	88
55	2	8	98	97	98	97	97
	1	2	90	86	82	61	86
60	0	8	100	100	97	97	100
	1	2	92	89	86	60	88
61	0	8	100	100	100	97	100
	1	2	94	89	92	56	89

VII. MINQE(U, I, 1)

Rao (1970, 1971a) suggests the use of priors ρ(A) = ρ(B) = 1 when independent information is not available a priori. Swallow (1978) suggests a ρ(A) = 1 and the use of MINQE(U, I, 1) over ANOVA for the

12.7

Points Where the Variance (MINQH)/Variance

All Unknown Components

$\frac{VB(M)}{VB(N)} \leq 1$	$\frac{VE(M)}{VE(N)} \leq 1$	$\frac{V(MH)}{V(M)} \leq 1$ over 3 grids	$\frac{VA(MH)}{VA(M)} \leq 1$	$\frac{VB(MH)}{VB(M)} \leq 1$	$\frac{VE(MH)}{VE(M)} \leq 1$
93	4	73	40	82	96
83	17	63	36	70	83
83	18	62	25	80	82
13	23	61	22	83	80
15	21	60	18	84	79
100	3	79	51	89	97
84	18	65	41	70	82
38	22	61	20	85	78
87	11	65	22	86	86
83	16	63	34	74	84
23	19	65	32	83	81
44	19	69	43	84	81
22	24	73	60	82	76
18	19	66	32	87	81
86	24	40	0	46	76
33	19	61	14	88	80
94	43	27	12	14	57
80	17	53	24	50	83
97	97	7	14	4	3
80	16	53	26	84	86
100	90	3	0	0	10
75	17	53	27	49	83
100	90	3	0	0	10
62	17	55	26	57	83

one-way classification if Rho(A) is felt to be greater than or equal to 1.

Table 12.8 like Tables 12.6 and 12.7 examines the percentages of unknowns where $V(M)/V(N) \leq 1$, $V(MH)/V(N) \leq 1$, and $V(MH)/V(M) \leq 1$ for the 15 designs of consideration when priors $\rho(A) = \rho(B) = 1.0$ are

TABLE

The Effects of Priors Equaling 1 on the Ratio

DESIGN	$\frac{V(M)}{V(N)} \leq 1$ over 3 grids	$\frac{VA(M)}{VA(N)} \leq 1$	$\frac{VB(M)}{VB(N)} \leq 1$	$\frac{VE(M)}{VE(N)} \leq 1$	$\frac{V(MH)}{V(N)} \leq 1$ over 3 grids
1	41	46	70	8	71
2	44	53	71	9	74
4	44	92	28	13	88
5	41	99	10	15	94
6	43	48	72	8	78
8	43	99	15	15	94
10	43	53	67	8	73
13	32	63	22	11	76
17	43	85	32	11	84
18	41	100	11	12	82
21	41	96	15	12	75
53	44	67	56	8	77
55	45	74	54	8	78
60	43	79	42	8	81
61	40	81	30	8	83

adopted. One should pay special attention to the fact that the proportions of unknowns where $V(MH)/V(N) \leq 1$ are invariably larger than the respective proportion $V(M)/V(N) \leq 1$ for all 15 designs. This undoubtedly can be attributed to the fact that the proportion of unknowns

12.8

of the Variance over All Unknown Components

$\frac{VA(MH)}{VA(N)} \le 1$	$\frac{VB(MH)}{VB(N)} \le 1$	$\frac{V(MH)}{V(M)} \le 1$ over 3 grids	$\frac{VA(MH)}{VA(M)} \le 1$	$\frac{VB(MH)}{VB(M)} \le 1$	$\frac{VE(MH)}{VE(M)} \le 1$
47	67	67	66	43	92
54	69	70	54	65	91
93	92	70	43	80	87
99	82	68	33	86	85
65	68	69	70	45	92
99	85	69	37	86	85
54	64	69	64	52	92
63	65	79	63	89	63
86	65	82	71	85	89
100	47	66	16	93	88
96	29	67	18	95	88
67	62	74	52	79	92
74	59	76	62	74	92
80	61	76	56	78	92
82	68	76	54	84	92

where VE(M)/VE(N) $\le$ 1 is very small and VE(MH) = VE(N). Also of interest is the large proportions in the column representing the percentages of V(MH)/V(M) $\le$ 1.

VIII. MINQE(U, I, 0)

MINQE(U, I, 0) has some appeal because it leads to tremendous simplification of the arithmetic, especially for large data sets. It is Rao's MINQE(U, I) with zero prior weights except for the error component. This is usually set equal to 1. This method was recently advocated by Hartley, Rao, and La Motte (1978) because of the follow-

TABLE

The Effects of Zero Priors on the Ratio

DESIGN	$\frac{V(M)}{V(N)} \leq 1$ over 3 grids	$\frac{VA(M)}{VA(N)} \leq 1$	$\frac{VB(M)}{VB(N)} \leq 1$	$\frac{VE(M)}{VE(N)} \leq 1$	$\frac{V(MH)}{V(N)} \leq 1$ over 3 grids
1	39	17	100	< 1	54
2	3	9	< 1	< 1	41
4	< 1	< 1	< 1	< 1	34
5	< 1	< 1	< 1	< 1	36
6	36	7	100	< 1	50
8	< 1	< 1	< 1	< 1	36
10	7	20	< 1	< 1	50
13	5	13	< 1	< 1	39
17	4	12	< 1	< 1	38
18	< 1	1	< 1	< 1	35
21	32	95	< 1	< 1	66
53	8	23	< 1	< 1	46
55	9	26	< 1	< 1	45
60	14	40	< 1	< 1	50
61	16	49	< 1	< 1	54

ing properties: unbiasedness, locally best (at zero), asymptotic consistency, admissibility, best unbiased if the design is balanced, and computationally efficient. It would seem that the latter point far outweighs consideration of the one underlined.

Goodnight (1979) suggests this procedure because of those optimal properties but seems to make no mention concerning the potentially hazardous effect when the true unknown components are not in fact close to zero.

12.9

of the Variances over All Unknown Components

$\frac{VA(MH)}{VA(N)} \leq 1$	$\frac{VB(MH)}{VB(N)} \leq 1$	$\frac{V(MH)}{V(M)} \leq 1$ over 3 grids	$\frac{VA(MH)}{VA(M)} \leq 1$	$\frac{VB(MH)}{VB(M)} \leq 1$	$\frac{VE(MH)}{VE(M)} \leq 1$
19	41	80	98	41	99
9	13	91	95	77	99
< 1	2	97	91	99	99
< 1	9	96	89	99	99
7	42	80	99	42	99
< 1	8	99	98	99	99
21	28	92	98	80	99
14	4	99	98	99	99
12	2	99	99	99	99
2	3	67	< 1	99	99
95	1	76	27	99	99
24	13	99	96	99	99
28	7	99	99	99	99
48	3	99	96	99	99
59	4	97	92	99	99

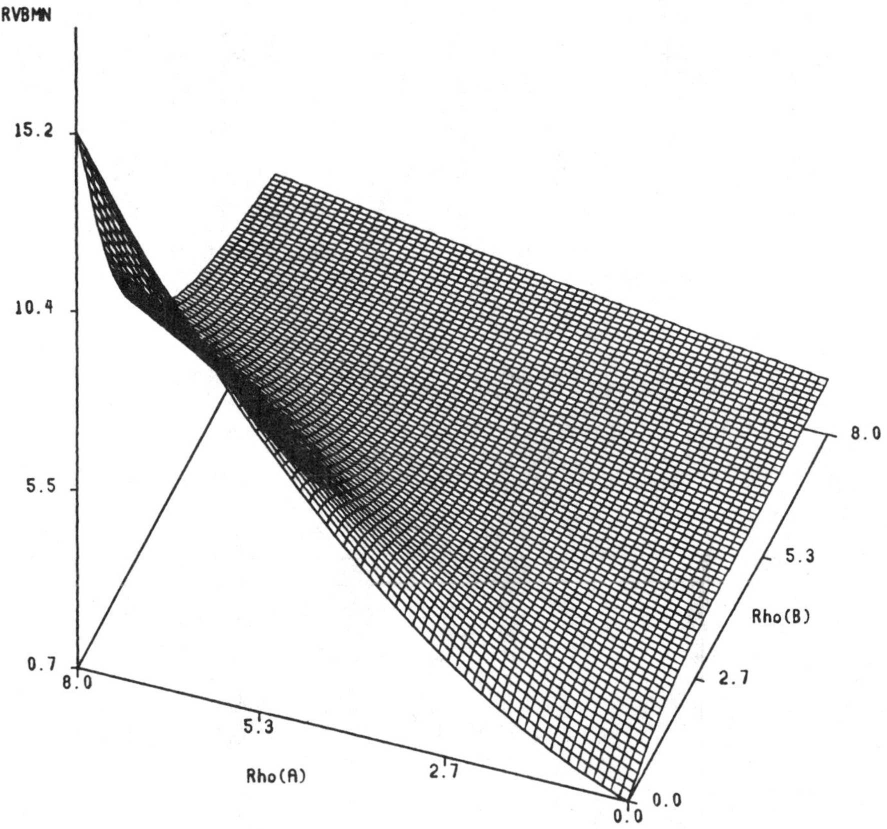

Figure 12.6 Ratio of the Variance (MINQE(U,I))/Variance (ANOVA) for Estimating the B Component (Using Design 17 the a priori Values Are $\rho(A) = 0$ and $\rho(B) = 0$ (.2 Percent of Region ≤ 1.0))

Table 12.9 examines the effects of using zero priors. The proportions $V(M)/V(N) \leq 1$ are consistently smaller for all designs compared with those considered previously. However, for designs 1 and 6 the variance of the MINQE(U, I, 0) of θ_b is uniformly less than the variance of the ANOVA estimate over the grid. Also, for 95 percent of the cases considered in design 21 the variance of the MINQE(U, I, 0) of θ_a is less than the variance of the ANOVA estimate.

The combined MINQH estimator minimizes the ratio of the variances with the ANOVA estimator for a larger proportion of unknowns

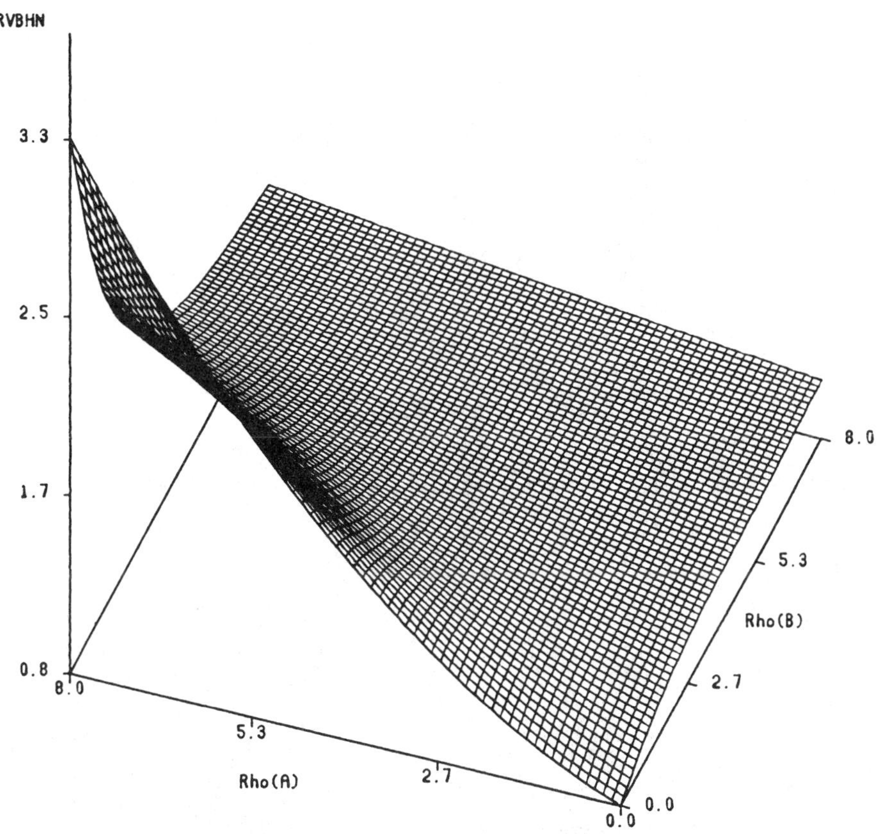

Figure 12.7 Ratio of the Variance (MINQH)/Variance (ANOVA) for Estimating the B Component (Using Design 17 the a priori Values are $\rho(A) = 0$ and $\rho(B) = 0$ (.2 Percent of Region $\leq$ 1.0))

than MINQE(U, I, 0) and with the exception of design 18 the variance of the combined estimators is less than the variance of the MINQE(U, I, 0) for a majority of those scaled unknowns considered. Figure 12.6 clearly shows that for $\rho(A) = 0 = \rho(B)$ and the true Rho(A) = 8 and Rho(B) = 0, the ratio of the variances denoted by VB(M)/VB(N) equals 15.2. On the other hand, Figure 12.7 shows the ratio VB(MH)/VB(N) = 3.34. Figure 12.8 examines the case when $\rho(A) = \rho(B) = 0$ and Rho(A) = Rho(B) = 8.0; then VE(M)/VE(N) $\doteq$ 118, a truly alarming feature when implementing MINQE(U, I, 0).

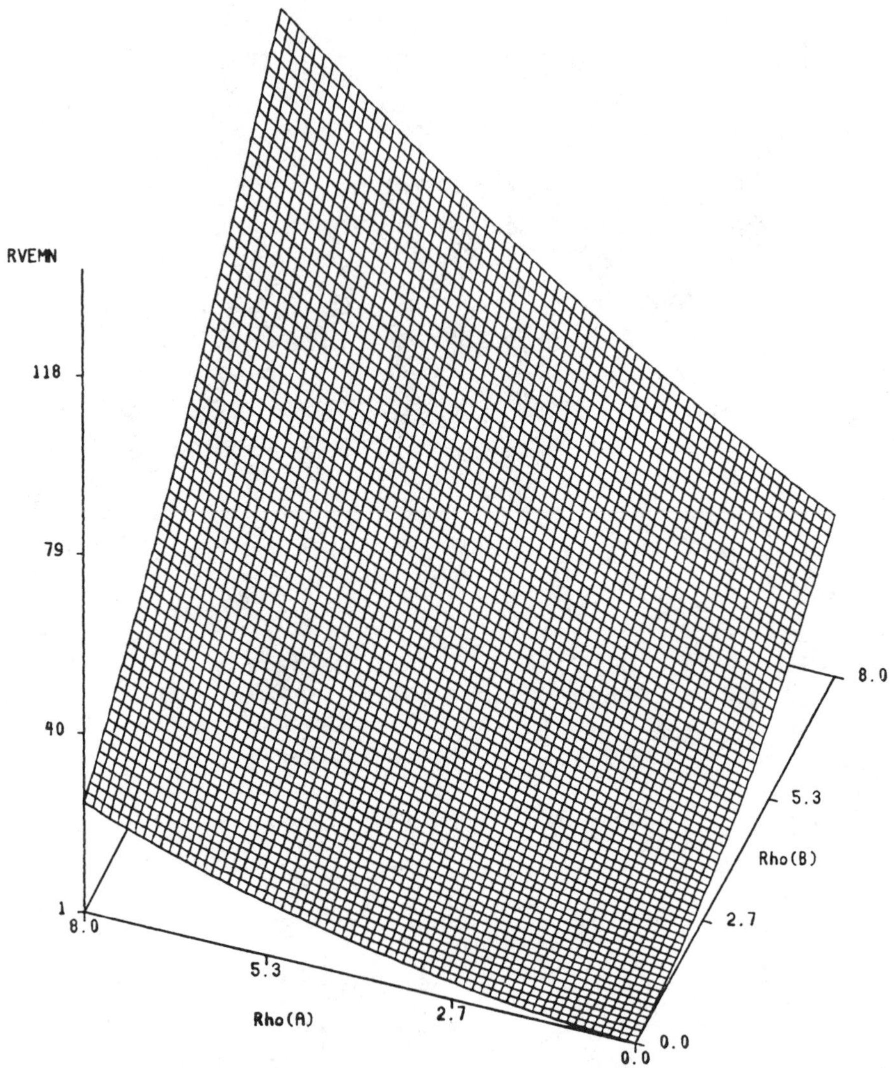

Figure 12.8 Ratio of the Variance (MINQE(U,I))/Variance (ANOVA) for Estimating the E Component (Using Design 17 the a priori Values Are $\rho(A) = 0$ and $\rho(B) = 0$ (.2 Percent of Region ≤ 1.0))

IX. SUMMARY OF THE RESULTS

The emphasis in this study has been on determining efficient expressions for calculating Minimum Norm Quadratic Estimators for the two-stage nested model. Three different procedures for estimating the variance components were developed.

Since simple expressions for the variance components and their variances cannot be obtained, analytic comparisons are not possible. For the two-stage nested design a number of unbalanced designs were considered and for each design the variances of the variance components for certain values of the variance components and pre-determined prior values were calculated.

From this study the following conclusions were obtained:

1. When a priori weights for the variance components are both less than one, a balanced design is suggested as being optimal for all three estimation procedures.
2. Prior weights can be selected which upon implementation in the MINQE(U, I, α) equations guarantee smaller variances than for ANOVA estimates for at least 78 percent of the grid areas considered.
3. Selecting large prior values (i.e., $\rho(A) = \rho(B) = 8.0$) gives smaller variances for MINQE(U, I, α) estimates than for ANOVA estimates over a large part of the grid for all designs considered.
4. Those prior values recommended when using the combined MINQH(U, I, α) estimator are of the order $\rho(A) = 2$ and $\rho(B) = 1$ as compared with $\rho(A) = \rho(B) = 8.0$ for the MINQE(U, I, α).
5. If values for priors are scarce and someone recommends using $\rho(A) = \rho(B) = 1$, the combined estimator MINQH appears to have some optimal properties over MINQE.
6. Zero priors should probably not be used.

REFERENCES

Ahrens, H. (1978). MINQE and ANOVA estimator for one-way classification - a risk comparison. *Biometrical Journal 20*, 535-556.

Anderson, R.L. (1975). Designs and estimators for variance components. In *Statistical Design and Linear Models* (Chapter 1). J. N. Srivastava (ed.), Amsterdam: North Holland.

Anderson, R.L. (1978). Recent developments in designs and estimators for variance components. Technical Report No. 118, University of Kentucky, Lexington, Kentucky.

Anderson, R.L. and Crump, P.P. (1967). Comparison of designs and estimation procedures for estimating parameters in a two-stage nested process. *Technometrics 9*, 499-516.

Bush, N. and Anderson, R.L. (1963). A comparison of three different procedures for estimating variance components. *Technometrics 5*, 421-440.

Crump, P.P. (1954). Optimal designs to estimate the parameters of a variance component model. Unpublished Ph.D. dissertation, North Carolina State University, Raleigh, North Carolina.

Crump, S.L. (1946). The estimation of variance components in analysis of variance. *Biometrics Bulletin 2*, 7-11.

Gaylor, D.W. (1960). The construction and evaluation of some designs for the estimation of parameters in random models. Unpublished Ph.D. dissertation, North Carolina State University, Raleigh, North Carolina.

Ginsburg, E.H. (1973). On the planning of the experiment on estimation of intraclass correlation. *Biom. Zeitschrift 15*, 47-52.

Goldsmith, C.H. (1969). Three stage nested designs for estimating variance components. Unpublished Ph.D. dissertation, North Carolina State University, Raleigh, North Carolina.

Goldsmith, C.H. and Gaylor, D.W. (1970). Three stage nested designs for estimating variance components. *Technometrics 12*, 487-498.

Goodnight, J. (1979). New features in GLM and VARCOMP. Proceedings of Fourth Annual SAS Users' Group International Conference, SAS Institute, Cary, North Carolina.

Graybill, F.A. (1954). On quadratic estimates of variance components. *Ann. Math. Stat. 25*, 367-372.

Graybill, F.A. and Wortham, A.W. (1956). A note on uniformly best unbiased estimators for variance components. *J. Am. Stat. Assoc. 51*, 266-268.

Hartley, H.O., Rao, J.N.K. and La Motte, L.R. (1978). A simple 'synthesis'-based method of variance component estimation. *Biometrics 34*, 233-242.

Hammersley, J.M. (1949). The unbiased estimate and standard error of the interclass variance. *Metron 15*, 189-205.

Harville, D.A. (1969). Quadratic unbiased estimation of variance components for the one-way classification. *Biometrika 56*, 313-326.

Harville, D.A. (1977). Maximum likelihood approaches to variance component estimation. *J. Am. Stat. Assoc. 72*, 320-340.

Henderson, C.R. (1953). Estimation of variance and covariance components. *Biometrics 9*, 226-252.

Herrendörfer, G. (1974). Experimental design for estimating variance component between groups for a simple analysis of variance. *Biom. Zeitschrift 16*, 406-415.

Herrendörfer, G. (1979). Optimal designs for estimating the variance components in a simple variance component model (Model II), I. *Biom. Journal 21*, 11-16.

Hess, J.L. (1979). Sensitivity of MINQUE with respect to a priori weights. *Biometrics 35*, 645-649.

Kleffe, J. (1977). Optimal estimation of variance components - a survey. *Sankhyā Series B 39*, 211-244.

Kleffe, J. (1978). On quadratic estimation of heteroscedastic variances. *Mathematische Operationsforschung und Statistik 9*, 27-44.

La Motte, L.R. (1973a). On non-negative quadratic unbiased estimation of variance components. *J. Am. Stat. Assoc. 68*, 728-730.

La Motte, L.R. (1973b). Quadratic estimation of variance components. *Biometrics 29*, 311-330.

Muse, H.D. (1974). Comparison of designs to estimate variance components in a two-way classification model. Unpublished Ph.D. dissertation, University of Kentucky, Lexington, Kentucky.

Muse, H.D. and Anderson, R.L. (1978). Comparison of designs to estimate variance components in a two-way classification model. *Technometrics 20*, 159-166.

Prairie, R.R. (1962). Optimal designs to estimate variance components and to reduce product variability for nested classifications. Unpublished Ph.D. dissertation, North Carolina State University, Raleigh, North Carolina.

Rao, C.R. (1970). Estimation of heteroscedastic variances in linear models. *J. Am. Stat. Assoc. 65*, 161-172.

Rao, C.R. (1971a). Estimation of variance and covariance components MINQUE theory. *J. Mult. Anal. 1*, 257-275.

Rao, C.R. (1971b). Minimum variance quadratic unbiased estimation of variance components. *J. Mult. Anal. 1*, 445-456.

Rao, P.S.R.S. (1979). Theory of the MINQUE - a review. *Sankhyā Series B 39*, 201-210.

Rao, C.R. and Mitra, S.K. (1979). *Generalized Inverse of Matrices and its Applications*. New York: John Wiley.

Searle, S.R. (1979). Notes on variance component estimation: A detailed account of maximum likelihood and kindred methodology. Paper BU-673-M in the Biometrics Unit, Cornell University, Ithaca, New York.

Steel, R.G.D. and Torrie, J.H. (1980). *Principles and Procedures of Statistics*. 2nd ed., New York: McGraw-Hill.

Swallow, W.H. and Searle, S.R. (1978). Minimum variance quadratic unbiased estimation (MIVQUE) of variance components. *Technometrics 20*, 265-272.

Swallow, W.H. (1981). Variances of locally minimum variance quadratic unbiased estimates ("MIVQUE's") of variance components. *Technometrics 23*, 271-284.

Thitakomal, B. (1977). Extension of previous results on properties of estimators of variance components. Unpublished Ph.D. dissertation, University of Kentucky, Lexington, Kentucky.

Thompson, W.O. and Anderson, R.L. (1975). A comparison of designs and estimators for the two-stage nested design. *Technometrics 17*, 37-44.

Townsend, E.C. and Searle, S.R. (1971). Best quadratic unbiased estimation of variance components from unbalanced data in the one-way classification. *Biometrics 29*, 643-657.

Chapter 13

PATH ANALYSIS, CORRELATION, AND THE ANALYSIS OF VARIANCE

Ted H. Emigh
North Carolina State University
Raleigh, North Carolina

I. INTRODUCTION

Path analysis, also called the method of path coefficients, was developed by Wright (1918, 1920, 1921) in order to make sense out of the correlational relationships among several variables and their effect on a target variable. In the more than sixty years since its introduction, path analysis has developed to a large extent, but many concerns about the validity of the method still remain. Among these concerns are the use of the coefficients of determination when the sources are correlated, and the validity of path analysis under nonadditivity or asymmetry of the system. In this paper, I shall explore path analysis from an analysis of variance viewpoint through a comparison of the analysis of commonality with path analysis.

Before we begin, it should be pointed out that the algebra for regression, correlation, the analysis of variance, and, to a lesser degree, path analysis is largely the same. For the purposes of this paper, I shall make a larger distinction among regression, correlation, and the analysis of variance than is done in practice. I shall refer to the regression viewpoint as the process of fitting a model to data with the purpose of prediction. I shall refer to correlation analysis or the correlation viewpoint to mean the process of determining correlational relationships among the variables, particularly when we are not primarily interested in the effects of the variables on one particular variable. I shall refer to the analysis of variance viewpoint as the process of attributing variation in a single variable to variation in the other variables through the analysis of variance. While there can be debate on this division, it will make the discussion that follows clearer.

II. REVIEW OF PATH ANALYSIS

In this section, I shall briefly summarize path analysis from the viewpoint of multiple linear regression. The reader may wish to refer to Li (1975) for a more complete development of path analysis. Li's summary is a very good description of path analysis as a method of multiple linear regression and as correlation analysis. The reader also may wish to refer to Wright's original sources (1920, 1921a,b,c,d,e,f, 1931, 1934) or to such texts as Kempthorne (1957), Li (1975), or Wright (1968, 1969).

Consider the multiple linear regression model

$$X_0 = \beta_0 + \beta_1 X_1 + \dots + \beta_m X_m + E$$

We may standardize the variables, so that $x_i = (X_i - \mu_i)/\sigma_i$, i=0,..., m, where $\mu_i = E[X_i]$ and $\sigma_i^2 = Var[X_i]$. Also, we may standardize the error term as $e = (E - \mu_E)/\sigma_E$. The standardized regression model is

$$x_0 = P_1 x_1 + P_2 x_2 + \dots + P_m x_m + P_e e \tag{2.1}$$

where $P_i = \beta_i \sigma_i/\sigma_0$ and $\beta_e = 1$. We may define $k = m + 1$, $P_k = P_e$ and $x_k = e$. Hence $x_0 = \Sigma P_i x_i$. Since the error term has the same form as the other variables, we do not need to define a formal error term.

However, we need to add the condition that this term is independent of the other model terms. The P_i term is called the path coefficient for the ith source and is the standardized partial regression coefficient of X_i on X_0 after $X_1, X_2, \ldots, X_{i-1}, X_{i+1}, \ldots, X_k$.

Using these relationships, the properties of path coefficients are obtained as

$$\begin{aligned} \mathrm{Corr}(x_{i'}, x_0) &= \mathrm{Corr}(x_{i'}, \Sigma P_i x_i) \\ &= \Sigma P_i \,\mathrm{Corr}(x_{i'}, x_i) \\ &= \Sigma P_i \rho_{i'i} \end{aligned} \tag{2.2}$$

and

$$\begin{aligned} 1 = \mathrm{Corr}(x_0, x_0) &= \Sigma P_i \rho_{i0} \\ &= \Sigma P_i \Sigma P_{i'} \rho_{ii'} \\ &= \Sigma P_i \rho_{ii'} P_{i'} \end{aligned} \tag{2.3}$$

Estimates for the path coefficients may be obtained through either multiple linear regression and equation (2.1) or by estimating the correlations in (2.2) and (2.3) and solving for the path coefficients.

This model may be described by the path diagram for the sources A, B, and C influencing Y given in Figure 13.1. The arrows pointing from sources A, B, and C to Y show an assumed causal relationship for A, B, and C with Y. The two-headed arrows among the A, B, and C sources indicate that there are no assumed causal relationships among the sources, but that they are correlated. The basic path equations become

$$y = P_A A + P_B B + P_C C \tag{2.1a}$$

and

$$1 = P_A^2 + P_B^2 + P_C^2 + 2P_A\rho_{AB}P_B + 2P_A\rho_{AC}P_C + 2P_B\rho_{BC}P_C \tag{2.3a}$$

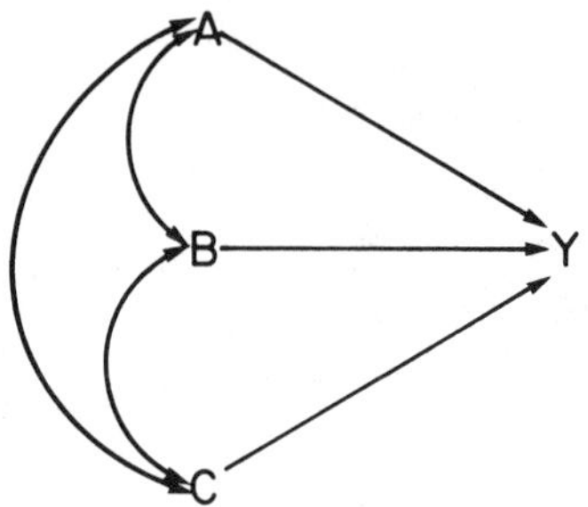

Figure 13.1 Path Diagram for Three Correlated Sources

The terms in (2.3a) divide the total standardized variation of the Y variable into what Wright calls coefficients of determination. Aside from the graphical convenience of the path diagrams, path analysis is mainly concerned with estimating and interpreting these quantities. Some comments about these coefficients are given by Kempthorne (1977, 1978), and an initial (and now dated) discussion of determination and correlation is given by Niles (1922, 1923) and Wright (1923). When the sources are independent, i.e., when $\rho_{AB} = \rho_{AC} = \rho_{BC} = 0$, there is little difficulty in calling the squares of the path coefficients the coefficients of determination, but even then the relationship between this and causality is quite obscure. The case when correlations are present is far more difficult to interpret.

III. PATH ANALYSIS AND CORRELATION ANALYSIS

While path analysis is often developed through multiple linear regression, path analysis is not primarily concerned with prediction from a model (as is regression analysis) nor is it used with experimental data (compared to observational data) as is the analysis of variance. Correlation analysis is concerned with observational data, and we now consider the correlational aspects of path analysis. In path analysis we are concerned with attributing the variability of the Y-variate (dependent variable) to one or more sources. On the other hand, the correlation viewpoint is usually concerned with the relationship of the variables, rather than with how they affect one particular variate.

We can see from the basic equations of path analysis, (2.2) and (2.3), that the correlation coefficient is of utmost importance in path analysis. As Tukey (1954) has pointed out, one of the basic path equations (2.3) is equivalent to the following equality of correlation coefficients

$$R^2 = \sum_i \rho_{i.}^2 + 2 \sum_{i<j} \rho_{ij}\rho_{i.}\rho_{j.}$$

where R^2 is the multiple correlation of the regression (with or without including the error term in the model) and $\rho_{i.}$ is the partial correlation of X_i with X_0 after $X_1,\ldots,X_{i-1}$, $X_{i+1},\ldots,X_m$.

The correlation viewpoint has been influenced strongly by the multivariate normal distribution (and generally assumes this distribution), and with the symmetry of this distribution. Although it is usually not stated, one of the assumptions for path analysis is that the sources act symmetrically (see, e.g., Wright, 1968, page 299). When there is a correlation among sources, it is difficult to attribute the determination of the Y-variable to each of the sources. This is due to the lack of a model of how the correlations arise. For example, source B might control the level or value of source C, in which case the coefficient of determination should be much larger for B than for C, with an appropriate reduction in the joint term, and possibly the unique term for C. On the other hand, both might be caused equally by another source, D. In this case, the coefficients of determination should be divided more evenly between B and C, or, better, D should be included in the model. For multivariate normal distributions and path analysis the ambiguity as to how to divide the variation is resolved through dividing the joint term evenly (proportionately) among the sources.

In some respects, path analysis is similar to analysis of variance problems, where we are interested in attributing variation to each of the sources. However, the usual use of the analysis of variance is with data from designed experiments and, hopefully, uncorrelated sources. Even when correlations are present, as with missing value data, the correlations are considered nuisances to be eliminated, whereas path analysis considers the correlations important parts of the data. In the next section, we shall consider how the analysis of variance can be used in a path analysis situation where the sources are correlated, and the correlations are meaningful.

IV. PATH ANALYSIS AND THE ANALYSIS OF VARIANCE

The division of the total sum of squares (or variance) can be accomplished through the analysis of variance. The regression model is the same as for path analysis, and, indeed, the algebra is largely the same.

We may start with the same model as path analysis:

$$X_0 = \beta_0 + \beta_1 X_1 + \ldots + \beta_m X_m + e$$

The analysis of variance gives an indication of how well the data fit the model or any of the various submodels. If regression is applied to a subset of the model, using $\{X_{(1)},\ldots,X_{(r)}\}$, where $X_{(i)}$ is some variable from the overall regression, we may define the subset regression sum of squares (corrected) as $R[X_{(1)},\ldots,X_{(r)}]$. This definition is consistent with Searle (1971) and Speed and Hocking (1976).

If our regression model includes two variables (X_1 and X_2) then the possible regression sums of squares are $R[X_1]$, $R[X_2]$, and $R[X_1,X_2]$. For the usual sequential analysis there are two possible arrangements, given in Table 13.1. The first arrangement is to fit the model in the order X_1, then X_2 after X_1. The second is to fit the model in the order X_2, then X_1 after X_2. The usual terminology is to call $R[X_1]$ the sum of squares for X_1 ignoring X_2, while $R[X_2|X_1] = R[X_1,X_2] - R[X_1]$ is called the sum of squares for X_2 eliminating X_1.

TABLE 13.1

Possible Sequential Sums of Squares and the Analysis of Commonality for Two Variables

Sequence 1		Sequence 2		Analysis of Commonality	
Source	Regression	Source	Regression	Source	Commonality
X_1	$R[X_1]$	X_2	$R[X_2]$	X_1 uniquely	$U(X_1)$
X_2 after X_1	$R[X_2 \mid X_1]$	X_1 after X_2	$R[X_1 \mid X_2]$	X_2 uniquely	$U(X_2)$
				X_1, X_2 jointly	$Com(X_1, X_2)$
Total	$R[X_1, X_2]$		$R[X_1, X_2]$		$R[X_1, X_2]$

Kempthorne (1957) was the first to suggest that path analysis might be defined through the use of the analysis of variance (see Tukey, 1954, for a discussion of path analysis and regression). The method Kempthorne described was developed independently by Mood (1971) and called the analysis of commonality. Since that time, Beaton (1973) has formalized the definition of the analysis of commonality, and Emigh (1974 and 1977) has used the analysis of commonality to study path analysis.

In the analysis of commonality, the additional sum of squares that is contributed by X_i after fitting every other term in the model, is the unique sum of squares for that source, denoted by $U(X_i)$. In the same way, the additional sum of squares contributed by X_i and X_j, with $i \neq j$, after fitting every other term in the model is the unique sum of squares for that pair, denoted by $U(X_i, X_j)$. If $U(X_1) + U(X_2) \neq U(X_1, X_2)$, as is usually the case, the difference is the contribution to the sum of squares which is common to both X_1 and X_2 and which is independent of the other terms in the model. This difference is the common term: $\mathrm{Com}(X_1, X_2) = U(X_1, X_2) - U(X_1) - U(X_2)$. While this term is usually positive, it is possible for the common term to be negative. The relationship of the terms in the analysis of commonality to the components of variance in an analysis of variance model has been given by Emigh (1977 and 1983).

The analysis of commonality for the model with two possible terms is given in Table 13.1. In this model, the common term has several equivalent forms:

$$\begin{aligned} \mathrm{Com}(X_1, X_2) &= R[X_1, X_2] - R[X_1 | X_2] - R[X_2 | X_1] \\ &= R[X_1] + R[X_2] - R[X_1, X_2] \\ &= R[X_1] - R[X_1 | X_2] \\ &= R[X_2] - R[X_2 | X_1] \end{aligned}$$

For three sources, X_1, X_2, and X_3, the seven possible regression sums of squares are: $R[X_1]$, $R[X_2]$, $R[X_1, X_2]$, $R[X_3]$, $R[X_1, X_3]$,

$R[X_2,X_3]$ and $R[X_1,X_2,X_3]$. The unique terms are defined as the partial regression sum of squares. For example,

$$U(X_1) = R[X_1|X_2,X_3] = R[X_1,X_2,X_3] - R[X_2,X_3]$$

The two-term joint sources are found as in the above example after eliminating the third variable, as

$$\text{Com}(X_1,X_2) = R[X_1,X_2|X_3] - U(X_1) - U(X_2)$$

With the analysis of commonality, up to m-term joint effects (common terms) are possible, where m is the number of model sources. These higher order joint effects are of particular interest in the analysis of commonality. The three-term joint source is found by obtaining the unique term for the three terms together, then subtracting out each unique term and each pair of two-term common terms. We may write the three-term common effect for this model as

$$\begin{aligned}\text{Com}(X_1,X_2,X_3) &= R[X_1,X_2,X_3] - U(X_1) - U(X_2) - \text{Com}(X_1,X_2) - \\ &\quad U(X_3) - \text{Com}(X_1,X_3) - \text{Com}(X_2,X_3) \\ &= R[X_1,X_2,X_3] - R[X_1,X_2] - R[X_1,X_3] - \\ &\quad R[X_2,X_3] + R[X_1] + R[X_2] + R[X_3]\end{aligned}$$

The common terms for a larger model are defined in an analogous manner. The unique term for a source, $U(X_i)$, is that part of the variation in Y which can be uniquely attributed to that source (at least within the context of the model under consideration). The common term between two sources, $\text{Com}(X_i,X_j)$, is that part of the variation in Y which can be uniquely attributed to the combination of X_i and X_j, after each of the unique terms for X_i and X_j. In general, the common term among r sources, $\text{Com}(X_{(1)},\ldots,X_{(r)})$, is that part of the variation in Y which can be uniquely attributed to $X_{(1)},\ldots,X_{(r)}$, but which cannot be uniquely attributed to any subset of these sources.

Emigh (1977) was able to determine the population parameters in the analysis of commonality for unbalanced random classification models. The population parameters for the analysis of commonality with three sources are given in Table 13.2. The sampling properties for this type of problem when the study is observational, rather than experimental have been worked out by Emigh (1983). The interpretation of the terms in the analysis of commonality for a classification model are apparent from Table 13.2. The unique term is the average variance of that term within each of the other treatment classes in the model. The common term between two of the terms, say A and B, is made up of two parts. The first part is the average covariance between, say, A and B effects within the C classes. The second part consists of the variance of each effect taken separately over the classes of the other effect, which are averaged over the classes of the third effect. The three-way effect is harder to describe, but is analogously made up of three sets of terms. The first set of terms consists of the covariances of the conditional means of two effects averaged over the classes of the third effect. The second set of terms consists of the variances of the conditional means of one effect given the other two effects. Finally, the third set of terms consists of the conditional means of each effect over a second effect, which are then averaged over the classes of the third effect. The extension to larger models is analogous to this, although the algebra becomes tedious.

While both path analysis and the analysis of variance have roots in the general linear model, they represent differing viewpoints in their interpretation of data. Li (1975) demonstrates how the results of a regression model can be derived through the use of path analysis. In particular, he shows how the results of a selection of variables problem can be derived through the use of path analysis. Even so, the procedures are quite different. A comparison of the terms of the analysis of commonality with the terms of a path analysis are given in Table 13.3. The population parameters for the analysis of commonality for a problem with three effects (Table 13.2) are written using the notation from path analysis. Compared with

TABLE 13.2

Population Parameters for the Analysis of Commonality with Three Classification Sources

Source	Commonality
U(A)	$E_{B,C}[Var(A\|B,C)]$
U(B)	$E_{A,C}[Var(B\|A,C)]$
Com(A,B)	$E_C[Cov(A,B\|C)] + E_C\{Var_B[E(A\|B,C)]\} + E_C\{Var_A[E(B\|A,C)]\}$
U(C)	$E_{A,B}[Var(C\|A,B)]$
Com(A,C)	$E_B[Cov(A,C\|B)] + E_B\{Var_C[E(A\|B,C)]\} + E_B\{Var_A[E(C\|A,B)]\}$
Com(B,C)	$E_A[Cov(B,C\|A)] + E_A\{Var_C[E(B\|A,C)]\} + E_A\{Var_B[E(C\|A,B)]\}$
Com(A,B,C)	$Cov_C[E(A\|C),E(B\|C)] + Cov_B[E(A\|B),E(C\|B)] + Cov_A[E(B\|A),E(C\|A)] +$
	$Var_{B,C}[E(A\|B,C)] + Var_{A,C}[E(B\|A,C)] + Var_{A,B}[E(C\|A,B)] -$
	$E_C\{Var_B[E(A\|B,C)]\} - E_C\{Var_A[E(B\|A,C)]\} - E_B\{Var_C[E(A\|B,C)]\} -$
	$E_B\{Var_A[E(C\|A,B)]\} - E_A\{Var_C[E(B\|A,C)]\} - E_A\{Var_C[E(B\|A,C)]\}$

TABLE 13.3

Path Analysis and Standardized Analysis of Commonality for Three Sources in Terms of Partial Correlation Coefficients and Path Coefficients

Source	Path Analysis	Commonality
A	P_A^2	$U(A) = P_A^2(1-\rho_{AB}^2)(1-\rho_{AC\cdot B}^2) = P_{A\cdot BC}^2$
B	P_B^2	$U(B) = P_B^2(1-\rho_{AB}^2)(1-\rho_{BC\cdot A}^2) = P_{B\cdot AC}^2$
A,B	$2P_A\rho_{AB}P_B$	$Com(A,B) = (1-\rho_{AC}^2)\rho_{AB\cdot C}^2P_A^2 + (1-\rho_{BC}^2)\rho_{AB\cdot C}^2P_B^2 + 2P_A(\rho_{AB}-\rho_{AC}\rho_{BC})P_B = P_{A(B)\cdot C}^2 + P_{B(A)\cdot C}^2 + 2P_{A\cdot C}\rho_{AB\cdot C}P_{B\cdot C}$
C	P_C^2	$U(C) = P_C^2(1-\rho_{AC}^2)(1-\rho_{BC\cdot A}^2) = P_{C\cdot AB}^2$
A,C	$2P_A\rho_{AC}P_C$	$Com(A,C) = (1-\rho_{AB}^2)\rho_{AC\cdot B}^2P_A^2 + (1-\rho_{BC}^2)\rho_{AC\cdot B}^2P_C^2 + 2P_A(\rho_{AC}-\rho_{AB}\rho_{BC})P_C = P_{A(C)\cdot B}^2 + P_{C(A)\cdot B}^2 + 2P_{A\cdot B}\rho_{AC\cdot B}P_{C\cdot B}$
B,C	$2P_B\rho_{BC}P_C$	$Com(B,C) = (1-\rho_{AB}^2)\rho_{BC\cdot A}^2P_B^2 + (1-\rho_{AC}^2)\rho_{BC\cdot A}^2P_C^2 + 2P_B(\rho_{BC}-\rho_{AB}\rho_{AC})P_C = P_{B(C)\cdot A}^2 + P_{C(B)\cdot A}^2 + 2P_{B\cdot A}\rho_{BC\cdot A}P_{C\cdot A}$
A,B,C	0	$Com(A,B,C) = [\rho_{AB}^2 - (1-\rho_{AC}^2)\rho_{AB\cdot C}^2]P_A^2 + [\rho_{BC}^2 - (1-\rho_{AB}^2)\rho_{BC\cdot A}^2]P_B^2 + [\rho_{AC}^2 - (1-\rho_{BC}^2)\rho_{AC\cdot B}^2]P_C^2 + 2P_A\rho_{AC}\rho_{BC}P_B + 2P_A\rho_{AB}\rho_{BC}P_C + 2P_B\rho_{AB}\rho_{AC}P_C = P_{A(B,C)}^2 + P_{B(A,C)}^2 + P_{C(A,B)}^2 + 2P_{A(C)}P_{B(C)} + 2P_{A(B)}P_{C(B)} + 2P_{B(A)}P_{C(A)}$
Total	1	1

path analysis, the analysis of commonality will always give a smaller value for the unique contribution of an effect (except in the case of independent factors). The two-way commonality terms can be larger or smaller than the joint effects in a path analysis, depending on the value and sign of Com(A,B,C), on the reduction in P_A^2 to U(A), and on the three-way commonality term which has no counterpart in path analysis.

A key assumption in path analysis, often not explicitly stated, is that the effects are assumed symmetric with respect to each other (Wright, 1968). In this case, the $E_C\{Var_B[E(A|B,C)]\}$ term, for example, is removed from the common term and placed in the unique term. We can see how the allocation of the portions of the path analysis is accomplished by comparing the path analysis column in Table 13.3 with the analysis of commonality column.

Li (1975, pages 121-125) uses an example from Draper and Smith (1966) to demonstrate the algebra of path coefficients. We can use the same example to compare path analysis with the analysis of commonality. If we simplify the problem by excluding X_3, and let A = X_1, B = X_2, and C = X_4, we have the problem as shown in Figure 13.2. The path analysis and the analysis of commonality for this problem are given in Table 13.4.

The analyses give quite different interpretations of the data. The values for A (U(A) vs. P_A^2) are similar, reflecting the relatively small correlations between A and B and between A and C. The values throughout the rest of the table are quite different. Since the correlation between B and C is very highly negative (-0.9730), it will be very hard to distinguish between the effects of each of these sources on the Y-variable. In the analysis of commonality, this leads to very small unique effects for these sources, U(B) = 0.0098 and U(C) = 0.0036, and a relatively large common term (Com(B,C) = 0.4349). On the other hand, the path analysis gives much larger values to the coefficients of determination for these sources and a relatively smaller common term. It is also obvious that the coefficient for the combination of B and C, $P_B^2 + P_C^2 + 2P_B\rho_{BC}P_C = 0.4750$ is very sim-

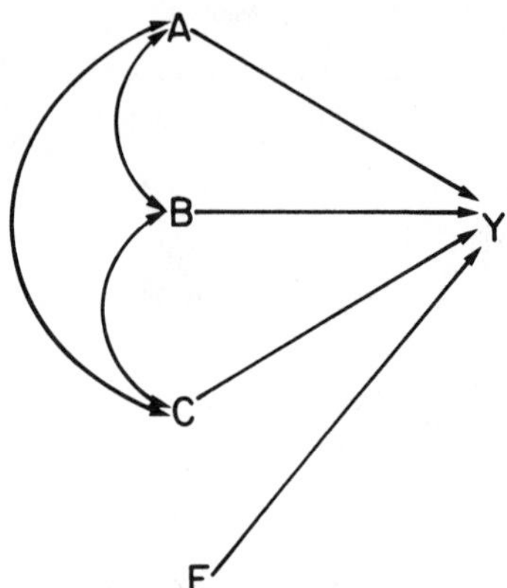

Figure 13.2 Path Diagram and Correlations for an Example by Draper and Smith (1966). Correlations for the Path Diagram Are: $\rho_{AB} = 0.2286$, $\rho_{AC} = -0.9730$, $\rho_{BC} = -0.2454$, $\rho_{AY} = 0.7307$, $\rho_{BY} = 0.8163$, and $\rho_{CY} = -0.8213$

TABLE 13.4

Path Analysis and Analysis of Commonality for the Problem Given in Figure 13.2

Source	Path Analysis	Commonality
A	0.3233	0.3023
B	0.1856	0.0098
A,B	0.1118	-0.0043
C	0.0691	0.0036
A,C	0.0732	0.0101
B,C	0.2203	0.4349
A,B,C	0	0.2259
Error	0.0177	0.0177
Total	1.0000	1.0000

TABLE 13.5

Path Analysis and Analysis of Commonality for the Problem Given in Figure 13.2 and Table 13.4, Ignoring Source C

Source	Path Analysis	Commonality
A	0.3296	0.3124
B	0.4693	0.4447
A,B	0.1798	0.2216
Error	0.0213	0.0213
Total	1.0000	1.0000

ilar to the unique term for the combination of B and C in the analysis of commonality, U(B) + U(C) + Com(B,C) = 0.4483. This indicates that the allocation to the unique terms is greater for path analysis than for the analysis of commonality (as we saw in Table 13.3), which is a reflection of the implicit symmetry assumption.

Because of the additive nature of the sums of squares, it is easy to consider submodels of the situation. For example, if we are to ignore source C in the model the analysis of commonality is obtained directly by adding terms in Table 13.4 to get Table 13.5, for example, the common term in Table 13.5 is obtained by adding Com(A,B,C) to Com(A,B) from Table 13.4. Path analysis requires a more extensive reworking of the model.

Since the analysis of commonality is based on the analysis of variance, it is possible to have classification models with general interactions. Path analysis has been extended to allow for classification models with general interactions (e.g., Wright, 1969) but, in general, the results have not been accepted.

V. CONCLUSIONS

The analysis of commonality gives us an opportunity to examine the manner in which path analysis divides the total variation in the dependent variable into coefficients of determination for the independent sources. We see that the analysis of commonality will give a minimal allocation to the individual sources compared to path analysis. This allows researchers to decide how they wish to allocate the joint terms among the sources, based on their knowledge of the system. Path analysis assumes a symmetry of source effects, and will give unique terms which are larger than given by the analysis of commonality. While this may be appropriate in some situations, it may give misleading conclusions if used indiscriminately.

While the analysis of commonality is not as well developed as path analysis, the method will allow the use of more general models than will path analysis. There is considerable debate as to whether or not path analysis is valid under non-additive models, e.g., when there are interaction terms present. The analysis of commonality is based on the general linear model and will allow the inclusion of interaction terms in the model. It is possible that the analysis of commonality can be used for nonlinear least squares models, although this has not been explored.

ACKNOWLEDGMENTS

Paper No. 8894 of the Journal Series of the North Carolina Agricultural Research Service, Raleigh, North Carolina.

REFERENCES

Beaton, Albert E. (1973). Commonality. Unpublished paper, Educational Testing Service, Princeton, New Jersey.

Draper, N.R. and H. Smith (1966). *Applied Regression Analysis*. New York: Wiley.

Emigh, T.H. (1974). Statistical methodology of the nature-nurture controversy in human intelligence. MS thesis, Iowa State University Library, Ames, Iowa.

Emigh, T.H. (1977). Partition of variance under unknown dependent association of genotypes and environments. *Biometrics 33*, 505-514.

Emigh, T.H. (1983). The analysis of variance when treatments are chosen at random. Unpublished paper, North Carolina State University, Raleigh, North Carolina.

Kempthorne, O. (1957). *An Introduction to Genetic Statistics*. New York: Wiley.

Kempthorne, O. (1977). Book Review: *Path Analysis - A Primer*, by C.C. Li. *Journal of Heredity 68*, 270-271.

Kempthorne, O. (1978). Logical, epistemological and statistical aspects of nature-nurture data interpretation. *Biometrics 34*, 1-23.

Li, C.C. (1975). *Path Analysis*. Pacific Grove: Boxwood Press.

Mood, A.M. (1971). Partitioning variance in multiple regression analysis as a tool for developing learning models. *American Educational Research Journal 8*, 191-202.

Niles, H.E. (1922). Correlation, causation and Wright's theory of 'path coefficients'. *Genetics 7*, 258-273.

Niles, H.E. (1923). The method of path coefficients: An answer to Wright. *Genetics 8*, 256-260.

Searle, S.R. (1971). *Linear Models*. New York: Wiley.

Speed, F.M. and R.R. Hocking (1976). The use of the R( )-notation with unbalanced data. *Am. Stat. 30*, 30-33.

Tukey, J.W. (1954). Causation, regression, and path analysis. In *Statistics and Mathematics in Biology*. Oscar Kempthorne, Theodore A. Bancroft, John W. Gowan and Jay L. Lush (eds.). Ames: Iowa State University Press.

Wright, S. (1918). On the nature of size factors. *Genetics 3*, 367-374.

Wright, S. (1920). The relative importance of heredity and environment in determining the piebald pattern of guinea-pigs. *Proceedings of the National Academy of Sciences, USA 6*, 320-332.

Wright, S. (1921a). Correlation and causation. *Journal of Agricultural Research 20*, 557-585.

Wright, S. (1921b). Systems of mating I. The biometric relations between parent and offspring. *Genetics 6*, 111-123.

Wright, S. (1921c). Systems of mating II. The effects of inbreeding on the genetic composition of a population. *Genetics 6*, 124-143.

Wright, S. (1921d). Systems of mating III. Assortative mating based on somatic resemblance. *Genetics 6*, 144-161.

Wright, S. (1921e). Systems of mating IV. The effects of selection. *Genetics 6*, 162-166.

Wright, S. (1921f). Systems of mating V. General considerations. *Genetics 6*, 167-178.

Wright, S. (1923). The theory of path coefficients: A reply to Nile's criticism. *Genetics 8*, 239-255.

Wright, S. (1934). The method of path coefficients. *Ann. Math. Stat. 5*, 161-215.

Wright, S. (1958). *Systems of Mating and Other Papers*. Ames: Iowa State University Press.

Wright, S. (1968). *Evolution and the Genetics of Populations I. Genetic and Biometric Foundations*. Chicago: University of Chicago Press.

Wright, S. (1969). *Evolution and the Genetics of Populations II. The Theory of Gene Frequencies*. Chicago: University of Chicago Press.

Chapter 14

A GENERALIZED VERSION OF ALBERT'S THEOREM, WITH APPLICATIONS TO THE MIXED LINEAR MODEL

David A. Harville
Iowa State University
Ames, Iowa

I. INTRODUCTION

The mixed model for an $n\times 1$ vector y of data is

$$y = W\alpha + Z_1 b_1 + \dots + Z_m b_m + b_{m+1} \tag{1.1}$$

where $W, Z_1, \dots, Z_m$ are given matrices, $b_1, \dots, b_{m+1}$ are independently distributed random column vectors with $b_i \sim N(0, \sigma_i^2 I)$ $(i=1,\dots,m+1)$, and α and $\sigma = (\sigma_1^2, \dots, \sigma_{m+1}^2)'$ are column vectors of unknown parameters. [We use the notation $x \sim N(\mu, \Sigma)$ to indicate that x is a random column vector having a multivariate normal distribution with mean vector μ and (possibly singular) covariance matrix Σ and the notation $q \sim \chi^2(s,\lambda)$ to indicate that q is a random variable whose distribution is noncentral chi-square with s degrees of freedom and noncentrality parameter λ.] Define $\Omega = \{\sigma\colon\ \sigma_1^2 \geq 0, \dots, \sigma_m^2 \geq 0, \sigma_{m+1}^2 > 0\}$.

If $A_1, \dots, A_k$ are $n\times n$ nonnull symmetric matrices of constants, and the matrix A, defined by $A = A_1 + \dots + A_k$, is idempotent, we call the collection $\{y'A_1y, \dots, y'A_ky\}$ of quadratic forms a *partition* of the quadratic form $y'Ay$. It will be called a *proper partition* if, in addition, there exist k linear functions $\gamma_1,\ldots,\gamma_k$ of $\sigma_1^2,\ldots,\sigma_{m+1}^2$ such that, for each $\sigma \in \Omega$, $y'A_1y/\gamma_1,\ldots,y'A_ky/\gamma_k$ are distributed independently as noncentral chi-square variates.

Let $P = W(W'W)^-W'$, and define conformal partitionings $W = (W_1,W_2)$ and $\alpha' = (\alpha_1',\alpha_2')$. Suppose that the model has been parameterized so that the null hypothesis $\alpha_2 = 0$ is of interest. Then, we may, as discussed by Brown (1979), wish to know if there exists a proper partition of $y'(I-P_1)y$, where $P_1 = W_1(W_1'W_1)^-W_1$, that includes $y'(P-P_1)y$ as one of its terms and, if so, to construct it. When our interest centers on the variance components, we may be satisfied with a proper partition of $y'(I-P)y$.

Brown (1979) gave necessary and sufficient conditions for the existence of proper partitions in these two cases and described an algorithm for constructing them. However, in many potential applications, the computations required by his algorithm would be prohibitive.

Albert (1976) introduced a general theorem on the distribution of quadratic forms that can be used to check whether a given partition of the total sum of squares $y'y$ is a proper partition. In the present paper, we derive a generalization of Albert's Theorem that can be used to determine whether any particular partition of $y'(I-P_1)y$ or of $y'(I-P)y$, or more generally of $y'Ay$, where A is an idempotent matrix, is a proper partition. Thus, if in a particular case, there is a partition [e.g., the partition of $y'(I-P)y$ generated by Henderson's (1953) Method 3 and described by Searle (1971, Sec. 10.4)] that is regarded as a good candidate for a proper partition, this result can be used to check whether it is in fact proper. If this check confirms that the trial partition is proper, the proper partition would be obtained with much less effort than required by Brown's algorithm.

Our approach to Albert's Theorem differs somewhat from that taken by Albert himself. We derive our generalized version essentially as a corollary to a generalization of Cochran's (1934) Theorem described by Rao and Mitra (1971). This approach has the virtue that it makes clear the precise nature of the relationship between Albert's and Cochran's Theorems and between their generalizations.

II. GENERALIZED VERSION OF COCHRAN'S THEOREM

The following theorem can be constructed, e.g., from parts of Rao and Mitra's (1971) Theorems 9.3.3 and 9.2.1 and Lemma 9.3.1 and is essentially a generalization of Cochran's (1934) Theorem.

Theorem 1. Let $x \sim N(\mu,\Sigma)$, take $A_1,\ldots,A_k$ to be nonrandom nonnull symmetric n×n matrices, and define $A = A_1+\ldots+A_k$. If

$$A\,\Sigma\,A = A \tag{2.1}$$

and

$$A_i\,\Sigma\,A_i = A_i \quad (i=1,\ldots,k) \tag{2.2}$$

then $x'A_1x,\ldots,x'A_kx$ are distributed independently as

$$\chi^2[\text{rank}(A_i\Sigma),\ \mu'A_i\mu/2] \qquad (i=1,\ldots,k)$$

Conversely (for nonsingular Σ), if $x'A_1x,\ldots,x'A_kx$ are distributed independently as $\chi^2(s_i,\lambda_i)$ for some s_i and λ_i $(i=1,\ldots,k)$, then conditions (2.1) and (2.2) are satisfied and $s_i = \text{rank}(A_i\Sigma)$ and $\lambda_i = \mu'A_i\mu/2$ $(i=1,\ldots,k)$.

If $\mu = 0$, $\Sigma = I$, and $A_1,\ldots,A_k$ are such that $A = I$, then condition (2.1) is trivial (it becomes $I = I$) and condition (2.2) becomes $A_i^2 = A_i$ (which is equivalent to the condition that $\text{rank}(A_1)+\ldots+\text{rank}(A_k) = n$), and Theorem 1 reduces essentially to Cochran's Theorem.

If in Theorem 1 we replace A_i by $(1/\gamma_i)A_i$, where γ_i is a nonzero scalar, we obtain the following result.

Corollary. Let $x \sim N(\mu,\Sigma)$, and take $A_1,\ldots,A_k$ to be nonrandom nonnull symmetric n×n matrices. If, for some nonzero nonrandom scalars $\gamma_1,\ldots,\gamma_k$,

$$H \Sigma H = H \tag{2.3}$$

where $H = (1/\gamma_1)A_1 + \ldots + (1/\gamma_k)A_k$, and

$$A_i \Sigma A_i = \gamma_i A_i \qquad (i=1,\ldots,k) \tag{2.4}$$

then $x'A_1x/\gamma_1, \ldots, x'A_kx/\gamma_k$ are distributed independently as

$$\chi^2[\text{rank}(A_i\Sigma),\ \mu'A_i\mu/(2\gamma_i)] \qquad (i=1,\ldots,k)$$

Conversely (for nonsingular Σ), if $x'A_1x/\gamma_1, \ldots, x'A_kx/\gamma_k$ are distributed independently as $\chi^2(s_i,\lambda_i)$ for some γ_i, s_i, and λ_i $(i=1,\ldots,k)$, then conditions (2.3) and (2.4) are satisfied, and $s_i = \text{rank}(A_i\Sigma)$ and $\lambda_i = \mu'A_i\mu/(2\gamma_i)$ $(i=1,\ldots,k)$.

III. MAIN RESULT

The corollary to Theorem 1 can be used to establish the following generalization of Albert's (1976) Theorem.

Theorem 2. Let $x \sim N(\mu,\Sigma)$, and take $A_1,\ldots,A_k$ to be nonrandom nonnull symmetric $n\times n$ matrices such that the matrix A, defined by $A = A_1+\ldots+A_k$, is idempotent. If

$$A_i^2 = A_i \qquad (i=1,\ldots,k) \tag{3.1}$$

and if, for some nonzero nonrandom scalars $\gamma_1,\ldots,\gamma_k$,

$$A \Sigma A_i = \gamma_i A_i \qquad (i=1,\ldots,k) \tag{3.2}$$

then γ_i is positive $(i=1,\ldots,k)$, and $x'A_1x/\gamma_1,\ldots,x'A_kx/\gamma_k$ are distributed independently as

$$\chi^2[\text{rank}(A_i\Sigma),\ \mu'A_i\mu/(2\gamma_i)] \qquad (i=1,\ldots,k)$$

Conversely (for nonsingular Σ), if $x'A_1x/\gamma_1,\ldots,x'A_kx/\gamma_k$ are distributed independently as $\chi^2(s_i,\lambda_i)$ for some γ_i, s_i, and λ_i $(i=1,\ldots,k)$, then conditions (3.1) and (3.2) are satisfied, and $s_i = \text{rank}(A_i\Sigma)$ and $\lambda_i = \mu'A_i\mu/(2\gamma_i)$ $(i=1,\ldots,k)$.

Proof. Suppose that (3.1) and (3.2) are satisfied. Then, $A_iA_j = 0$ for $j\neq i=1,\ldots,k$ (see, e.g., Graybill's (1969) Theorem 12.3.6), so that

$$A_i\ \Sigma\ A_i = A_iA\ \Sigma\ A_i = \gamma_iA_i^2 = \gamma_iA_i \tag{3.3}$$

Further, defining $H = (1/\gamma_1)A_1+\ldots+(1/\gamma_k)A_k$, we find that $HA = H$ and $A\ \Sigma\ H = A$, so that

$$H\ \Sigma\ H = HA\ \Sigma\ H = HA = H$$

Now, applying the corollary to Theorem 1, we conclude that $x'A_1x/\gamma_1$, $\ldots,x'A_kx/\gamma_k$ are distributed independently as $\chi^2[\text{rank}(A_i\Sigma),\ \mu'A_i\mu/(2\gamma_i)]$ $(i=1,\ldots,k)$. Moreover, (3.3) implies that $\gamma_i\text{tr}(A_i^2) = \text{tr}(A_i'\ \Sigma\ A_i)$ and thus (since $\gamma_i \neq 0$, $\text{tr}(A_i^2) = \text{tr}(A_i'A_i) > 0$, and $\text{tr}(A_i'\ \Sigma\ A_i) \geq 0$) that $\gamma_i > 0$.

Conversely, suppose that Σ is nonsingular and that $x'A_1x/\gamma_1,\ldots,$ $x'A_kx/\gamma_k$ are distributed independently as $\chi^2(s_i,\lambda_i)$ for some γ_i, s_i, and λ_i $(i=1,\ldots,k)$. Then, from the corollary to Theorem 1, we have that $s_i = \text{rank}(A_i\Sigma)$ and $\lambda_i = \mu'A_i\mu/(2\gamma_i)$ $(i=1,\ldots,k)$ and that conditions (2.3) and (2.4), or equivalently the conditions $\Sigma\ H\ \Sigma\ H\ \Sigma = \Sigma\ H\ \Sigma$ and

$$\Sigma[(1/\gamma_i)A_i]\Sigma[(1/\gamma_i)A_i]\Sigma = \Sigma[(1/\gamma_i)A_i]\Sigma \qquad (i=1,\ldots,k) \tag{3.4}$$

are satisfied. Note that (since Σ is nonsingular), condition (3.4) implies that

$$A_i\ \Sigma\ A_i = \gamma_iA_i \qquad (i=1,\ldots,k)$$

Further, according to Rao and Mitra's (1971) Lemma 9.3.1,

$$[1/(\gamma_i\gamma_j)]\ \Sigma\ A_i\ \Sigma\ A_j\ \Sigma = 0$$

and hence $A_i\ \Sigma\ A_j = 0$ ($j \neq i = 1,\ldots,k$).

Combining various of these results gives

$$A\ \Sigma\ A_i = A_i\ \Sigma\ A_i = \gamma_i A_i$$

and

$$A_i^2 = A_i'A_i = (1/\gamma_i^2)A_i\ \Sigma\ A^2\ \Sigma\ A_i = (1/\gamma_i^2)A_i\ \Sigma\ A\ \Sigma\ A_i = (1/\gamma_i)A_i\ \Sigma\ A_i = A_i$$

($k=1,\ldots,k$), i.e., conditions (3.2) and (3.1), which completes the proof. □

The special case of Theorem 2 where $A_1,\ldots,A_k$ are such that $A = I$ (and where Σ is nonsingular) includes Albert's Theorem. Note that [in light of condition (3.1)] the restriction to nonzero γ_i's in (3.2) is redundant if Σ is nonsingular.

IV. APPLICATION TO THE MIXED MODEL

Theorem 2 can be applied to the mixed model (1.1) to determine whether any particular partition of a quadratic form $y'Ay$, where A is idempotent, is a proper partition. The specifics are included in the following theorem, which generalizes Albert's (1976) corollary to his theorem.

Theorem 3. Assume that y follows the mixed model (1.1), take $A_1,\ldots,A_k$ to be nonnull symmetric n×n matrices of constants such that the matrix A, defined by $A = A_1+\ldots+A_k$, is idempotent. If

$$A_i^2 = A_i \qquad (i=1,\ldots,k) \tag{4.1}$$

and, if for some scalar constants $\nu_{11},\ldots,\nu_{1m},\ldots,\nu_{k1},\ldots,\nu_{km}$,

$$AZ_jZ_j'A_i = \nu_{ij}A_i \qquad (i=1,\ldots,k;\ j=1,\ldots,m) \tag{4.2}$$

then ν_{ij} is nonnegative ($i=1,\ldots,k$; $j=1,\ldots,m$), and, for each $\sigma \in \Omega$, $y'A_1y/\gamma_1,\ldots,y'A_ky/\gamma_k$ are distributed independently as

$$\chi^2[\text{rank}(A_i),\ \alpha'W'A_iW\alpha/(2\gamma_i)]$$

with $\gamma_i = \sigma^2_{m+1} + \Sigma^m_{j=1}\nu_{ij}\sigma^2_j$ ($i=1,\ldots,k$). Conversely, if for each $\sigma \in \Omega$, $y'A_1y/\gamma_1,\ldots,y'A_ky/\gamma_k$ are distributed independently as $\chi^2(s_i,\lambda_i)$ for some γ_i, s_i, and λ_i ($i=1,\ldots,k$), then condition (4.1) is satisfied, condition (4.2) is satisfied for some scalar constants $\nu_{11},\ldots,\nu_{1m},\ldots,\nu_{k1},\ldots,\nu_{km}$, and $\gamma_i = \sigma^2_{m+1} + \Sigma^m_{j=1}\nu_{ij}\sigma^2_j$, $s_i = \text{rank}(A_i)$, and $\lambda_i = \alpha'W'A_iW\alpha/(2\gamma_i)$ ($i=1,\ldots,k$).

An argument similar to that outlined by Albert in establishing his corollary to his theorem can be used to show that Theorem 3 follows from Theorem 2. (To see that (4.1) and (4.2) imply that $\nu_{ij} \geq 0$, observe that (4.1) and (4.2) imply that $\text{tr}(A_i) > 0$ and that $\nu_{ij}\text{tr}(A_i) = \text{tr}(A_i'Z_jZ_j'A_i) \geq 0$.)

Note that, if condition (4.2) of Theorem 3 is satisfied, then ν_{ij} is a characteristic root of $Z_j'AZ_j$ or ν_{ij} equals zero ($i=1,\ldots,k$; $j=1,\ldots,m$).

REFERENCES

Albert, A. (1976). When is a sum of squares an analysis of variance? *Ann. Stat. 4,* 775-778.

Brown, K.G. (1979). Partitioning the sum of squares for error in a general linear model. Technical Report 79-2, University of Maine at Orono.

Cochran, W.G. (1934). The distribution of quadratic forms in a normal system, with applications to the analysis of covariance. *Proc. Cambridge Phil. Soc. 30,* 178-191.

Graybill, F.A. (1969). *Introduction to Matrices with Applications in Statistics.* Belmont: Wadsworth.

Henderson, C.R. (1953). Estimation of variance and covariance components. *Biometrics 9,* 226-252.

Rao, C.R. and Mitra, S.K. (1971). *Generalized Inverse of Matrices and its Applications*. New York: Wiley.

Searle, S.R. (1971). *Linear Models*. New York: Wiley.

Chapter 15

SOME NEW SIMULTANEOUS CONFIDENCE INTERVALS IN MANOVA AND THEIR GEOMETRIC REPRESENTATION AND GRAPHICAL DISPLAY*

K. Ruben Gabriel and David Gheva
University of Rochester
Rochester, New York

I. THE MANOVA SET-UP AND ITS CONFIDENCE BOUNDS

We consider a one-way MANOVA set-up, with n samples observed on m variables. We denote the sample sizes by N_i, i=1,...,n, and write $\underline{N}' = (N_1,\ldots,N_n)$ and $N = \Sigma_{i=1}^{n} N_i$, so that N-n is the number of "within samples" degrees of freedom. We write $s_{v,w}$ for the "within" estimate of the covariance $\sigma_{v,w}$ of variables v and w and define (m×m) matrices $S = ((s_{v,w}))$ and $\Sigma = ((\sigma_{v,w}))$. We let $\sigma = \text{rank}(\Sigma)$ and limit our treatment to the case $N\text{-}n \geq \sigma$, which ensures that rank(S) = σ a.e. In the following, write S^{-1} to indicate any generalized inverse of S. For $\sigma = m$, the inverse is unique; for $\sigma < m$, the results presented in this paper hold for any choice of generalized inverse. (Gheva, 1983, Chapter 2).

Let $y_{i,v}$ be the i-th sample mean of the v-th variable and let its expectation be $\psi_{i,v}$. In matrix form, let the means be Y = $((y_{i,v}))$ and their expectations $\Psi = ((\psi_{i,v}))$, both (n×m) matrices.

*This work was supported by ONR contract N00014-80-0387.

MANOVA inferences usually relate to bilinear functions $\underline{a}'\,\Psi\,\underline{b} = \Sigma_i\Sigma_v\, a_i\psi_{i,v}b_v$, where $\underline{a}$ defines any contrast between the samples, i.e., $\Sigma_i\, a_i = 0$, and $\underline{b}$ defines any linear combination of the variables. In the following, both Y and Ψ will consist of the means and expectations, respectively, expressed as deviations from the overall means of all samples. Taking deviations in that way does not affect either $\underline{a}'Y$ or $\underline{a}'\Psi$ since $\Sigma_i\, a_i = 0$.

The classical MANOVA 1-α simultaneous confidence intervals on all contrasts $\underline{a}$ and all combinations $\underline{b}$ are Roy and Bose's (1953, (4.2.6))

$$\underline{a}'\,\Psi\,\underline{b} \in [\underline{a}'Y\underline{b} \pm \sqrt{(n-1)}\,\sqrt{\lambda_\alpha}\sqrt{(\Sigma_i a_i^2/N_i)}\,\sqrt{\underline{b}'S\underline{b}}],$$

$$\forall\, \underline{a}\ (\Sigma_i a_i = 0);\ \forall\, \underline{b} \qquad (1)$$

where the notation $y \in [x \pm z]$ signifies that $x-z \leq y \leq x+z$. The critical point used in these intervals satisfies

$$(n-1)\lambda_\alpha = (N-n)\theta_\alpha/(1-\theta_\alpha) \qquad (2)$$

where θ_α is obtained from Heck's (1960) charts entered with "s" = min(σ,n-1), "m" = ($|n-1-\sigma|$ - 1)/2 and "n" = (N-n-σ-1)/2.

The new intervals we propose are narrowest when the critical value they use is the upper α point τ_α^2 of the maximum of the Hotelling T^2's for all $\tilde{n} = \binom{n}{2}$ pairwise comparisons. Tables of this distribution (Siotani, 1952) are not adequate for all applications, but conservative inferences may always be obtained by using the Bonferroni bound

$$\tau_\alpha^2 \leq \sigma(N-n)(N-n-\sigma+1)^{-1}F_{(\sigma,N-n-\sigma+1),\ \tilde{\alpha}} \qquad (3)$$

where $\tilde{\alpha} = \alpha/\tilde{n}$. Another conservative bound is

$$\tau_\alpha^2 \leq (n-1)\lambda_\alpha \tag{4}$$

where the right hand side is as in (2), above.

Construction of the new intervals uses the functions W introduced by Hochberg, Weiss and Hart (1982) which ensure that

$$W_i + W_e \geq (N_i^{-1} + N_e^{-1})^{\frac{1}{2}} \quad \forall\ i,e \tag{5}$$

In particular, we use Gabriel and Gheva's (1982) function

$$W_i = \sqrt{(1 + r_i)}/\{(1 + \sqrt{r_i})\sqrt{N_i}\} \tag{6}$$

where

$$r_i = (\max_e N_e/N_i) \vee (N_i/\min_e N_e) \tag{7}$$

This is easier to compute than the optimal function proposed by Hochberg, Weiss and Hart (1982) which requires linear programming.

The intervals we propose are

$$\underline{a}'\ \Psi\ \underline{b}\ \varepsilon\ [\underline{a}'Y\underline{b} \pm \tau_\alpha(\Sigma_i|a_i|W_i)\sqrt{\underline{b}'S\underline{b}}],\ \forall\ \underline{a}\ (\Sigma_i a_i = 0);\ \forall\ \underline{b} \tag{8}$$

They include the following intervals for all pairwise comparisons,

$$(\underline{\Psi}_i' - \underline{\Psi}_e')\underline{b}\ \varepsilon\ [(\underline{y}_i' - \underline{y}_e')\underline{b} \pm \tau_\alpha(W_i + W_e)\sqrt{\underline{b}'S\underline{b}}],\ \forall\ i,e;\ \forall\ \underline{b} \tag{9}$$

where $\underline{\Psi}_i'$ and $\underline{y}_i'$, i=1,...,n, are the i-th rows of Ψ and Y, respectively. In particular, they include the following bounds on pairwise comparisons on single variables,

$$(\psi_{i,v} - \psi_{e,v})\ \varepsilon\ [(y_{i,v} - y_{e,v}) \pm \tau_\alpha(W_i + W_e)\sqrt{s_{v,v}}], \tag{10}$$

$$\forall\ i,e;\ \forall\ v$$

Statements (8), which include (9) and (10), are simultaneously true with probability at least $1-\alpha$. (For a proof, see Appendix A). Thus, the intervals of (8) are conservative confidence intervals. Since they are of confidence at least $1-\alpha$ when they use the upper α point of the maximum T^2 distribution, they are a fortiori conservative when they use the upper bounds (3) or (4) instead. Their centers are the same as those of the classical intervals (1). However, for pairwise comparisons they often are narrower than the latter. For many other contrasts, they will be found to be wider. (The relation is analogous to that between Tukey's and Scheffé's bounds (Scheffé, 1953, Chapter 3). See also Appendix B.)

II. A SIMPLE EXAMPLE AND ITS GEOMETRIC REPRESENTATION

The new multivariate confidence intervals and test decision rules and their geometric representation will be introduced by means of a simple bivariate example. Morrison's (1976, p. 190) data on two weeks' weight losses of 4 rats in each of six groups are presented in Table 15.1. (The factorial structure of these groups as 2 sexes by 3 drugs is ignored in the present one-way analysis). Note that in this example all sample sizes are $N_i = 4$, $i=1,2,\ldots,6$.

Some examples of the new confidence intervals calculated from equation (8) are shown in Table 15.2, along with the corresponding Roy-Bose intervals (1). As expected, for pairwise comparisons, the new intervals are somewhat narrower than the old ones, whereas for all other comparisons shown in Table 15.2, the new intervals are wider. Indeed, for some intervals, they are appreciably wider. In particular, this is so for the comparison of the average of 4 groups with the average of the other 2 groups. This results also in different test decisions: Thus, $\psi_{1,1} = \psi_{2,1}$ would be rejected by the new procedure but accepted by the old; on the other hand, hypothesis

$$\tfrac{1}{4}(\psi_{1,1} - \psi_{1,2} + \psi_{2,1} - \psi_{2,2} + \psi_{3,1} - \psi_{3,2} + \psi_{4,1} - \psi_{4,2}) = \tfrac{1}{2}(\psi_{5,1} - \psi_{5,2} + \psi_{6,1} - \psi_{6,2})$$

would be accepted by the new procedure, though rejected by the old.

TABLE 15.1

Weight Loss Experiment: Data (Morrison, 1976, p. 190)

Variable			Means		Deviations from Overall Mean	
			1	2	1	2
A.	Male	Drug U	6.50	6.25	-3.25	-2.4167
B.	Fem.	Drug U	7.50	8.25	-2.25	-0.4167
C.	Male	Drug V	7.25	8.25	-2.50	-0.4167
D.	Fem.	Drug V	7.75	8.75	-2.00	0.0833
E.	Male	Drug W	16.00	12.00	6.25	3.3333
F.	Fem.	Drug W	13.50	8.50	3.75	-0.1667
Overall			9.75	8.67	0.00	0.0000
			Within Sums of Squares and Products		Variance Estimate	
			94.5	76.5	5.25	4.25
			76.5	114.00	4.25	6.3333
Eigenvectors of Variance Estimate					.660898	.750476
					.750476	-.660898
Eigenvalues					10.0760	1.5073

The geometric representation of the new intervals uses "uncertainty ellipses" about the sample means $\underline{y}_i$, i=1,...,6. (These ellipses and their connection with "uncertainty intervals" are discussed in Appendix C.) The i-th uncertainty ellipse is defined as

$$\{Y\}_i = \{\underline{y} | (\underline{y} - \underline{y}_i)' S^{-1} (\underline{y} - \underline{y}_i) \leq \tau_\alpha^2 W_i^2\} \tag{11}$$

It uses the conservative critical value $\tau^2_{.05} = 4.285$, so that $\tau_{.05} W_i = 1.515$ for all i. The resulting uncertainty ellipses are displayed in Figure 15.1.

TABLE 15.2

Weight Loss Experiment: Simultaneous 95% Confidence Intervals

Contrast a	Combination b	Roy-Bose Intervals (1)		New Intervals (8)	
1, 0, 0, 0, - 1, 0	1, 0	-16.87	-2.13	-16.44	-2.56
1, 0, 0, 0, -.5, -.5	1, 0	-14.64	-1.86	-15.19	-1.31
.25, .25, .25, .25, -.5, -.5	1, 0	-12.02	-2.98	-14.44	-0.56
1, 0, 0, 0, 0, - 1	1, 0	-14.37	0.37	-13.94	-0.06
1, 0, 0, 0, - 1, 0	0, 1	-13.85	2.35	-13.38	1.88
1, 0, 0, 0, -.5, -.5	0, 1	-11.01	3.01	-11.62	3.63
.25, .25, .25, .25, -.5, -.5	0, 1	- 7.33	2.58	-10.00	5.25
1, 0, 0, 0, 0, - 1	0, 1	-10.35	5.85	- 9.88	5.38

1, 0, 0, 0, - 1, 0	2,-1	-23.59	-2.91	-22.99	-3.51
1, 0, 0, 0, -.5, -.5	2,-1	-21.46	-3.54	-22.24	-2.76
.25, .25, .25, .25, -.5, -.5	2,-1	-18.96	-6.29	-22.36	-2.89
1, 0, 0, 0, 0, - 1	2,-1	-22.09	-1.41	-21.49	-2.01
1, 0, 0, 0, - 1, 0	1,-1	- 9.40	1.90	- 9.07	1.57
1, 0, 0, 0, -.5, -.5	1,-1	- 9.14	0.64	- 9.57	1.07
.25, .25, .25, .25, -.5, -.5	1,-1	- 8.59	-1.66	-10.45	0.20
1, 0, 0, 0, 0, - 1	1,-1	-10.40	0.90	-10.07	0.57
0, 0, 0, 0, 1, - 1	1,-1	- 6.65	4.65	- 6.32	4.32

Note: The critical value $\sqrt{(4-1)}\sqrt{\lambda_{.05}}$ for (1) is obtained by reading $\theta_{.05}(2,1,7.5) = .535$ from Heck's (1960) charts and computing $\sqrt{(18 \times .535/(1-.535))} = 4.551$. The critical value $\tau_{.05} = 4.285$ for (8) is obtained by interpolating for $\tilde{\alpha} = .05/15$ in the F distribution with 2 and 17 d.f.

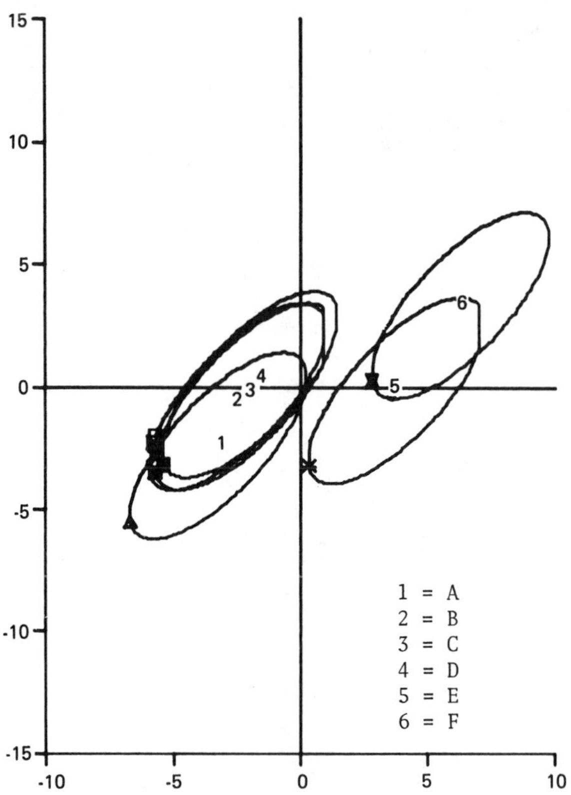

Figure 15.1 Weight Loss Experiment: 95% Uncertainty Ellipses {Y}

Simultaneous tests on pairwise comparisons of samples are carried out by checking whether the corresponding pair of ellipses has any point in common, i.e., whether

$$\{Y\}_i \cap \{Y\}_e \neq \phi \qquad (12)$$

This test is related to Hotelling's T^2 test of $\underline{\Psi}_i' = \underline{\Psi}_e'$ as the new intervals are related to those of Roy and Bose. (See Appendix D.)

Figure 15.1 shows that the last two uncertainty ellipses (E, F) are disjoint from the first four (A, B, C and D). Within each set, the ellipses overlap. Hence the only significant differences are those between the rats of both sexes receiving drug W and the rest of the rats.

Inferences on individual variables can be obtained by projecting the uncertainty ellipses onto the axes representing those variables. (See Appendix C.) If there is an overlap of the "shadows" that the uncertainty ellipses $\{Y\}_i$ and $\{Y\}_e$ of samples i and e throw upon the direction of a variable, then one may conclude that samples i and e do not differ significantly on that variable. To describe this more precisely, define $P\langle\underline{b}\rangle$ as the orthogonal projection operator onto $V\langle\underline{b}\rangle$, the direction through $\underline{b}$; now write the shadow of the uncertainty ellipse $\{Y\}_i$ onto $V\langle\underline{b}\rangle$ as $P\langle\underline{b}\rangle\{Y\}_i$. The graphical test then consists of rejecting $\underline{\Psi}_i'\underline{b} = \underline{\Psi}_e'\underline{b}$ if, and only if,

$$P\langle\underline{b}\rangle\{Y\}_i \cap P\langle\underline{b}\rangle\{Y\}_e \neq \phi \tag{13}$$

For the present example, Figure 15.1 shows that there are significant differences on the first variable - horizontal axis - since the shadow of the fifth ellipse (males on Drug W) is disjoint from the shadows of the first four ellipses. Also, the shadow of the sixth ellipse is disjoint from that of the first. However, there are no significant differences on the second variable - since all six ellipses' shadows overlap on the vertical axis.

Inferences on linear combinations of variables can also be represented geometrically. Thus, inspection of the plot of Figure 15.1 shows that the shadows of $\{Y\}_5$ and $\{Y\}_6$ (marked as E and F) are most clearly separated from those of $\{Y\}_1$, $\{Y\}_2$, $\{Y\}_3$ and $\{Y\}_4$ (marked as A, B, C and D) in the direction of roughly -40° w.r.t. the horizontal axis. This direction represents a linear combination with coefficients $\underline{b}' = (\cos(-40°), \sin(-40°)) = (.7660, -.6428)$.

Geometric construction of the new confidence intervals (8) for various contrasts $\underline{a}$ and combinations $\underline{b}$ is illustrated next. The shadow of $\{Y\}_i$ onto $V\langle\underline{b}\rangle$ can be rewritten as

$$P\langle\underline{b}\rangle\{Y\}_i = \underline{b}^*[1_i\langle\underline{b}\rangle/||\underline{b}||,\ u_i\langle\underline{b}\rangle/||\underline{b}||] \tag{14}$$

where $\underline{b}^*$ is the normalized form of $\underline{b}$ and

$$l_i\langle\underline{b}\rangle = \underline{y}_i'\underline{b} - \tau_\alpha W_i\sqrt{\underline{b}'S\underline{b}} \tag{15}$$

and

$$u_i\langle\underline{b}\rangle = \underline{y}_i'\underline{b} + \tau_\alpha W_i\sqrt{\underline{b}'S\underline{b}} \tag{16}$$

are, respectively, the lower and upper bounds of the i-th sample mean's uncertainty interval for linear combination $\underline{b}$ of variables. (Gabriel and Gheva, 1982. See Appendix B.) Differencing these bounds for the i-th and the e-th sample gives the new confidence intervals (9) in the form

$$(\underline{\Psi}_i' - \underline{\Psi}_e')\underline{b} \in [l_i\langle\underline{b}\rangle - u_e\langle\underline{b}\rangle,\ u_i\langle\underline{b}\rangle - l_e\langle\underline{b}\rangle] \tag{17}$$

Figure 15.2 shows the comparison of the first sample with the sixth sample on variable $2y_1 - y_2$. (This linear combination is not far from the (cos(-40°), sin(-40°) noted above.) The shadows of the two ellipses $\{Y\}_1$ and $\{Y\}_5$ onto $V\langle\underline{b}\rangle$, the direction through $\underline{b}' = (2,-1)$ have coordinates approximately 0.0 and -4.0 for $\{Y\}_1$ and 6.0 and 1.5 for $\{Y\}_5$. The length of $\underline{b}$ is $||\underline{b}|| = \sqrt{5}$, and so

$$l_1\langle\underline{b}\rangle = -4.5 \times \sqrt{5} = -10.1 \qquad u_1\langle\underline{b}\rangle = 0 \times \sqrt{5} = 0.0$$

$$l_5\langle\underline{b}\rangle = 1.5 \times \sqrt{5} = 3.4 \qquad u_5\langle\underline{b}\rangle = 6 \times \sqrt{5} = 13.4$$

Applying (17), one obtains a graphical approximation of (-23.5; -3.5) to the new confidence interval on $(\underline{\Psi}_1' - \underline{\Psi}_5')\underline{b}$. This is close to the interval of (-22.99; -3.51) calculated directly from (9). (See Table 15.2.)

Inference on any other contrast can also be represented graphically. This is done by normalizing its coefficients $\underline{a}$ so that $\Sigma_i|a_i| = 2$ and thereby making it into a comparison of the weighted

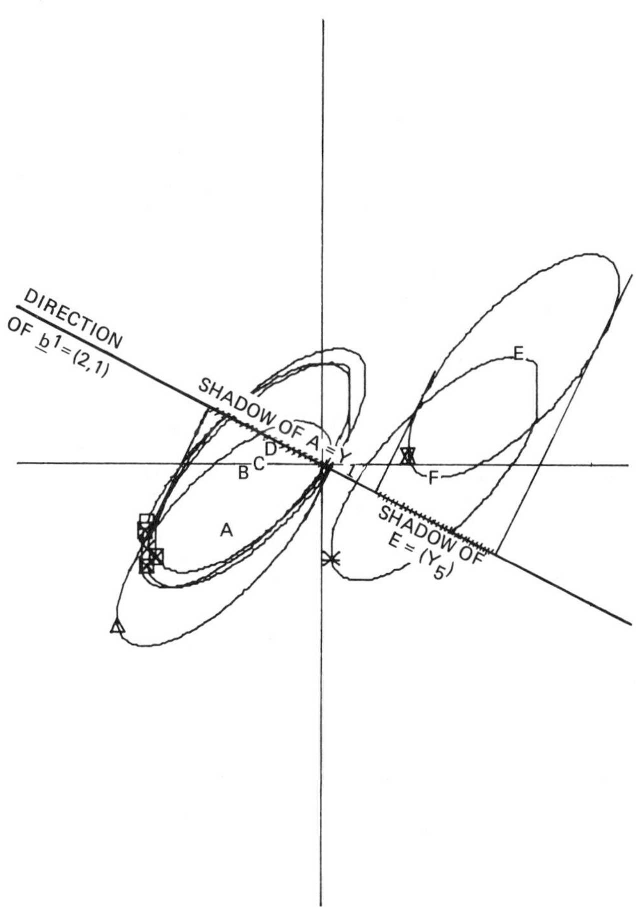

Figure 15.2 Weight Loss Experiment: Uncertainty Ellipses and Shadows onto $\underline{b}' = (2,-1)$

means of two separate subsets of samples. (Note that pairwise comparisons can also be viewed in this way. See Appendix B.) Then one may consider the following two "average uncertainty ellipses",

$$\{Y\}_{\underline{a}+} = \{\underline{y} | (\underline{y} - \Sigma_i^+ a_i \underline{y}_i)' S^{-1} (\underline{y} - \Sigma_i^+ a_i \underline{y}_i) \leq (\tau_\alpha \Sigma_i^+ a_i W_i)\} \quad (18)$$

and

$$\{Y\}_{\underline{a}-} = \{\underline{y} | (\underline{y} - \Sigma_e^- |a_e| \underline{y}_e)' S^{-1} (\underline{y} - \Sigma_e^- |a_e| \underline{y}_e) \leq (\tau_\alpha \Sigma_e^- |a_e| W_e)\} \quad (19)$$

where Σ_i^+ indicates a sum over all indices i such that $a_i \geq 0$ and Σ_e^- indicates a sum over all indices e such that $a_e < 0$. Intersection of these two ellipses indicates that the contrast hypothesis $\underline{a}' \Psi = \underline{0}'$ is acceptable. Similarly, intersection of the shadows of these two ellipses onto any direction $V\langle\underline{b}\rangle$ indicates the acceptance of univariate hypothesis $\underline{a}' \Psi \underline{b} = 0$, and conversely, that hypothesis will be rejected if

$$P\langle\underline{b}\rangle\{Y\}_{\underline{a}+} \cap P\langle\underline{b}\rangle\{Y\}_{\underline{a}-} \neq \phi \quad (20)$$

Differencing the coefficients of the end points of these two shadows, as in (17), yields the new confidence intervals (8) in the form

$$\underline{a}' \Psi \underline{b} \in [l_{\underline{a}+}\langle\underline{b}\rangle - u_{\underline{a}-}\langle\underline{b}\rangle,\ u_{\underline{a}+}\langle\underline{b}\rangle - l_{\underline{a}-}\langle\underline{b}\rangle] \quad (21)$$

(For an illustration of such averaging, see Section VIII and Figure 15.7, below.)

III. STANDARDIZING THE VARIABILITY

The geometric representation of the new confidence intervals on bilinear functions $\underline{a}' \Psi \underline{b}$ and of the corresponding tests of hypotheses $\underline{\Psi}_i' = \underline{\Psi}_e'$, or $\underline{a}' \Psi = \underline{0}'$, can be greatly simplified by standardizing the variables so their estimated variances and covariances become an identity matrix. To this purpose, define

$$Z = YS^{-\frac{1}{2}} \tag{22}$$

for a symmetric nonnegative definite (m×m) matrix $S^{-\frac{1}{2}}$ which is the inverse of a matrix $S^{\frac{1}{2}}$ satisfying

$$S^{\frac{1}{2}}S^{\frac{1}{2}} = S \tag{23}$$

Also define

$$\underline{d} = S^{\frac{1}{2}}\underline{b} \tag{24}$$

so that

$$\underline{a}'Z\underline{d} = \underline{a}'Y\underline{b} \tag{25}$$

and

$$||\underline{d}|| = \sqrt{\underline{b}'S\underline{b}} \tag{26}$$

With this standardized geometric representation, an uncertainty interval for a combination with coefficient vector $\underline{b}$ is obtained by projecting the appropriate uncertainty circle

$$\{Z\}_i = \{\underline{z} \mid ||\underline{z} - \underline{z}_i|| \leq \tau_\alpha W_i\} \tag{27}$$

onto $V\langle S^{\frac{1}{2}}\underline{b}\rangle$ and multiplying the bounds by $\sqrt{\underline{b}'S\underline{b}}$. (Details are given in Appendix E.)

IV. THE SIMPLE EXAMPLE IN STANDARDIZED FORM

For the weight loss data of Table 15.1, the standardized matrix of means is

$$Z = \begin{bmatrix} -1.339 & -0.483 \\ -1.239 & -0.335 \\ -1.388 & 0.397 \\ -1.213 & 0.540 \\ 2.901 & 0.229 \\ 2.276 & -1.018 \end{bmatrix}$$

obtained from (22) by use of

$$S = \begin{bmatrix} .5964 & -.2477 \\ -.2477 & .5332 \end{bmatrix}$$

The radii of all six uncertainty circles are

$$\tau_\alpha W_i = 4.285/(\sqrt{2}\sqrt{4}) = 1.515$$

since $W_i = 1/\sqrt{2N_i}$ in the case of equal sample sizes (see (6), (7) above).

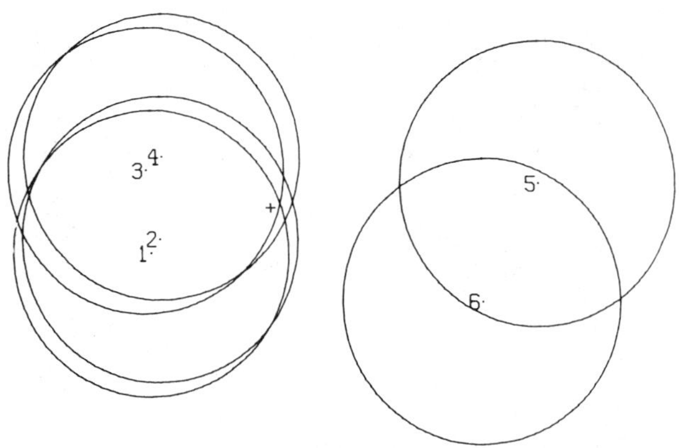

Figure 15.3 Weight Loss Experiment: Plot of $Z = YS^{-1/2}$ with 95% Uncertainty Circles {Z}

Figure 15.3 shows the plot of the resulting uncertainty circles. The separation and overlaps are the same as in Figure 15.1 but may be easier to look at because of the more symmetric representation. The significant differences are evident from this figure. However, the individual variables cannot be identified on this display.

V. BIPLOT REPRESENTATION

Another standardized geometric representation of this type is the biplot. This does allow identification of the individual variables and their linear combinations (Gabriel, 1971, 1972, 1981).

The biplot uses the singular value decomposition (S.V.D.) of the weighted matrix of means $D_N^{\frac{1}{2}}YS^{-\frac{1}{2}}$, where

$$D_N = \text{Diag}(N_1,\ldots,N_n) \tag{28}$$

This singular value decomposition may be written

$$D_N^{\frac{1}{2}}YS^{-\frac{1}{2}} = P'\,\Lambda\,Q \tag{29}$$

where P and Q are subject to $PP' = Q'Q = QQ' = I_m$ and $\Lambda = \text{Diag}(\lambda_1,\ldots,\lambda_m)$ is subject to $\lambda_1 \geq \ldots \geq \lambda_m \geq 0$. The biplot factorization then is

$$Y = JK' \tag{30}$$

where

$$J = D^{-\frac{1}{2}}P'\Lambda \tag{31}$$

and

$$K = S^{\frac{1}{2}}Q' \tag{32}$$

which satisfies

$$K'S^{-1}K = I_m \tag{33}$$

provided rank(Y) = m.

Biplot representation consists of markers $\underline{j}_1,\ldots,\underline{j}_n$ for the n sample means and markers $\underline{k}_1,\ldots,\underline{k}_m$ for the m variables. These are obtained from the corresponding rows of J and K, respectively, and therefore the i-th sample mean on variable v is represented by the inner product

$$y_{i,v} = \underline{j}_i'\underline{k}_v \tag{34}$$

which can be visualized as the product of the signed length of the projection of $\underline{j}_i$ onto $V\langle\underline{k}_v\rangle$, times the length of $\underline{k}_v$.

Inferences on sample comparisons are made by means of uncertainty circles

$$\{J\}_i = \{\underline{j} \mid \|\underline{j} - \underline{j}_i\| \leq \tau_\alpha W_i\} \tag{35}$$

about the $\underline{j}$-markers. Uncertainty intervals for individual variables are then obtained by projecting the uncertainty circles onto the directions representing the variables. Thus, the uncertainty interval of sample i on the combination of variables with coefficients $\underline{b}$ is $P\langle K'\underline{b}\rangle\{J\}_i$. (See (F.7) of Appendix F.) This construction permits inferences on individual variables and linear combinations of variables in a manner analogous to that of the unstandardized representation, except that circles are easier to work with (see the example in Section VI, below). Also it permits approximate representation of higher dimensional data by means of lower rank approximation. (See Sections VII and X, below.)

VI. BIPLOT OF THE SIMPLE EXAMPLE

The computations for the biplot of the weight loss data use

$$D_N^{\frac{1}{2}} Y S^{-\frac{1}{2}} = \begin{bmatrix} -2.679 & -0.967 \\ -2.477 & 0.670 \\ -2.775 & 0.794 \\ -2.427 & 1.080 \\ 5.903 & 0.498 \\ 4.555 & -2.036 \end{bmatrix}$$

whose S.V.D. yields, by (31) and (32),

$$J = \begin{bmatrix} -1.260 & -.664 \\ -1.273 & .161 \\ -1.429 & .201 \\ -1.276 & .367 \\ 2.842 & .628 \\ 2.397 & -.693 \end{bmatrix}$$

and

$$K = \begin{bmatrix} 1.924 & 1.244 \\ 0.634 & 2.435 \end{bmatrix}$$

The uncertainty circles' radii are 1.515, just as in the previous construction.

The biplot is shown in Figure 15.4, in which the directions for the two variables are indicated by arrows from the origin, and the six samples by uncertainty circles. It is evident that the configuration of these circles is merely a rotation of that of the circles in Figure 15.4. However, the biplot displays the variables as well and

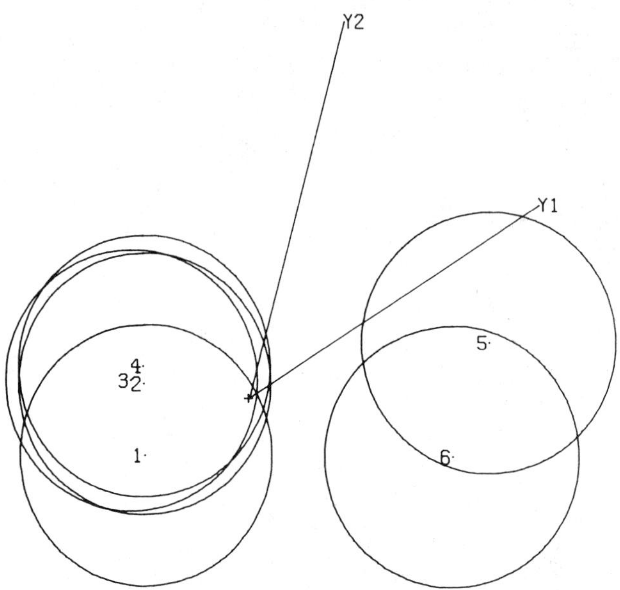

Figure 15.4 Weight Loss Experiment: Biplot of Means with 95% Uncertainty Circles {J}

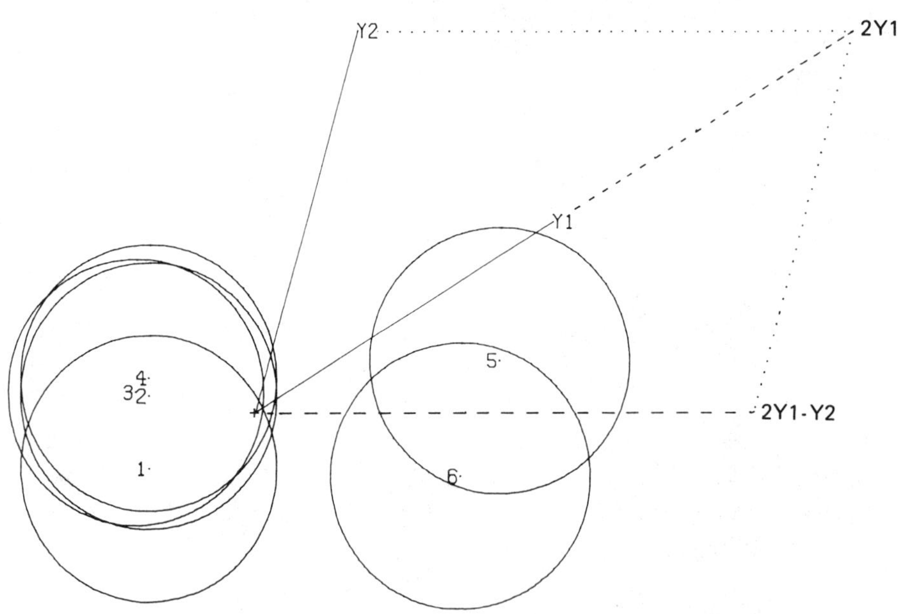

Figure 15.5 Weight Loss Experiment: Biplot with Construction of $\underline{k}$ Marker for Combination of Variables $2y_1 - y_2$

makes it possible to see in which directions, i.e., on what variables, the samples differ. From the circles' shadows onto the two lines it is immediately evident that there are no significant differences on the second variable. It is also obvious that the fifth sample differs from the first four on the first variable and perhaps -- the figure is not clear enough -- the sixth sample also differs from the first. Also, by vector addition, one may construct k-markers for combinations of variables. Thus, Figure 15.5 shows the construction of the vector for the combination with $\underline{b}' = (2,-1)$; it is very clear that both samples 5 and 6 differ from the first four samples on that combination.

The conclusions from this display are clearly the same as those for the original display (Figure 15.1) of the unstandardized data. Indeed, that must always be so.

VII. BIPLOTS OF HIGHER DIMENSIONAL DATA

Geometric visualization and graphic representation of the uncertainty ellipses and circles is physically feasible only for m = 2 and perhaps m = 3. However, for higher dimensional data, projections onto a plane (or 3D space) may give useful approximations. In view of the Householder-Young (1938) Theorem on lower rank approximation, the least squares fit in the plane (or 3D) can be calculated by means of the singular value decomposition. Thus, the biplot planar approximation is obtained by using only the first two coordinates of J and K (and the 3D bimodel approximation, by using their first three coordinates).

On the plane, then, one would construct the biplot by plotting

$$\tilde{\underline{j}}_i' = (j_{i,1}, j_{i,2}) \tag{36}$$

and

$$\tilde{\underline{k}}_v' = (k_{v,1}, k_{v,2}) \tag{37}$$

as well as uncertainty circles

$$\{\tilde{J}\}_i = \{\underline{j} \mid \|\underline{j} - \tilde{\underline{j}}_i\| \leq \tau_\alpha W_i\} \tag{38}$$

for sample i (=1,...,n), and correspondingly $\{\tilde{J}\}_{\underline{a}+}$ and $\{\tilde{J}\}_{\underline{a}-}$ for any contrast with coefficients $\underline{a}$ ($\Sigma_i a_i = 0$). In 3D, one would similarly construct the bimodel by using three columns of J and of K.

The use and interpretation of such an approximate biplot (or bimodel) is similar to that of the exact biplot described above (Sections V and VI) for bivariate data. Thus, one would accept $\underline{\Psi}_i' = \underline{\Psi}_e'$ if $\{\tilde{J}\}_i$ and $\{\tilde{J}\}_e$ overlapped, and one would similarly accept $\underline{\Psi}_i'\underline{b} = \underline{\Psi}_e'\underline{b}$ if $P\langle K'\underline{b}\rangle\{\tilde{J}\}_i$ and $P\langle K'\underline{b}\rangle\{\tilde{J}\}_e$ overlapped. Also, one would obtain approximations $[\tilde{l}_i\langle\tilde{K}'\underline{b}\rangle, \tilde{u}_i\langle\tilde{K}'\underline{b}\rangle]$ to uncertainty intervals $[l_i\langle K'\underline{b}\rangle, u_i\langle K'\underline{b}\rangle]$ by observing the shadow

$$P\langle\tilde{K}'\underline{b}\rangle\{J\}_i = (\tilde{K}'\underline{b})*[l_i\langle\tilde{K}'\underline{b}\rangle/\|K'\underline{b}\| , \tilde{u}_i\langle\tilde{K}'\underline{b}\rangle/\|\tilde{K}'\underline{b}\|] \tag{39}$$

of $\{\tilde{J}\}_i$ onto $V\langle\tilde{K}'\underline{b}\rangle$. This is the analog of (13) in the reduced space.

The closeness of the biplot (or 3D bimodel) approximation would evidently depend primarily on how close $\tilde{K}'\underline{b}$ was to $K'\underline{b}$. Secondarily, even if $\tilde{K}'\underline{b}$ was not close to $K'\underline{b}$, but $\tilde{\underline{j}}_i - \tilde{\underline{j}}_e$ was close to $\underline{j}_i - \underline{j}_e$, the approximation would be off by a factor $\|\tilde{K}'\underline{b}\|/\|K'\underline{b}\|$. This would not affect the test of $\underline{\Psi}_i'\underline{b} = \underline{\Psi}_e'\underline{b}$ which checks the intersection of $P\langle\tilde{K}'\underline{b}\rangle\{\tilde{J}\}_i$ and $P\langle\tilde{K}'\underline{b}\rangle\{\tilde{J}\}_e$. And since this is so for any $\underline{b}$, it follows that if $\tilde{\underline{j}}_i - \tilde{\underline{j}}_e$ is close to $\underline{j}_i - \underline{j}_e$, the test of $\underline{\Psi}_i' = \underline{\Psi}_e'$ by intersection of $\{\tilde{J}\}_i$ and $\{\tilde{J}\}_e$ closely approximates the full rank test (12) which checks the intersection of $\{J\}_i$ and $\{J\}_e$.

VIII. A THREE-DIMENSIONAL EXAMPLE

An example of three dimensional data relates to 4 groups' observations on three maternal attitude scales (Morrison, 1976, p. 210). Table 15.3 presents the sample sizes, means and sums of squares and products within samples.

TABLE 15.3

Maternal Attitude Scales: Data (Morrison, 1976, p. 210)

Group	Sample Sizes	Means on Scales 1	2	3
A	8.00	18.00	20.00	19.75
B	5.00	13.80	15.20	14.20
C	4.00	13.00	14.00	15.00
D	4.00	10.00	9.00	11.00
Average	—	14.524	15.619	15.857

Scales	Within Sums of Squares and Products 1	2	3
1	2.5176	.6588	1.0706
2	.6588	1.8118	.2235
3	1.0706	.2235	2.2529

The biplot of the maternal attitude means, in the form of deviations from the average of the means, is obtained from the S.V.D. of

$$D_N^{\frac{1}{2}}YS^{-\frac{1}{2}} =$$

$$\begin{bmatrix} \sqrt{8} & 0 & 0 & 0 \\ 0 & \sqrt{5} & 0 & 0 \\ 0 & 0 & \sqrt{4} & 0 \\ 0 & 0 & 0 & \sqrt{4} \end{bmatrix} \begin{bmatrix} 3.4762 & 4.3810 & 3.8929 \\ -.7238 & -.4190 & -1.6571 \\ -1.5238 & -1.6910 & -.8571 \\ -4.5238 & -6.6190 & -4.8571 \end{bmatrix} \begin{bmatrix} .7091 & -.1133 & -.1684 \\ -.1133 & .7732 & .0001 \\ -.1684 & .0001 & .7270 \end{bmatrix}$$

the right hand side is $S^{-\frac{1}{2}}$, where S is the estimated variance-covariance matrix. The S.V.D. is $D_N^{\frac{1}{2}}YS^{-\frac{1}{2}} = P' \Lambda Q$, with

$$P' = \begin{bmatrix} -.6931 & -.2246 & .2970 \\ .1145 & .8599 & -.0967 \\ .1584 & -.3598 & -.8093 \\ .6938 & -.2840 & .4974 \end{bmatrix}$$

$$\Lambda = \text{Diag}(16.1672,\ 1.9805,\ 0.7790)$$

$$Q = \begin{bmatrix} -.3177 & -.7833 & -.5343 \\ .1415 & .5180 & -.8436 \\ .9367 & -.3436 & -.0537 \end{bmatrix}$$

The 2-dimensional biplot coordinates are obtained from the first two components as

$$J = \begin{bmatrix} 1/\sqrt{8} & 0 & 0 & 0 \\ 0 & 1/\sqrt{5} & 0 & 0 \\ 0 & 0 & 1/\sqrt{4} & 0 \\ 0 & 0 & 0 & 1/\sqrt{4} \end{bmatrix} \begin{bmatrix} -.6931 & -.2246 \\ .1145 & .8599 \\ .1548 & -.3589 \\ .6938 & -.2840 \end{bmatrix} \begin{bmatrix} 16.1672 & 0 \\ 0 & 1.9805 \end{bmatrix}$$

$$= \begin{bmatrix} 3.961 & -.157 \\ .828 & .762 \\ 1.280 & -.356 \\ 5.609 & -.281 \end{bmatrix}$$

and

$$K = S^{-\frac{1}{2}} \begin{bmatrix} -.3177 & .1415 \\ -.7833 & .5180 \\ -.5343 & -.8436 \end{bmatrix} = \begin{bmatrix} -\ .851 & .034 \\ -1.138 & .675 \\ -\ .932 & -1.153 \end{bmatrix}$$

The radii of the uncertainty circles are:

i	1	2	3	4
$\tau_{.05}W_i$	1.120	1.406	1.586	1.586

The resulting biplot is displayed in Figure 15.6.

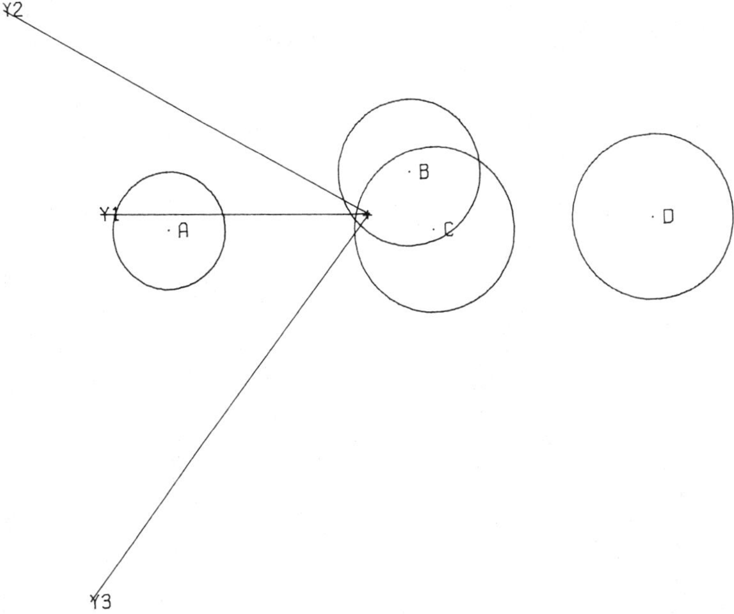

Figure 15.6 Maternal Attitude Scales: Biplot with 95% Uncertainty Circles

Inspection of the biplot on Figure 15.6 shows strikingly that groups A and D differ very significantly. Also, they are both significantly different from groups B and C which are intermediate between them. The latter two groups do not differ significantly from one another. This is immediately evident from the uncertainty circles. The differences are most obvious along the horizontal axis of the biplot, which corresponds closely to scale 1. Projections onto the vector for scale 2, Figure 15.7, also show separation between A and D. Thus, the outer and inner distances between the shadows of $\{J\}_1$ and $\{J\}_4$ are measured as about 11.3 and 5.6, respectively. The length of the $\underline{k}_2$ vector for scale 2 is about 1.3, so the confidence interval for $\psi_{1,2} - \psi_{4,2}$ is about $(5.6 \times 1.3, 11.3 \times 5.6) = (7.3, 14.7)$, clearly this difference is highly significant. This graphical determination of the interval compares well with the exact cal-

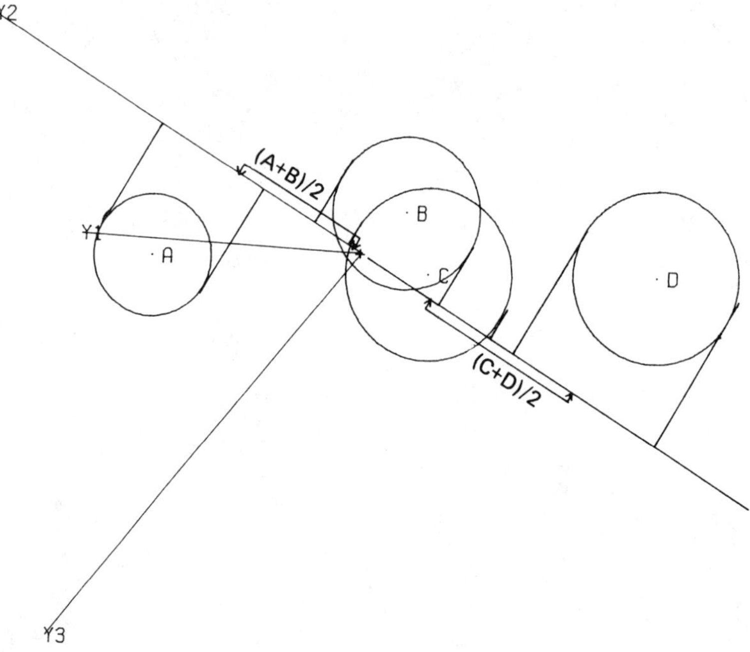

Figure 15.7 Maternal Attitude Scales: Biplot with Construction of Confidence Intervals on $\psi_{1,2} - \psi_{4,2}$ and on $(\psi_{1,2} + \psi_{2,2})/2 - (\psi_{3,2} + \psi_{4,2})/2$

culation of (7.36, 14.64) in Table 15.4. (Note that Table 15.4 also shows the calculation of the intervals for the rank two approximation used in the biplot, i.e., (7.40, 14.55). The graphical determination really approximates the latter, but for this particular bilinear function the two are very close together.)

Another comparison that could be made, though inspection of the biplot does not indicate that it would be particularly revealing, is that of the average of groups A and B versus the average of groups C and D. The "average" shadows are shown in Figure 15.7 and indicate the confidence interval of

TABLE 15.4

Maternal Attitude Data: Simultaneous 95% Confidence Intervals

Contrast A, B, C, D	Combination 1, 2, 3	Roy-Bose Intervals (1)	New Intervals (8) On Data	New Intervals (8) On Rank 2 Approx.
1, 0, 0, -1	1, 0, 0	3.57, 12.43	3.71, 12.29	5.85, 10.45
	0, 1, 0	7.24, 14.76	7.36, 14.64	7.40, 14.55
	0, 0, 1	4.56, 12.94	4.69, 12.81	4.77, 12.78
	0, 1, -1	-3.06, 7.56	-2.89, 7.39	-2.78, 7.17
.5, .5, -.5, -.5	1, 0, 0	1.12, 7.68	-0.12, 8.92	1.86, 6.71
	0, 1, 0	3.31, 8.89	2.27, 9.93	2.36, 9.89
	0, 0, 1	0.87, 7.08	-0.30, 8.25	-0.26, 8.17
	0, 1, -1	-1.81, 7.54	-3.29, 7.54	-3.07, 7.40

Note: The critical value $\sqrt{(6-1)}\sqrt{\lambda_{.05}}$ is obtained by reading $\theta_{.05}(3,-\frac{1}{2},6.5) = 0.550$ from Heck's (1960) charts and computing $\sqrt{(17 \times .550/(1-.550))} = 4.558$. The critical value $\tau_{.05}$ is obtained by interpolating for $\tilde{\alpha} = .05/6 = .00833$ in the F distribution with 3 and $21 - 4 - 3 + 1 = 15$ d.f. and computing $\sqrt{(3(21-4)/(21-4-3+1))F_{(3,21-4-3+1),.00833}} = 4.415$.

$$2.1 \leq (\psi_{1,2} + \psi_{2,2})/2 - (\psi_{3,2} + \psi_{4,2})/2 \leq 9.2$$

This compares closely with the exact and the rank two approximate intervals in Table 15.4.

A different type of comparison relates to non-trivial linear combinations of the variables. Thus, the "scale 2 minus scale 3" difference would be represented on Figure 15.7 by $\underline{k}_2' = \underline{k}_3'$, a vector from the vertex of $\underline{k}_3'$ to the vertex of $\underline{k}_2'$. This new vector would be almost exactly vertical on Figure 15.7 and it is clear that the projections of all the four uncertainty circles on it would intersect. This is interpreted as acceptance of the hypothesis $(0, 1, -1)\Psi' = \underline{0}'$. Indeed, in Table 15.4 both the A vs. D and the A + B vs. C + D contrasts on the "scale 2 minus scale 3" combination are found to have confidence intervals which include zero.

In the above examples, the graphical construction of confidence intervals closely approximates the intervals constructed from (8). However, this is not necessarily so for all bilinear functions $\underline{a}' \Psi \underline{b}$. Table 15.4 shows that for $\underline{b}' = (1,0,0)$, i.e., on the first scale y_1, the rank two approximate confidence intervals are quite different from the exact ones of (8). Construction of the corresponding intervals on the biplot provides graphical estimates of the rank 2 approximations, rather than of the exact intervals. Thus, the use of these graphical estimates may be misleading if some vectors lie mainly outside the biplot plane (or bimodel 3D space). It would seem that one should follow up graphical inferences made from the biplot (or bimodel) by checks on the fit of the $\underline{k}_v$ vectors onto the biplot plane (or bimodel 3D space), i.e., by the size of $(\|\underline{k}_v\|^2 - \|\tilde{\underline{k}}_v\|^2)/\|\underline{k}_v\|^2$. For variables for which this is large, biplot (bimodel) inspection is not adequate.

Table 15.4 shows simultaneous 95% confidence intervals for selected contrasts and combinations, by both Roy-Bose and the new method. That table also shows the intervals that would be calculated from the rank two approximation to the data, which was used for the biplot.

It is again evident that the new intervals are slightly narrower than the classical ones for the pairwise comparison. On the other hand, for the other contrast which compares the average of the first two groups with the average of the last two, the new intervals are appreciably wider than those due to Roy and Bose. Indeed, in two of the cases illustrated i.e., on scales 1 and 3, the new intervals would not have shown the A,B vs. C,D difference to be significant, whereas the Roy-Bose intervals would have shown it.

Though the classical method may be more sensitive for many contrasts, its practical application suffers from lack of a method of exploring which particular contrast-combination pairs, out of the infinitely many which exist, are the significant ones. They cannot all be calculated, and there is no indication which ones should be tested. The advantage of the new method is that it permits approximate graphical display and thereby allows the investigator's eye to guide the search for significant comparisons. This has been illustrated by using the biplot on two and three dimensional data. For data of higher dimensions, the biplot approximation may not always be adequate. Imaginative use of rotations and projections on a graphical display device may then be helpful (Tsianco et al., 1981).

IX. SUMMARY

When data from a MANOVA set-up are to be explored by simultaneous inference techniques, one may use either the classical Roy-Bose intervals or the new intervals proposed here. The new intervals have the advantage of being narrower for pairwise comparisons but suffer from being wider for many other contrasts. However, the new intervals allow for graphical display, exact in bivariate situations, approximate when more variables are involved, and this permits the investigator to be guided by the data in searching for interesting contrasts and combinations. That is an important asset when the data involve many comparisons and are difficult to disentangle.

REFERENCES

Gabriel, K.R. (1971). The biplot - graphic display of matrices with application to principal component analysis. *Biometrika 58*, 453-467.

Gabriel, K.R. (1972). Analysis of meteorological data by means of canonical decomposition and the biplot. *J. Appl. Meteor. 11*, 1071-1077.

Gabriel, K.R. (1978). A simple method of multiple comparison of means. *J. Amer. Statist. Assoc. 73*, 724-729.

Gabriel, K.R. (1980). Biplot. In *Encyclopedia of Statistical Sciences Vol. 1*, N.L. Johnson and S. Kotz (eds.). New York: Wiley.

Gabriel, K.R. (1981). Display of Multivariate Matrices for Inspection of Data and Diagnosis, Chapter 8 of *Interpreting Multivariate Data*, V. Barnett, (ed.), London: Wiley.

Gabriel, K.R. and Gheva, D. (1982). An improved graphical procedure for multiple comparisons. University of Rochester, Statistics Technical Report 82/02.

Gheva, D. (1983). Biplot Approximate Display of Contingency Table Analysis. University of Rochester Ph.D. Dissertation.

Heck, D.L. (1960). Charts of some upper percentage points of the largest characteristic root. *Ann. Math. Statist. 31*, 625-642.

Hochberg, Y. (1974). Some generalizations of the T-method in simultaneous inference. *J. Multiv. Anal. 4*, 224-234.

Hochberg, Y., Weiss, G., and Hart, S. (1982). On graphical procedures for multiple comparisons. *J. Amer. Statist. Assoc. 78*, 767-772.

Householder, A.S. and Young G. (1938). Matrix approximation and latent roots. *Am. Math. Monthly 45*, 165-171.

Morrison, D.F. (1976). *Multivariate Statistical Methods (2nd ed.)*. New York: McGraw-Hill.

Roy, S.M., and Bose, R.C. (1953). Simultaneous confidence interval estimation. *Ann. Math. Statist. 24*, 513-536.

Scheffé, H. (1959). *The Analysis of Variance*. New York: Wiley.

Siotani, M. (1960). Notes on multivariate confidence bounds. *Ann. Inst. Statist. Math. 11,* 167-182.

Tsianco, M.C., Odoroff, C.L., Plumb, S. and Gabriel, K.R. (1981). BGRAPH - A Program for Biplot Multivariate Graphics. Version 1: User's Guide. University of Rochester, Statistics Technical Report 81/20.

APPENDIX A: PROOF OF THE CONSERVATISM OF THE PROPOSED NEW SIMULTANEOUS CONFIDENCE INTERVALS.

Recall that Roy and Bose (1953, (4.3.1)) also proposed simultaneous $(1-\alpha)$-confidence intervals associated with the maximum T^2 test. These were of the form

$$(\underline{\Psi}_i' - \underline{\Psi}_e')\underline{b} \in [(\underline{y}_i' - \underline{y}_e')\underline{b} \pm \tau_\alpha(1/N_i + 1/N_e)\sqrt{\underline{b}'S\underline{b}}],$$

$$\forall\ i,e;\ \forall\ \underline{b} \qquad \text{(A.1)}$$

In view of (5), the new intervals (9) on pairwise comparisons contain these. The new intervals on pairwise comparisons are therefore also of confidence at least $1-\alpha$.

Next, recall that Hochberg (1974, Lemma 3.1) proved that for any $z_1,\ldots,z_n$ and $x_1,\ldots,x_n$, the following two statements are equivalent:

$$|z_i - z_e| \leq x_i^2 + x_e^2,\ \forall\ i,e = 1,\ldots,n \qquad \text{(A.2)}$$

and

$$|\Sigma_i a_i z_i| \leq \Sigma_i |a_i| x_i^2,\ \forall\ a_1,\ldots,a_n\ (\Sigma_i a_i = 0) \qquad \text{(A.3)}$$

Apply this with $z_i = (\underline{y}_i' - \underline{\Psi}_i')\underline{b}$ and $x_i = \tau_\alpha W_i \sqrt{\underline{b}'S\underline{b}}$ to see that the statements (8) are equivalent to the statements (9).

It therefore follows that the statements (8) are also simultaneously true with probability at least $1-\alpha$.

APPENDIX B: ON THE RELATION BETWEEN THE NEW CONFIDENCE INTERVALS AND THE UNCERTAINTY INTERVALS - THE UNIVARIATE CASE.

For any single variable, the new confidence intervals (10) can be represented, as shown by Gabriel (1979), Hochberg, Weiss and Hart (1982) and Gabriel and Gheva (1982), by a smaller number of "uncertainty intervals" defined as follows. For the v-th variable, the i-th sample uncertainty interval on $\psi_{i,v}$ is defined as

$$[l_i\langle v\rangle,\ u_i\langle v\rangle] = [y_{i,v} \pm \tau_\alpha W_i \sqrt{s_{v,v}}] \tag{B.1}$$

These are not confidence intervals on the $\psi_{i,v}$'s, but they permit the simple construction of confidence intervals on differences as

$$\psi_{i,v} - \psi_{e,v} \in [l_i\langle v\rangle - u_e\langle v\rangle,\ u_i\langle v\rangle - l_e\langle v\rangle] \tag{B.2}$$

which is readily checked to be a re-expression of (10) by means of (B.1).

For arbitrary linear combinations of variables, with coefficients $\underline{b}$, the uncertainty interval on $\underline{\Psi}_i'\underline{b}$ is similarly defined as

$$[l_i\langle\underline{b}\rangle,\ u_i\langle\underline{b}\rangle] = [\underline{y}_i'\underline{b} \pm \tau_\alpha W_i \sqrt{\underline{b}'S\underline{b}}] \tag{B.3}$$

for the i-th sample. The confidence intervals of (9) can in that case be constructed as in (17) in the form

$$(\underline{\Psi}_i' - \underline{\Psi}_e')\underline{b} \in [l_i\langle\underline{b}\rangle - u_e\langle\underline{b}\rangle,\ u_i\langle\underline{b}\rangle - l_e\langle\underline{b}\rangle] \tag{B.4}$$

To make inferences on an arbitrary contrast with coefficients $\underline{a}$, it is convenient to standardize it to have sum of absolute coefficients $\Sigma_i|a_i| = 2$. It then becomes a difference of two weighted means. (This argument is at the basis of the well known proof of the extension of Tukey's method of multiple comparisons to arbitrary contrasts (Scheffé, 1959, Chapter 3).) Corresponding to (B.3) and

(B.4) one may construct the following two "average uncertainty intervals" on the positive and negative components of $\underline{a}'\ \Psi\ \underline{b}'$, i.e., on $\Sigma_i^+ \Sigma_v a_i \psi_{i,v} b_v$ and $\Sigma_e^- \Sigma_v |a_e| \psi_{e,v} b_v$, respectively. These are

$$[l_{\underline{a}+}\langle\underline{b}\rangle,\ u_{\underline{a}+}\langle\underline{b}\rangle] = [\Sigma_i^+ a_i l_i\langle\underline{b}\rangle,\ \Sigma_i^+ a_i u_i\langle\underline{b}\rangle] \tag{B.5}$$

and

$$[l_{\underline{a}-}\langle\underline{b}\rangle,\ u_{\underline{a}-}\langle\underline{b}\rangle] = [\Sigma_e^- |a_e| l_e\langle\underline{b}\rangle,\ \Sigma_e^- |a_e| u_e\langle\underline{b}\rangle] \tag{B.6}$$

Differencing these two uncertainty intervals will be seen to yield

$$\underline{a}'\ \Psi\ \underline{b} \in [l_{\underline{a}+}\langle\underline{b}\rangle - u_{\underline{a}-}\langle\underline{b}\rangle,\ u_{\underline{a}+}\langle\underline{b}\rangle - l_{\underline{a}-}\langle\underline{b}\rangle] \tag{B.7}$$

the new confidence intervals (8) in the form (21).

For any one variable, or one linear combination of variables, the uncertainty intervals are thus seen to have a ready graphical representation on the real line and the construction (B.2), (B.4) and (B.7) of confidence intervals is readily visualized (Gabriel, 1978). Since acceptance/rejection of a null hypothesis on a difference depends on inclusion/exclusion of zero in its confidence interval, it is seen to depend on the overlap/non-overlap of the corresponding uncertainty intervals. Thus, in view of (17) and (B.4), $\underline{\Psi}_i'\underline{b} = \underline{\Psi}_e'\underline{b}$ is acceptable if, and only if,

$$[l_i\langle\underline{b}\rangle,\ u_i\langle\underline{b}\rangle] \cap [l_e\langle\underline{b}\rangle,\ u_e\langle\underline{b}\rangle] \neq \phi \tag{B.8}$$

Also, in view of (21) and (B.7), $\underline{a}'\ \Psi\ \underline{b} = 0$ is acceptable if, and only if

$$[l_{\underline{a}+}\langle\underline{b}\rangle,\ u_{\underline{a}+}\langle\underline{b}\rangle] \cap [l_{\underline{a}-}\langle\underline{b}\rangle,\ u_{\underline{a}-}\langle\underline{b}\rangle] \neq \phi \tag{B.9}$$

Criterion (B.8) will be shown (Appendix C) to be equivalent to (13) and, analogously, criterion (B.9) is equivalent to (20).

APPENDIX C: MULTIVARIATE UNCERTAINTY ELLIPSES AND THEIR RELATION TO UNIVARIATE UNCERTAINTY INTERVALS.

It is possible to represent the uncertainty intervals simultaneously for all variables by means of the uncertainty ellipses $\{Y\}_i$ of (11) about the centroids $\underline{y}_i$, i=1,...,n. Some straightforward algebra shows that the envelope of such an ellipse can be written

$$[\{Y\}_i] = \{\underline{y} | \underline{y} = \underline{y}_i + \tau_\alpha W_i R' \Lambda \underline{t}; \ ||\underline{t}|| = 1\} \tag{C.1}$$

where $S = R' \Lambda^2 R$ is the singular value decomposition of S.

To relate these ellipses to the univariate confidence intervals, we first standardize the linear combinations of variables to coefficients $\underline{b}^* = \underline{b}/||\underline{b}||$, so that

$$P\langle\underline{b}\rangle\underline{y} = \underline{b}^*(\underline{b}'\underline{y})/||\underline{b}|| \tag{C.2}$$

We next denote

$$||P\langle\underline{b}\rangle\{Y\}_i||^{min} = \min(\underline{b}'\underline{y}/||\underline{b}||: y \in \{Y\}_i) \tag{C.3}$$

and

$$||P\langle\underline{b}\rangle\{Y\}_i||^{max} = \max(\underline{b}'\underline{y}/||\underline{b}||: \underline{y} \in \{Y\}_i) \tag{C.4}$$

so that the "shadow", or projection, of $\{Y\}_i$ onto $V\langle\underline{b}\rangle$ is

$$P\langle\underline{b}\rangle\{Y\}_i = \underline{b}^*[||P\langle\underline{b}\rangle\{Y\}_i||^{min}, ||P\langle\underline{b}\rangle\{Y\}_i||^{max}] \tag{C.5}$$

Comparing (C.3) and (C.4) with (C.1) and using (C.2), we find

$$||\underline{b}|| \, ||P\langle\underline{b}\rangle\{Y\}_i||^{min} = \underline{b}'\underline{y}_i - \tau_\alpha W_i \sqrt{\underline{b}' S \underline{b}} \tag{C.6}$$

and

$$||\underline{b}||\,||P\langle\underline{b}\rangle\{Y\}_i||^{max} = \underline{b}'\underline{y}_i + \tau_\alpha W_i\sqrt{\underline{b}'S\underline{b}} \qquad (C.7)$$

Introducing (15) and (16) or (B.1) yields

$$||P\langle\underline{b}\rangle\{Y\}_i||^{min} = 1_i\langle\underline{b}\rangle/||\underline{b}|| \qquad (C.8)$$

and

$$||P\langle\underline{b}\rangle\{Y\}_i||^{max} = u_i\langle\underline{b}\rangle/||\underline{b}|| \qquad (C.9)$$

This establishes that

$$P\langle\underline{b}\rangle\{Y\}_i = \underline{b}^*[1_i\langle\underline{b}\rangle/||\underline{b}||,\ u_i\langle\underline{b}\rangle/||\underline{b}||] \qquad (C.10)$$

as in (14). Hence criterion (B.8) for non-significance of the test of $\underline{\Psi}_i'\underline{b} = \underline{\Psi}_e'\underline{b}$ is equivalent to

$$P\langle\underline{b}\rangle\{Y\}_i \cap P\langle\underline{b}\rangle\{Y\}_e \neq \phi \qquad (C.11)$$

which corresponds to (13).

Analogously to this use of uncertainty ellipses (11) for inferences on pairwise comparisons, one may use the average uncertainty ellipses of (18) and (19) for inferences on arbitrary contrasts $\underline{a}'\ \Psi\ \underline{b}$.

APPENDIX D: THE ELLIPSE OVERLAP TESTS OF THE EQUALITY OF TWO SAMPLES OR THE NULLITY OF A CONTRAST.

To infer on the overall hypothesis $\underline{\Psi}_i' = \underline{\Psi}_e'$ of equality of samples i and e on all variables' expectations, one may consider this hypothesis as the intersection of the univariate hypotheses $\underline{\Psi}_i'\underline{b} = \underline{\Psi}_e'\underline{b}$ for all $\underline{b}$. Thus, $\underline{\Psi}_i' = \underline{\Psi}_e'$ is accepted if and only if all the above univariate hypotheses are accepted. Geometrically, this means that

$\underline{\Psi}_i' = \underline{\Psi}_e'$ is accepted if and only if (13) does not hold for any $\underline{b}$, i.e., the shadows of $\{Y\}_i$ and $\{Y\}_e$ in all directions overlap. But that is equivalent to saying that $\underline{\Psi}_i' = \underline{\Psi}_e'$ is rejected unless the two uncertainty ellipses overlap, as stated in (12).

The new bounds for two sample differences are noted (see (5)) to be wider than the classical ones if both used the same critical value. Hence the implied new tests are conservative. Also, the ellipse overlap test was seen (Appendix C) to be built from the new tests by the Union-Intersection principle, just as Hotelling's T^2 test is known to be the Union-Intersection test based on the univariate t tests. Hence we may regard the ellipse overlap test (12) as a conservative analog of Hotelling's T^2 test.

The geometric, and conservative, analog of the generalized Hotelling T^2 test of $\underline{a}'\Psi = \underline{0}'$ for arbitrary contrast $\underline{a}$ would be to accept if

$$\{Y\}_{\underline{a}+} \cap \{Y\}_{\underline{a}-} \neq \phi \tag{D.1}$$

i.e., if the two average uncertainty ellipses defined for that contrast overlap.

APPENDIX E: THE STANDARDIZED REPRESENTATION.

Confidence intervals (8) can be rewritten

$$\underline{a}'\ \Psi\ \underline{b} \ \varepsilon\ [\underline{a}'Z\underline{d} \pm \tau_\alpha \|\underline{d}\| \Sigma_i |a_i| W_i] \tag{E.1}$$

with corresponding special form

$$(\underline{\Psi}_i' - \underline{\Psi}_e')\underline{b}\ \varepsilon\ [(\underline{z}_i' - \underline{z}_e')\underline{d} \pm \tau_\alpha \|\underline{d}\|(W_i + W_e)] \tag{E.2}$$

for (9). The uncertainty intervals (14) then become

$$[l_i\langle\underline{b}\rangle,\ u_i\langle\underline{b}\rangle] = [\underline{z}_i'\underline{d} \pm \tau_\alpha \|\underline{d}\| W_i] \tag{E.3}$$

and the ellipses (11) are replaced by circles (27). The projections of $\{Z\}_i$ onto $V\langle\underline{d}\rangle$ thus become, in view of (24),

$$P\langle\underline{d}\rangle\{Z\}_i = \underline{d}^*[1_i\langle\underline{b}\rangle/\sqrt{\underline{b}'S\underline{b}},\ u_i\langle\underline{b}\rangle/\sqrt{\underline{b}'S\underline{b}}] \quad (E.4)$$

where $\underline{d}^* = \underline{d}/\|\underline{d}\|$.

Inferences on pairwise comparisons for all variables are now made by testing for intersection of circles instead of ellipses. Inferences on single variables are made by projections of the circles, just as projections of ellipses were used in the unstandardized case.

Similarly, average uncertainty circles

$$\{Z\}_{\underline{a}+} = \{\underline{z} \mid \|\underline{z}\| - \Sigma_i^+ a_i z_i\| \le \tau_\alpha \Sigma_i^+ a_i W_i\} \quad (E.5)$$

and

$$\{Z\}_{\underline{a}-} = \{\underline{z} \mid \|\underline{z}\| - \Sigma_e^- |a_e| z_e\| \le \tau_\alpha \Sigma_e^- |a_e| W_e\} \quad (E.6)$$

are defined analogously to (18) and (19) and used for inferences on arbitrary contrasts.

APPENDIX F: THE STANDARDIZED REPRESENTATION AND THE BIPLOT.

In view of (22), (30) and (32),

$$J = ZQ' \quad (F.1)$$

i.e.,

$$\underline{j}_i' = \underline{z}_i' Q' \quad (F.2)$$

The rows of J are thus seen to be orthogonal transformations of the rows of Z. Similarly, the uncertainty circles $\{Z\}_i$ of (27) rotate into biplot uncertainty circles $\{J\}_i$ of (35). The latter circles have the same radii and relative positions as the former $\{Z\}_i$'s. Because of the rotation by means of Q', the coefficients $\underline{d} = S^{\frac{1}{2}}\underline{b}$ of (24) need to be replaced by

$$\underline{t} = Q'\underline{d} = Q'S^{\frac{1}{2}}\underline{b} \tag{F.3}$$

i.e., in view of (32), by

$$\underline{t} = K'\underline{b} \tag{F.4}$$

for which

$$||\underline{t}|| = \sqrt{\underline{b}'S\underline{b}} \tag{F.5}$$

Thus, we define the standardized vector

$$\underline{t}^* = K'\underline{b}/||K'\underline{b}|| \tag{F.6}$$

and the shadow of the circle $\{J\}_i$ onto $V\langle K'\underline{b}\rangle$ is readily seen to be

$$P\langle K'\underline{b}\rangle\{J\}_i = \underline{t}^*[1_i\langle\underline{b}\rangle/\sqrt{\underline{b}'S\underline{b}},\ u_i\langle\underline{b}\rangle/\sqrt{\underline{b}'S\underline{b}}] \tag{F.7}$$

Here, again, is a geometric representation of $1/\sqrt{\underline{b}'S\underline{b}}$ times the uncertainty interval (14) for $\underline{\Psi}_i'\underline{b}$. Note, however, that this is now represented in the direction of $K'\underline{b}$ rather than of $\underline{b}$, as in (14).

The conservative analogs of Hotelling's T^2 tests again check the intersections of the uncertainty circles. Thus, one accepts $\underline{\Psi}_i' = \underline{\Psi}_e'$ if and only if

$$\{J\}_i \cap \{J\}_e \neq \phi \tag{F.8}$$

Mutatis mutandis, for any $\underline{a}$ such that $\Sigma_i a_i = 0$, one accepts $\underline{a}'\Psi = \underline{0}'$ if, and only if,

$$\{J\}_{\underline{a}+} \cap \{J\}_{\underline{a}-} \neq \phi \qquad \text{(F.9)}$$

where $\{J\}_{\underline{a}+}$ and $\{J\}_{\underline{a}-}$ are defined analogously to $\{Z\}_{\underline{a}+}$ and $\{Z\}_{\underline{a}-}$ in (E.5) and (E.6), respectively.

Inferences on sample (or contrast) comparisons on the biplot are carried out in the same manner as in the standardized representation, except that the biplot further allows projection onto directions for individual variables and linear combinations of variables, as illustrated in Figures 15.4 and 15.5, above.

Chapter 16

INFERENCE ON PARAMETERS IN A LINEAR MODEL: A REVIEW OF RECENT RESULTS

Jochen Müller, C. Radhakrishna Rao, and Bimal Kumar Sinha
University of Pittsburgh
Pittsburgh, Pennsylvania

I. INTRODUCTION

We consider the general Gauss-Markoff model

$$Y = X\beta + \varepsilon \tag{1.1}$$

where $E(\varepsilon) = 0$, $D(\varepsilon) = \sigma^2 V$, and the matrices X and V may be singular, and discuss problems of inference on the unknown parameters β and σ^2. We refer to the model (1.1) by the triplet $(Y, X\beta, \sigma^2 V)$. The paper is in three parts. In the first part, the Gauss-Markoff theory is extended to the case when V is singular. In the second, robustness of inference procedures for departures in the design matrix X, the dispersion matrix V and distributional assumptions on Y is considered. The third part introduces the concepts of linear sufficiency and completeness in linear models, without making any distributional assumptions.

The following notations are used throughout the paper.

(i) $\rho(A)$ denotes the rank of a matrix A and R(A), the range of A, i.e., the vector space generated by the columns of A.

(ii) A^- denotes a generalized inverse of A, satisfying the only condition $AA^-A = A$ (see Rao, 1973, p. 34).

(iii) Z denotes a matrix of full rank satisfying the condition $Z'X = 0$, where X is the design matrix.

(iv) The projection operators on R(A) are denoted by (see Rao, 1973, p. 48)

$$P_{A,M} = A(A'MA)^-A'M \text{ where M is p.d. (positive definite)}$$

$$P_A = A(A'A)^-A'$$

(v) E denotes the expectation operator and D the dispersion operator (providing the variance-covariance matrix of a vector variable).

(vi) For any matrix L, ker L' consists of all vectors a with $L'a = 0$.

(vii) $Y: n\times 1$; $X: n\times m$ with $\rho(X) = r \leq m$, $\beta: m\times 1$.

II. UNIFIED APPROACH TO LINEAR ESTIMATION

In this section, we consider some methods of estimating the unknown parameters β and σ^2 in the general model (1.1).

2.1 Inverse Partitioned Matrix Approach

Let

$$\begin{pmatrix} V & X \\ X' & 0 \end{pmatrix}^- = \begin{pmatrix} C_1 & C_2 \\ C_3 & C_4 \end{pmatrix}$$

for any g-inverse. Then the following proposition is proved in Rao (1971).

Proposition 2.1. (i) In the class of linear estimators L'Y such that $X'L = p$, the minimum variance linear unbiased estimator (MVLUE) of $p'\beta$ is $p'\hat{\beta}$ where

$$\hat{\beta} = C_3Y \text{ or } C_2'Y$$

(ii) If $p'\hat{\beta}$ and $q'\hat{\beta}$ are MVLUE's of $p'\beta$ and $q'\beta$ respectively, then

$$\text{Var}(p'\hat{\beta}) = \sigma^2 p'C_4p$$

$$\text{Cov}(p'\hat{\beta}, q'\hat{\beta}) = \sigma^2 p'C_4q = \sigma^2 q'C_4p$$

(iii) An unbiased estimator of σ^2 is

$$\hat{\sigma}^2 = f^{-1}Y'C_4Y, \; f = \rho(V:X) - \rho(X)$$

2.2 Unified Theory of Least Squares

When V is nonsingular and Y has multivariate normal distribution, we have the following well known results.

(1) Let $\hat{\beta}$ be such that

$$\min_{\beta} (Y-X\beta)'V^{-1}(Y-X\beta) = (Y-X\hat{\beta})'V^{-1}(Y-X\hat{\beta})$$

Then the MVLUE of $p'\beta$, $p \in R(X')$, is $p'\hat{\beta}$.

(2) $R_0^2 = (Y-X\hat{\beta})'V^{-1}(Y-X\hat{\beta}) \sim \sigma^2\chi^2(f)$ i.e., distributed as χ^2 on f d.f., where $f = \rho(V:X) - \rho(X)$.

(3) Let $K'\beta = \omega$ be a linear hypothesis where $R(K) \subset R(X')$ and $\rho(K) = h$, and

$$R_1^2 = \min_{K'\beta=\omega} (Y-X\beta)'V^{-1}(Y-X\beta)$$

Then

$$R_1^2 - R_0^2 \sim \sigma^2\chi^2(h)$$

If V is singular, the above statements are not applicable and the following question arises. Does there exist a symmetric matrix

M which takes the place of V^{-1} for which the above properties (1)-(3) hold? The answer is contained in Proposition 2.2 proved in Rao (1973).

Proposition 2.2. Let $M = (V+XUX')^-$ for any symmetric g-inverse and U be any symmetric matrix such that $\rho(V:X) = \rho(V+XUX')$.

(i) If $\hat{\beta}$ is such that

$$\min_{\beta} (Y-X\beta)'M(Y-X\beta) = (Y-X\hat{\beta})'M(Y-X\hat{\beta})$$

then the MVLUE of $p'\beta$, $p \in R(X')$, is $p'\hat{\beta}$.

(ii) $R_0^2 = (Y-X\hat{\beta})'M(Y-X\hat{\beta}) \sim \sigma^2\chi^2(f)$, $f = \rho(V:X) - \rho(X)$.

(iii) There is no choice of M for which the property (3) also holds for all testable hypotheses.

Contrary to what is stated in (iii), claims have been made about the existence of M for which the property (3) also holds. This is shown to be not true in Rao (1978).

2.3 Least Squares Theory with Derived Restrictions

If V is $n \times n$ and singular, then there exists a matrix N of rank $s = n-\rho(V)$ such that $N'V = 0$ which implies that

$$N'Y - N'X\beta = 0 \text{ w.p.1} \tag{2.3.1}$$

This stochastic relationship may be considered as a restriction on the parameter β, which is known when Y is observed. In such a case, the following proposition is proved by Goldman and Zelen (1964) and Mitra and Rao (1968).

Proposition 2.3. Let V^- be any g-inverse of V and $\hat{\beta}$ be such that

$$\min_{N'Y=N'X\beta} (Y-X\beta)'V^-(Y-X\beta) = (Y-X\hat{\beta})'V^-(Y-X\hat{\beta}) = R_0^2$$

Then

(i) $p'\hat{\beta}$ is the MVLUE of $p'\beta$, $p \in R(X')$.

(ii) $R_0^2 \sim \sigma^2\chi^2(f)$, $f = \rho(V:X) - \rho(X)$.

(iii) If

$$R_1^2 = \min_{\substack{N'Y=N'X\beta \\ K'\beta = \omega}} (Y-X\beta)'V^-(Y-X\beta)$$

then $R_1^2 - R_0^2 \sim \sigma^2\chi^2(h)$, where h denotes the degrees of freedom of the hypothesis $K'\beta = \omega$ to be tested. (Note that h is the rank of the variance covariance matrix of $K'\hat{\beta}$ and not necessarily the rank of K.)

2.4 Optimal Estimators in a Wider Class

In Sections 2.1 and 2.2, we considered the class of linear functions of Y as estimators of $p'\beta$, $p \in R(X')$. Now we consider a wider class of functions

$$T(Y) = f(N'Y) + Y'g(N'Y), \tag{2.4.1}$$

where N is as defined in (2.3.1), f is a scalar and g is a vector function, as possible estimators of $p'\beta$. The following proposition is proved in Rao (1979).

Proposition 2.4. (i) $p'\beta$ has an unbiased estimator in the class (2.4.1) iff $p \in R(X')$.

(ii) If $p'\beta$ is unbiasedly estimable, then the MVLUE of $p'\beta$ in the class of (2.4.1) is equivalent w.p. 1 to the MVLUE of $p'\beta$ in the class of linear functions $L'Y$, as considered in Sections 2.1 and 2.2.

(iii) If $L_*'Y$ is the MVLUE of $p'\beta$ in the class $L'Y$, then a general representation of the MVLUE in the wider class (2.4.1) is

$$L_*'Y + f(N'Y) + Y'g(N'Y)$$

where the functions f and g are such that they can be expressed in terms of a function h as

$$f(\xi) = -\xi' h(\xi)$$

$$g(\xi) = N\ h(\xi)$$

for all $\xi \in R(N'X)$ and arbitrary outside $R(N'X)$. A similar approach was given in a paper by Harville (1981).

2.5 Generalized Projection Operator

Consider the general linear model $(Y, X\beta, \sigma^2 V)$, where V may be singular. It is easily seen that

$$Y \in R(V:X) \text{ w.p.1}$$

The following proposition is established in Rao (1974).

Proposition 2.5. Let Z be a matrix of full rank such that $Z'X = 0$. Then:

(i) $R(X)$ and $R(VZ)$ are disjoint, and $R(X:VZ) = R(X:V)$.

(ii) The projection of $Y \in R(X:VZ)$ on $R(X)$ along $R(VZ)$ can be expressed as PY where P is any matrix satisfying the conditions

$$PX = X, \quad PVZ = 0$$

[Such a matrix P is called a generalized projection operator which reduces to the usual projection operator when $\rho(X:VZ) = n$, where n is the order of V. Note that P is not unique when $\rho(X:VZ) < n$].

From the above proposition we deduce:

Proposition 2.6. Let P be the projection operator on $R(X)$ along $R(VZ)$ as defined in Proposition 2.5, and CY be an unbiased estimator of $X\beta$ (i.e., $CX = X$). Then

$$D(CY) - D(PY)$$

is non-negative definite, where D denotes the dispersion (variance-covariance) operator, so that PY is the minimum dispersion unbiased estimator of $X\beta$ in the class of linear unbiased estimators.

Note that

$$D(CY) = D(CY-PY+PY)$$
$$= D(CY-PY) + D(PY) + \sigma^2(C-P)VP' + \sigma^2 PV(C-P)' \qquad (2.5.1)$$

Since $PVZ = 0 \Rightarrow PV = AX'$ for some A, we have

$$(C-P)VP' = (C-P)XA = 0$$

using the conditions $CX = X$ and $PX = X$. Thus from (2.5.1)

$$D(CY) = D(PY) + D(CY-PY)$$

which proves the Proposition 2.6.

Proposition 2.6 answers a question raised by Kempthorne (1976) on the construction of a projection operator when V is singular, and provides a general method for coordinate free estimation through the concept of a projection operator.

From the Proposition 2.6, we have

Proposition 2.7. Let P be the projection operator on $R(X)$ along $R(VZ)$ and $(X'X)^-$ be any g-inverse of $X'X$. Then

(i) $p'\hat{\beta}$ is the MVLUE of $p'\beta$, $p \in R(X')$, where

$$\hat{\beta} = (X'X)^- X'PY$$

(ii) An unbiased estimator of σ^2 is

$$f^{-1}Y'(I-P')V^-(I-P)Y, \quad f = \rho(V:X) - \rho(X)$$

Reference may also be made to Example 4, Rao (1973, p. 309), where an approach to linear estimation is given without appealing to concepts of linearity, unbiasedness and minimum variance. This is similar to the methods discussed in Section IV of this chapter.

III. ROBUSTNESS IN THE LINEAR MODEL

In this Section we will discuss robustness of some statistical procedures in linear models. To be specific, we will be concerned with the robustness of best linear unbiased estimators (BLUEs) in the context of estimation, and likelihood ratio tests in the context of tests of hypotheses when there is specification error in the design matrix and/or in the dispersion matrix. The consequences of deviations from the assumption of normality on tests will also be discussed.

We assume the same set up as in Section I. Let $(Y, X\beta, \sigma^2 I)$ be the assumed model while $(Y, X\beta, \sigma^2 V)$ is the correct model, resulting in specification error in the dispersion matrix. Throughout this section we assume that V is p.d. Then the BLUE of an estimable linear parametric function $A\beta$ is the same under both models if and only if

$$A(X'X)^{-}X'VZ = 0 \text{ for all } Z,\ Z'X = 0 \tag{3.1}$$

This follows from the condition that a BLUE must have zero covariance with every error function. Characterization of matrices V satisfying (3.1) is well known (Rao and Mitra, 1971; Rao, 1967; Zyskind, 1967). Generally, (3.1) is equivalent to the following representation of V:

$$V = I + X\Lambda_1X' + Z\Lambda_2Z' + X\Lambda_4Z' + Z\Lambda_4'X' \tag{3.2}$$

where Λ_1, Λ_2 and Λ_4 are arbitrary except that $A\Lambda_4Z' = 0$ and V is p.d. An equivalent representation of V is the following:

$$V = I + X\Lambda_1X' + Z\Lambda_2Z' + X_0\Lambda_3Z' + Z\Lambda_3'X_0' \tag{3.3}$$

where $X_0 = X(I - A^{-}A)$, Λ_1, Λ_2 and Λ_3 are arbitrary except that V is p.d. Some further necessary and sufficient conditions (i.e., equivalent conditions) for the representation (3.3) to hold are given in the following.

Proposition 3.1. The representation (3.3) is equivalent to any one of the following conditions:

(a) $Z'VZ_1 = 0$

(b) $P_X V^{-1}(I - P_{X_0,V^{-1}})$ is symmetric

(c) $(I - P_{X_0,V^{-1}})(I - P_{X,V^{-1}})$ is symmetric

(d) There exists an orthogonal matrix T such that $T'(I-P_X)T$, $T'(I-P_{X_0})T$, $T'V^{-1}(I-P_{X,V^{-1}})T$ and $T'(I-P_{X_0,V^{-1}})T$ are diagonal matrices.

In the above A: k×m with $\rho(A) = k$, Z_1: n×k is such that $Z_1'Z_1 = I_k$ and $Z_1'Z = 0$ (k×n-r). For a proof of the above proposition, see Mathew and Bhimasankaram (1982). Incidentally, if we demand (3.1) to hold for all A such that $R(A') \subset R(X')$, which means that for every estimable linear parametric function the BLUE is the same under both models, then we get the following.

Proposition 3.2. (3.1) holds for all A such that $R(A') \subset R(X')$ under any one of the following equivalent conditions:

(a) $X'VZ = 0$

(b) $VX = XQ$ for some Q

(c) VP_X is symmetric

(d) $P_{X,V^{-1}}$ is symmetric.

The result (a) which implies (b) was proved by Rao (1967), and (c) is due to Zyskind (1967). The result (d) appears in Mathew and Bhimansankaram (1982).

Consider next the problem of testing H_0: $A\beta = 0$ assuming normality of the errors. It is well known that the F-test based on $\lambda(X,I) = Y'(I-P_X)Y/Y'(I-P_{X_0})Y$ is both LRT and UMPI (under a suitable group of transformations) under the normal model $N(Y,X\beta,\sigma^2 I)$ (see, for example, Lehmann, 1959). We would like to study the robustness properties of this test in so far as whether the properties of its being LRT (criterion robustness) and UMPI (inference robustness) remain valid under deviations from the assumption of normality and the presence of specification errors in the design matrix and/or the dispersion matrix.

To begin with, note that if the correct model is $N(Y, X_1\beta, \sigma^2 V)$, denoting by X_1^0 the matrix $X_1(I-A^-A)$, the LRT testing H_0: $A\beta = 0$ is based on

$$\lambda(X_1, V) = Y'V^{-1}(I-P_{X_1^0, V^{-1}})Y/Y'V^{-1}(I-P_{X_1, V^{-1}})Y \tag{3.4}$$

Therefore, the F-test based on $\lambda(X,I)$ under $N(Y, X\beta, \sigma^2 I)$ is LRT under $N(Y, X_1\beta, \sigma^2 V)$ if and only if

$$\lambda(X,I) \equiv \lambda(X_1, V) \text{ for all } Y \tag{3.5}$$

Under the same design matrix $X_1 = X$ but a different dispersion matrix, the condition on the representation of V is the following:

$$V = I + X\Lambda_1 X' + (s-1)ZZ' + X_0\Lambda_3 Z' + Z\Lambda_3' X_0' \tag{3.6}$$

where Λ_1, Λ_3 are arbitrary and s is an arbitrary positive real number subject to (i) V is p.d. and (ii) $Z_1'X\Lambda_1 X'Z_1 = (s-1)I_k$. The following proposition provides other equivalent conditions.

<u>Proposition 3.3</u>. V has the representation (3.6) if and only if any one of the following equivalent conditions holds:

(a) $(I-P_{X_0})V(I-P_{X_0}) = a(I-P_{X_0})$ for some $a > 0$

(b) $V^{-1}(I-P_{X_0, V^{-1}}) = a(I-P_{X_0})$ for some $a > 0$

(c) $\begin{pmatrix} I-P_X \\ LP_X \end{pmatrix} (V-aI)(I-P_X : P_X L') = 0$, for some $a > 0$, with $A = LX$.

Part (c) of this proposition is due to Khatri (1980) and parts (a) and (b) are due to Mathew and Bhimasankaram (1982) as well as the representation (3.6) above. When V has the intraclass covariance structure, $V = (1-\rho)I + \rho \underset{\sim}{1}\underset{\sim}{1}'$, proceeding directly Ghosh and Sinha (1980) noted that $\lambda(X,I) \equiv \lambda(X,V)$ if and only if $\underset{\sim}{1} \in R(X_0)$. Some generalizations of (3.2) and (3.6) are reported in Chikuse (1981).

Under the same dispersion matrix I but a different design matrix X_1, the F-test remains LRT if and only if $\lambda(X,I) \equiv \lambda(X_1,I)$. This leads to the following.

Proposition 3.4. $\lambda(X,I) \equiv \lambda(X_1,I)$ for all Y if and only if

$$R(X) = R(X_1) \text{ and } R(X_0) = R(X_1^0) \tag{3.7}$$

Finally, the following proposition provides conditions under which (3.5) holds for arbitrary V and X_1.

Proposition 3.5. (3.5) holds if and only if (3.6) and (3.7) hold.

Propositions (3.4) and (3.5) are due to Mathew and Bhimasankaram (1982). The key to all these results, noted earlier by Sinha and Mukhopadhyay (1980), can be stated in the following most general form with a different simpler proof due to Müller and Sinha.

Lemma 3.6. Let A,B,C,D symmetric be such that

$$\frac{y'Ay}{y'By} = \frac{y'Cy}{y'Dy} \quad \text{for almost all } y$$

If there is an x such that $Ax = 0$ and $x'Bx \neq 0$ then

$$C = \gamma A \text{ for some } \gamma \in \mathbb{R}$$

Also

$$D = \gamma B$$

provided $A \neq 0$.

Proof. From the assumption it follows immediately that

$$y'Ay\ y'Dy = y'Cy\ y'By \text{ for all } y$$

Especially for $y = x$ this results in $x'Cx = 0$. Now insert $(y + \lambda x)$ to obtain

$$y'Ay[y'Dy + 2\lambda\ y'Dx + \lambda^2\ x'Dx]$$
$$= [y'Cy + 2\lambda\ y'Cx][y'By + 2\lambda\ y'Bx + \lambda^2\ x'Bx]$$

Comparison of the coefficients of λ^3 yields

$$0 = y'Cx\ (x'Bx) \text{ for all } y$$

whence $Cx = 0$. Therefore the coefficients of λ^2 become

$$y'Ay\ x'Dx = y'Cy\ x'Bx$$

from which

$$C = \frac{x'Dx}{x'Bx} A$$

follows. The remainder is evident. □

We now turn our attention to the robustness properties of the F-test for non-normal errors. The following result was proved by Ghosh and Sinha (1980).

Proposition 3.7. Let $Y = X\beta + \sigma\varepsilon$ with ε distributed according to a density $f(\varepsilon)$ given by

$$f(\varepsilon) = \int_0^\infty \frac{e^{-\frac{\tau}{2}\varepsilon'\varepsilon}}{(\sqrt{2\pi})^n} \tau^{n/2}\, dL(\tau)$$

Then for testing H_0: $A\beta = 0$, the F-test based on $\lambda(X,I)$ is both LRT and UMPI.

Recently Sinha and Drygas (1982) generalized this result to the following.

Proposition 3.8. Let $Y = X\beta + \sigma\varepsilon$ with ε distributed according to a density $q(\varepsilon'\varepsilon)$, $q \uparrow$, convex. Then the F-test based on $\lambda(X,I)$ is both LRT and UMPI.

This result is similar to a robustness property of the Hotelling's T^2-test proved by Kariya (1981) and is based on an application of a representation theorem due to Wijsman (1967). Under a slightly more general distribution of the errors, the following property of a BLUE holds (see Sinha and Drygas, 1982).

Proposition 3.9. Let $Y = X\beta + \sigma\varepsilon$ with ε having a spherically symmetric distribution. Then for any $c \in \mathbb{R}$ and any n.n.d. matrix C of appropriate order

$$P\{(GY-A\beta)'C(GY-A\beta) \leq c^2\} \geq P\{(LY-A\beta)'C(LY-A\beta) \leq c^2\}$$

where GY is any BLUE of estimable $A\beta$ and LY is any unbiased estimator of $A\beta$.

IV. SUFFICIENCY AND COMPLETENESS IN THE LINEAR MODEL

The well-tried principle of sufficiency has features some of which give rise to a similar concept in the linear model when no distributional assumptions are made. Suppose, for instance, s is a sufficient statistic for some parameter θ and t is independent of s. In this case the expected value of any integrable function $h(t)$ can be written as

$$E_\theta\, h(t) = E(h(t)\,|\,s) = \phi(s) \quad \text{a.s.}$$

which is a function independent of θ. Note that $\phi(s)$ must be constant, i.e., t is ancillary, if all underlying distributions share the same null sets (cf. Basu, 1958). It might have been the above equation that led Barnard (1963) to his notion of linear sufficiency. Adjusted to our model $(Y, X\beta, \sigma^2 V)$ it is as follows.

Definition 4.1. A linear statistic LY is called linearly sufficient if for all linear functions c'Y uncorrelated with LY there is a b such that $E_\beta(c'Y) = b'LY$ a.s.

If V is regular this simply means that the expected value of c'Y does not depend on β. Another approach to the idea of linear sufficiency arises from the fact that uniformly minimum variance unbiased estimators are functions of each sufficient statistic. According to this is a definition of Baksalary and Kala (1981), although they originally used a different terminology.

Definition 4.2. A linear statistic LY is called linearly sufficient if for each linear estimable function $p'\beta$ the BLUE is a linear function b'LY of LY.

On the other hand one may consider that the best prediction of Y given any statistic s is the conditional expectation $E_\theta(Y|s)$, which is independent of θ if s is sufficient. Reduced to linear terms this property results in a definition that is due to Drygas (1983).

Definition 4.3. A linear statistic LY is called linearly sufficient if the best linear predictor of Y given LY (written BLP(Y|LY)) is independent of β.

If the distribution of Y has a density p_θ then, under certain regularity conditions, Fisher's information matrix I is well defined. In this case a statistic s is sufficient if and only if its information matrix I_s equals the original I. In the linear model the assumptions above are met when Y is normally distributed and R(X) is contained in R(V). Then the information matrix for the parameter β reads

$$I = \frac{1}{\sigma^2} X'V^-X$$

This may be regarded as an information measure as well without the normal supposition. One can define therefore:

Definition 4.4. If $R(X) \subset R(V)$ a linear statistic LY is called linearly sufficient if $I_L = I$.

Each of these definitions can be transformed into algebraic terms, which all turn out to be equivalent. We present two of them, where $W = (V + XUX')^-$.

Proposition 4.5. LY is linearly sufficient if and only if $R(X) \subset R(XL')$ or $\ker L \cap R(W) \subset V(\ker X')$. (See Baksalary and Kala, 1981, Müller, 1982.)

If Y is normal with known variance and, in addition, R(X) is a subspace of R(V) then it follows immediately from Definition 3.4 that sufficiency and linear sufficiency are equivalent notions. This attractive property can likewise be confirmed without the regularity condition as was shown by Drygas (1983) and Müller (1982).

Proposition 4.6. If Y is normal with known variance then LY is linearly sufficient if and only if it is sufficient.

But the concept of linear sufficiency also makes some sense without the normal supposition as becomes evident from Definition 4.2 and from the following formulation which might be called a linear version of the Rao-Blackwell theorem (see Rao, 1973): Let LY be linearly sufficient and $a'\beta$ be any parametric function estimated by c'Y, say. Then $BLP(c'Y|LY)$ has the same bias as c'Y but smaller mean squared error. That means not only BLUEs but all admissible linear estimators are linear functions of LY. (As for admissibility see Rao, 1976.)

Sufficient statistics are especially useful when they are complete. The linear analogue of this concept arises quite naturally.

Definition 4.7. A linear statistic LY is called linearly complete if each linear function a'LY that has expected value 0 for all $\beta \in \mathbb{R}^m$ vanishes a.s.

Again the definition can easily be translated into an algebraic expression. Combined with this the above conditions for linear sufficiency turn from inclusions into equations. For normal variables Drygas (1983) showed the accordance with ordinary completeness.

Proposition 4.8. LY is linearly complete if and only if $R(LV) \subset R(LX)$.

If Y is normally distributed this is equivalent to completeness. LY is linearly sufficient and linearly complete, i.e. sufficient and complete in the normal case with known variance, if and only if $R(X) = R(WL')$ or $\ker L \cap R(W) = V(\ker X')$.

Generally a linearly sufficient LY does not provide all the information about σ^2 contained in the sample. This deficiency can be compensated when, in addition to LY, one or more quadratic forms are considered. One would like to extend now the idea of linear sufficiency to this situation, but among the four definitions above only Definition 4.2 can serve this purpose satisfactorily.

Definition 4.9. (LY, Y'TY) is called quadratically sufficient if LY is linearly sufficient and the residual sum of squares can be expressed as $Y'L'ALY + \alpha\, Y'TY$ for some symmetric A and real α.

Note that the residual sum of squares is a minimum variance unbiased estimator of $\sigma^2 \times$ (degrees of freedom) if Y is normal. Things become rather more complicated when one allows for more than one quadratic form while V is singular. With the above definition, however, the following can be proved (cf. Seely, 1978, Müller, 1982).

Proposition 4.10. (a) (LY, Y'TY) is quadratically sufficient if and only if for some $\alpha \in \mathbb{R}$, $\ker L \cap R(W) \subset V(\ker X') \cap \ker X'T \cap \ker(I-\alpha VT)$.

(b) If Y is normal a quadratically sufficient (LY,Y'TY) is sufficient. It is complete if LY is complete.

(c) If Y is normal a sufficient (LY,Y'TY) is quadratically sufficient provided one of the following two conditions holds:

(i) $Y'TY \geq 0$ a.s. (i.e., WTW is positive semidefinite).

(ii) Y'TY is invariant a.s. (i.e., WTX = 0) and LY is complete [i.e., $R(LV) \subset R(LX)$].

This work is sponsored by the Air Force Office of Scientific Research under Contract F49629-82-K-001. Reproduction in whole or in part is permitted for any purpose of the United States Government.

REFERENCES

Baksalary, J.K. and Kala, R. (1981). Linear transformations preserving the best linear unbiased estimator in a general Gauss-Markov model. *Ann. Stat. 9,* 913-916.

Barnard, G.A. (1963). The logic of least squares. *J. Roy. Stat. Soc. B 25,* 124-127.

Basu, D. (1958). On statistics independent of a complete sufficient statistic. *Sankhyā 15,* 377-380 and *Sankhyā 20,* 223-226.

Chikuse, Y. (1981). Representations of the covariance matrix for robustness in the Gauss-Markov model. *Comm. Statist. A10,* No. 19, 1997-2004.

Drygas, H. (1983). Sufficiency and completeness in the general Gauss-Markov model. Submitted to *Sankhyā*.

Ghosh, M. and Sinha, Bimal Kumar (1980). On the robustness of least squares procedures in regression models. *J. Mult. Analysis 10,* 332-342.

Goldman, A.J. and Zelen, M. (1964). Weak generalized inverse and minimum variance unbiased estimation. *J. Research Nat. Bureau of Standards 68 B,* 151-172.

Harville, D.A. (1981). Unbiased and minimum variance unbiased estimation of estimable functions for fixed linear models with arbitrary covariance structure. *Ann. Statist. 9,* 633-637.

Kariya, T. (1981). A robustness property of Hotelling's T^2-Test. *Ann. Statist. 9,* 211-214.

Kempthorne, O. (1976). Best linear unbiased estimation with arbitrary covariance matrix. Chapter 14 (pp. 203-225) in *Essays in Probability and Statistics*. Tokyo: Sinko Tsusho.

Khatri, C.G. (1980). Study of F-tests under dependent model. *Sankhyā Series A 43,* 107-110.

Lehmann, E.L. (1959). *Testing Statistical Hypotheses*. New York: Wiley.

Mathew, T. and Bhimasankaram, P. (1982). On the robustness of the LRT with respect to specification errors in a linear model. Indian Statistical Institute Tech. Report. To appear in *Sankhyā*.

Mitra, S.K. and Rao, C.R. (1968). Some results in estimation and tests of linear hypotheses under the Gauss-Markoff model. *Sankhyā A 30,* 313-322.

Müller, J. (1982). Sufficiency and completeness in the linear model. Tech. Rept. #82-32, Center for Multivariate Analysis, University of Pittsburgh.

Rao, C.R. (1967). Least squares theory using an estimated dispersion matrix and its application to measurement of signals. *Proc. Fifth Berkeley Symp. on Math. Stat. and Prob. 1,* 355-372.

Rao, C.R. (1971). Unified theory of linear estimation. *Sankhyā A33,* 370-396 and *Sankhyā A36,* 447.

Rao, C.R. (1973). *Linear Statistical Inference and its Applications*. New York: Wiley.

Rao, C.R. (1973). Unified theory of least squares. *Communications in Statistics 1,* 1-8.

Rao, C.R. (1974). Projectors, generalized inverses and the BLUE's. *J. Roy. Statist. Soc. 36 B,* 442-448.

Rao, C.R. (1976). Estimation of parameters in a linear model. *Ann. Statist. 4,* 1023-1037.

Rao, C.R. (1978). Least squares theory for possibly singular models. *Canad. J. Statist. 3,* 105-110.

Rao, C.R. (1979). Estimation of parameters in the singular Gauss-Markoff model. *Comm. Statist. A8,* 1353-1358.

Rao, C.R. and Mitra, S.K. (1971). *Generalized Inverse of Matrices and its Applications*. New York: Wiley.

Rao, C.R. and Yanai, H. (1979). General definition and decomposition of projectors and some applications to statistical problems. *J. Stat. Planning and Inference 3*, 1-17.

Seely, J. (1978). A complete sufficient statistic for the linear model under normality and a singular covariance matrix. *Comm. Stat. A7*, 1465-1473.

Sinha, Bikas K. and Mukhopadhyay, B.B. (1980). A note on a result of Ghosh and Sinha. *Calcutta Statist. Assoc. Bull. 29*, 169-171.

Sinha, Bimal K. and Drygas, H. (1982). Robustness properties of the F-test and best linear unbiased estimators in linear models. Tech. Rept. 82-28, Center for Multivariate Analysis, University of Pittsburgh.

Wijsman, R.A. (1967). Cross section of orbits and their application to densities of maximal invariants. *Fifth Berkeley Symp. Math. Statist. Prob. 1*, 389-400.

Zyskind, G. (1967). On canonical forms, nonnegative covariance matrices and best and simple least squares linear estimators in linear models. *Ann. Math. Statist. 38*, 1092-1109.

Chapter 17

CONSTRAINED NONLINEAR LEAST SQUARES

George E.P. Box
University of Wisconsin
Madison, Wisconsin

Hiromitsu Kanemasu
World Bank
Washington, D.C.

I. NONLINEAR LEAST SQUARES

Suppose an observation y_u is described in the model

$$y_u = f(\underset{\sim}{\xi}_u, \underset{\sim}{\theta}) + \varepsilon_u, \quad u = 1,2,\ldots,n \tag{1.1}$$

where $\underset{\sim}{\xi}_u = (\xi_{1u}, \xi_{2u}, \ldots, \xi_{ku})'$ are the levels of k independent variables, $\underset{\sim}{\theta} = (\theta_1, \theta_2, \ldots, \theta_p)'$ are p unknown parameters, $f(\underset{\sim}{\xi}_u, \underset{\sim}{\theta})$ is a known function of $\underset{\sim}{\xi}_u$ and $\underset{\sim}{\theta}$, and ε_u is a random error. The method of least squares obtains an estimate $\hat{\underset{\sim}{\theta}}$ of the parameters which minimizes the sum of squares

$$S(\underset{\sim}{\theta}) = \sum_{u=1}^{n} (y_u - f(\underset{\sim}{\xi}_u, \underset{\sim}{\theta}))^2 \tag{1.2}$$

or in vector notation

$$S(\underset{\sim}{\theta}) = (\underset{\sim}{y} - \underset{\sim}{f}_\theta)'(\underset{\sim}{y} - \underset{\sim}{f}_\theta) \tag{1.3}$$

where $\tilde{y}$ is the n×1 vector of y_u, u = 1,2,...,n and $\tilde{f}_\theta$ is the n×1 vector whose u^{th} element is $f(\tilde{\xi}_u,\tilde{\theta})$. Under the assumption that the errors are independently and normally distributed with equal variance, $\hat{\tilde{\theta}}$ is the maximum likelihood estimator of $\tilde{\theta}$.

The function $f(\tilde{\xi}_u,\tilde{\theta})$ is said to be linear in the parameters if the derivatives $\partial f(\tilde{\xi}_u,\tilde{\theta})/\partial\theta_j$ are independent of $\tilde{\theta}$ for all j. Otherwise, it is called nonlinear. When the model is linear, the least squares estimates $\hat{\tilde{\theta}}$ are simply the solution of the normal equation $X'X\tilde{\theta} = X'\tilde{y}$ where X is the n×p matrix of derivatives $\partial f(\tilde{\xi}_u,\tilde{\theta})/\partial\theta_j$. If the model is nonlinear Gauss (1821) suggested that $\hat{\tilde{\theta}}$ could be obtained by linearizing $f(\tilde{\xi}_u,\tilde{\theta})$ around the current estimates $\tilde{\theta}^{(0)}$ in successive iterations.

II. GAUSS METHOD AND OVERSHOOTING

A local approximation to the nonlinear function $f(\tilde{\xi}_u,\tilde{\theta})$ may be obtained by performing a Taylor series expansion at some guessed values $\tilde{\theta} = \tilde{\theta}^{(0)}$, and truncating after first order terms to obtain

$$f(\tilde{\xi}_u,\tilde{\theta}) \cong f(\tilde{\xi}_u,\tilde{\theta}^{(0)}) + \sum_{i=1}^{p}\left[\frac{\partial f(\tilde{\xi}_u,\tilde{\theta})}{\partial\theta_i}\right]_{\tilde{\theta}=\tilde{\theta}^{(0)}}(\theta_i - \theta_i^{(0)}) \quad (2.1)$$

$$u = 1,2,\ldots,n$$

In vector notation

$$\tilde{f}_\theta \cong \tilde{f}_0 + X_0(\tilde{\theta} - \tilde{\theta}^{(0)}) \quad (2.2)$$

where $\tilde{f}_0$ is the n×1 vector of $f(\tilde{\xi}_u,\tilde{\theta}^{(0)})$; u = 1,2,...,n and X_0 is the n×p matrix whose (u,j) element is $[\partial f(\tilde{\xi}_u,\tilde{\theta})/\partial\theta_j]_{\tilde{\theta}=\tilde{\theta}^{(0)}}$. This local linearization gives the approximate sum of squares

$$\begin{aligned}\bar{S}(\tilde{\theta}) &= (\tilde{y} - \tilde{f}_0 - X_0(\tilde{\theta} - \tilde{\theta}^{(0)}))'(\tilde{y} - \tilde{f}_0 - X_0(\tilde{\theta} - \tilde{\theta}^{(0)})) \quad (2.3)\\ &= \tilde{\varepsilon}_0'\tilde{\varepsilon}_0 - 2\tilde{\varepsilon}_0'X_0(\tilde{\theta} - \tilde{\theta}^{(0)}) + (\tilde{\theta} - \tilde{\theta}^{(0)})'X_0'X_0(\tilde{\theta} - \tilde{\theta}^{(0)})\end{aligned}$$

where $\tilde{\varepsilon}_0 = \tilde{y} - \tilde{f}_0$. Setting the derivatives $\partial\bar{S}(\tilde{\theta})/\partial\theta_j$ to zero gives the normal equations

$$X_0'X_0(\underset{\sim}{\theta} - \underset{\sim}{\theta}^{(0)}) = X_0'\underset{\sim}{\varepsilon}_0 \tag{2.4}$$

Thus provided that $X_0'X_0$ is nonsingular, new estimates $\underset{\sim}{\theta}^{(1)}$ of the parameters are given by

$$\underset{\sim}{\theta}^{(1)} - \underset{\sim}{\theta}^{(0)} = (X_0'X_0)^{-1}X_0'\underset{\sim}{\varepsilon}_0 \tag{2.5}$$

By linearizing now about $\underset{\sim}{\theta}^{(1)}$, a second approximation $\underset{\sim}{\theta}^{(2)}$ may now be obtained, and so on. Figure 17.1 shows the parameter space repre-

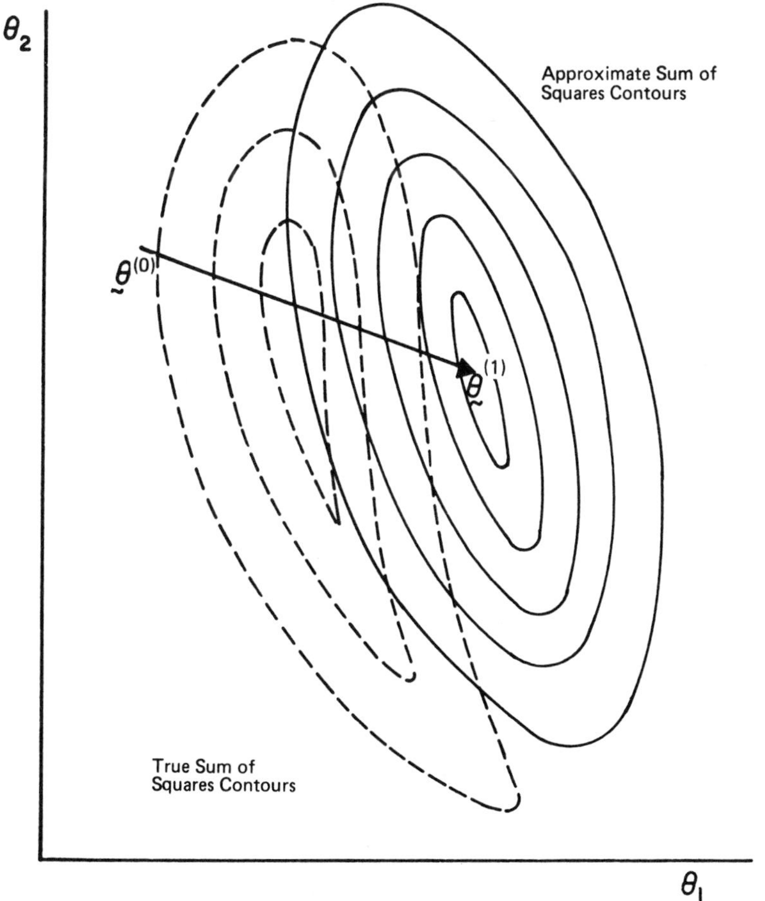

Figure 17.1 Parameter Space Representation of the Gauss Iteration

sentation of the first stage of this procedure for a two parameter model. Contours for true and approximate sums of squares $S(\underset{\sim}{\theta})$ and $\overline{S}(\underset{\sim}{\theta})$ are shown. Contours of $\overline{S}(\underset{\sim}{\theta})$ are necessarily elliptical while those for $S(\underset{\sim}{\theta})$ have the appearance of distorted ellipses.

In favorable circumstances the iteration will converge to the least squares estimates but in some examples wild oscillations can occur from one iteration to another and the process may not converge. Figure 17.1 shows a typical case of initial divergence, where the Gauss solution vector (2.5) "overshoots" and the true sum of squares given by the new estimates $\theta^{(1)}$ is larger than that given by the initial estimates $\theta^{(0)}$. A frequent cause of divergence is that the adjustment $\underset{\sim}{\theta}^{(1)} - \underset{\sim}{\theta}^{(0)}$ is too large and so invalidates the linear approximation (2.1).

III. STEEPEST DESCENT APPROACH AND APPLICATION OF RESPONSE SURFACE METHOD

Another iterative approach to nonlinear least squares uses steepest descent where initially the iteration moves from the current point $\underset{\sim}{\theta}^{(0)}$ in the direction of steepest descent vector, that is

$$-\left(\frac{\partial S(\underset{\sim}{\theta})}{\partial \theta_1}, \ldots, \frac{\partial S(\underset{\sim}{\theta})}{\partial \theta_p}\right)'$$

evaluated at $\underset{\sim}{\theta} = \underset{\sim}{\theta}^{(0)}$. This procedure is often effective in getting near the region of a minimum quickly. However, it may have an extremely slow rate of convergence after that, particularly in the common case of a ridgy minimum.

Box and Coutie (1956) proposed a method based on response surface methodology which employed steepest descent at the early stages (Box, 1954; Box and Wilson, 1951). If the initial point is remote from the minimum, the sum of squares surface $S(\underset{\sim}{\theta})$ might be capable of local approximation by a polynomial in $\underset{\sim}{\theta}$ of the first degree. The sum of squares is therefore determined at a series of points in the parameter space from which a planar approximation may be fitted and the direction of steepest descent calculated. This direction is followed until an increase in the sum of squares is encountered. The whole process is repeated until the need for a second degree approximation becomes manifest.

Sums of squares are then computed at a set of points arranged to allow a second degree polynomial to be locally fitted and an approximate minimum determined. This method is (see also Koshal, 1933) inefficient because it makes use of only the information contained in the sum of squares of residuals but not of that in the individual residuals themselves. It was shown by Box (1956) that when this missing information is included we are brought back to the Gauss method.

IV. MODIFIED GAUSS ITERATION

One way to overcome the difficulty of overshooting in the Gauss iteration is to determine direction only (but not distance) by the Gauss solution vector $\underset{\sim}{\theta}^{(g)} - \underset{\sim}{\theta}^{(0)} = (X_0'X_0)^{-1}X_0'\underset{\sim}{\varepsilon}_0$. Thus, the adjustment vector $\underset{\sim}{\theta} - \underset{\sim}{\theta}^{(0)}$ is given by

$$\underset{\sim}{\theta} - \underset{\sim}{\theta}^{(0)} = v(X_0'X_0)^{-1}X_0'\underset{\sim}{\varepsilon}_0 \tag{4.1}$$

where v is a positive quantity usually less than unity. This modified Gauss iteration was suggested by Box (1958) and incorporated into a computer program described by Booth, Box, Muller and Peterson (1959). In order to determine the value of v that approximately minimizes the sum of squares along the Gauss solution vector, they used the "halving and doubling" method in which, starting from v=1, the value of v is successively halved (or doubled) until the sum of squares finally starts to increase and then a quadratic curve is fitted to the last three points to locate an approximate local minimum. Hartley (1961) proved that, under a set of mild regularity conditions, the modified Gauss iteration as described above converges to the solution of $\partial S(\underset{\sim}{\theta})/\partial\theta_j = 0$; $j = 1,2,\ldots,p$ and also proposed a similar method to determine the value of v.

V. LEVENBERG-MARQUARDT'S CONSTRAINED ITERATION

5.1 Levenberg's Damped Least Squares

Levenberg (1944) tried to overcome the difficulty of "overshooting" in the Gauss iteration by introducing a constraint into the minimization of the sum of squares. Instead of minimizing the approximate sum of squares $\bar{S}(\underset{\sim}{\theta})$ itself he proposed to minimize

$$F(\underset{\sim}{\theta}) = \bar{S}(\underset{\sim}{\theta}) + \lambda \sum_{i=1}^{p} \omega_i(\theta_i - \theta_i^{(0)})^2 \tag{5.1}$$

where $\lambda\omega_1, \lambda\omega_2, \ldots, \lambda\omega_p$ are positive weighting factors expressing the relative importance of damping the different increments. Substituting (2.3) into (5.1),

$$F(\underset{\sim}{\theta}) = \underset{\sim}{\varepsilon}_0'\underset{\sim}{\varepsilon}_0 - 2\underset{\sim}{\varepsilon}_0'X_0(\underset{\sim}{\theta} - \underset{\sim}{\theta}^{(0)}) + (\underset{\sim}{\theta} - \underset{\sim}{\theta}^{(0)})'X_0'X_0(\underset{\sim}{\theta} - \underset{\sim}{\theta}^{(0)})$$
$$+ \lambda(\underset{\sim}{\theta} - \underset{\sim}{\theta}^{(0)})'\Omega(\underset{\sim}{\theta} - \underset{\sim}{\theta}^{(0)}) \tag{5.2}$$

where Ω is a p×p diagonal matrix whose i^{th} diagonal element is ω_i. Setting the derivatives $\partial F(\underset{\sim}{\theta})/\partial\theta_i$ to zero

$$-X_0'\underset{\sim}{\varepsilon}_0 + X_0'X_0(\underset{\sim}{\theta} - \underset{\sim}{\theta}^{(0)}) + \lambda\Omega(\underset{\sim}{\theta} - \underset{\sim}{\theta}^{(0)}) = \underset{\sim}{0} \tag{5.3}$$

Solving for $\underset{\sim}{\theta}$, we have

$$\underset{\sim}{\theta} - \underset{\sim}{\theta}^{(0)} = (X_0'X_0 + \lambda\Omega)^{-1}X_0'\underset{\sim}{\varepsilon}_0 \tag{5.4}$$

Geometrically, in the parameter space representation, this amounts to minimizing the approximate sum of squares $\bar{S}(\underset{\sim}{\theta})$ on the elliptical constraint whose principal axes are parallel to the axes of $\theta_1, \theta_2, \ldots, \theta_p$. This is illustrated in Figure 17.2. The dotted curve represents the locus of the solution of (5.4) for various values of λ. Levenberg proved that, provided the true sum of squares $S(\underset{\sim}{\theta})$

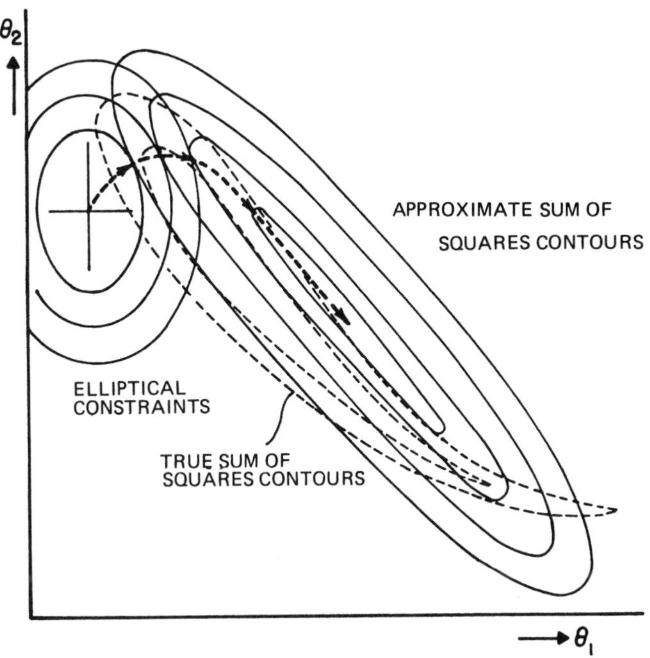

Figure 17.2 Parameter Space Representation of Levenberg's Constrained Minimization

does not already have a stationary value at the current estimate $\tilde{\theta}^{(0)}$, the sum of squares initially decreases as we move off the point $\tilde{\theta}^{(0)}$. He recommended the spherical constraint $\Omega = I_p$, where I_p is the p×p identity matrix.

An alternative choice for Ω was diagonal matrix D whose j^{th} diagonal element was the jth diagonal element [jj] of the matrix $X_0'X_0$.

5.2 Marquardt's Algorithm

Marquardt (1963) also considered the constrained minimization. He examined geometrical aspects of the case $\Omega = I_p$ in Levenberg's formulation. He noted a result of Morrison (1960) that the radius of the constraining sphere is a monotone decreasing function of λ that tends to zero as $\lambda \to \infty$, and then proved that as $\lambda \to \infty$ the solution vector of the constrained minimization rotates toward the steepest descent vector that remains fixed. Since the Gauss solution is of course a special case of the constrained minimization when $\lambda = 0$, a choice of some intermediate value for λ would provide a compromise between the two classical methods. The algorithm proposed by Marquardt, which may be termed (λ,ν) algorithm, reduces the level of λ by a factor ν from a relatively large initial value λ_0 as the iteration proceeds. This amounts to making use of the steepest descent-like procedure early in the iterative process and then later shifting to a solution that approximates the Gauss solution. Marquardt's method thus appears to possess the virtue of using each classical procedure in the circumstances where it is most effective.

Noting that the steepest descent procedure is dependent on the choice of parameter scales, Marquardt proposed the constrained minimization in the scale invariant metric, i.e., the minimization with a spherical constraint in $\theta_j^* = [jj]^{\frac{1}{2}}\theta_j (j = 1,2,\ldots,p)$. This is equivalent to using the constraint $\Omega = D$ of Levenberg. Marquardt's recommendations were followed by Meeter (1966) in the program GAUSHAUS at the University of Wisconsin and similar programs have been used elsewhere.

5.3 Constrained Minimization in the Transformed Space

The Levenberg-Marquardt procedure may be criticized as follows. The use of constant spherical or elliptical constraints implies that it is logical to solve the problem in some parameterization decided in advance by the investigator. However, usually there are many ways in which a problem might be parameterized. Instead of considering $\theta_1,\theta_2,\ldots,\theta_p$, we might with equal reason consider $\psi_1,\psi_2,\ldots,\psi_p$ where $\psi_j = \psi_j(\tilde{\theta})$ is some 1:1 transformation of $\tilde{\theta}$. Clearly, the nature of the constraints applied would differ depending on which parameterization was adopted.

Let us then consider the problem in a parameter metric that is not only scale invariant but also invariant under <u>linear transfor-</u>

mation. Such parameter metric $\underset{\sim}{\psi}$ is provided by $\underset{\sim}{\psi} = H\underset{\sim}{\theta}$ where H is any p×p nonsingular matrix that satisfies

$$H'H = X_0'X_0 \tag{5.5}$$

To see this, make an arbitrary linear transformation $\overline{\underset{\sim}{\theta}} = L\underset{\sim}{\theta}$. Correspondingly the derivative matrix X_0 will be transformed to $\overline{X}_0 = X_0L^{-1}$. The transformation of $\overline{\underset{\sim}{\theta}}$ corresponding to $\underset{\sim}{\theta} \to \underset{\sim}{\psi}$ will be $\overline{\underset{\sim}{\psi}} = \overline{H}\,\overline{\underset{\sim}{\theta}}$ with $\overline{H}'\overline{H} = \overline{X}_0'\overline{X}_0$. However, the requirement on $\overline{H}$ will be satisfied by $\overline{H} = HL^{-1}$ since

$$\overline{H}'\overline{H} = (HL^{-1})'(HL^{-1}) = L^{-1'}X_0'X_0L^{-1} = \overline{X}_0'\overline{X}_0 \tag{5.6}$$

Therefore

$$\overline{\underset{\sim}{\psi}} - \overline{H}\,\overline{\underset{\sim}{\theta}} = (HL^{-1})(L\underset{\sim}{\theta}) = H\underset{\sim}{\theta} = \underset{\sim}{\psi} \tag{5.7}$$

establishing that $\underset{\sim}{\psi}$ is invariant under linear transformation.

In the new space $\underset{\sim}{\psi}$ the linearized model will be

$$\underset{\sim}{f}_{\psi} = \underset{\sim}{f}_0 + Z_0(\underset{\sim}{\psi} - \underset{\sim}{\psi}^{(0)}) \tag{5.8}$$

where $\underset{\sim}{f}_{\psi}$ is the n×1 vector of $f(\underset{\sim}{\xi}_u, \underset{\sim}{\theta}(\underset{\sim}{\psi}))$; u = 1,2,...,n, $\underset{\sim}{\psi}^{(0)} = H\underset{\sim}{\theta}^{(0)}$, and Z_0 is the n×p matrix whose (u,j) element is

$$[\partial f(\underset{\sim}{\xi}_u, \underset{\sim}{\theta}(\underset{\sim}{\psi}))/\partial\psi_j]_{\underset{\sim}{\psi}=\underset{\sim}{\psi}^{(0)}}$$

Notice that Z_0 is related to X_0 by $Z_0 = X_0H^{-1}$. Therefore, the sum of squares contours for the linearized model given by

$$(\underset{\sim}{y} - \underset{\sim}{f}_0 - Z_0(\underset{\sim}{\psi} - \underset{\sim}{\psi}^{(0)}))'(\underset{\sim}{y} - \underset{\sim}{f}_0 - Z_0(\underset{\sim}{\psi} - \underset{\sim}{\psi}^{(0)})) = \text{constant} \tag{5.9}$$

will be spherical because

$$Z_0'Z_0 = (X_0H^{-1})'(X_0H^{-1}) = H^{-1'}X_0'X_0H^{-1} = I_p \tag{5.10}$$

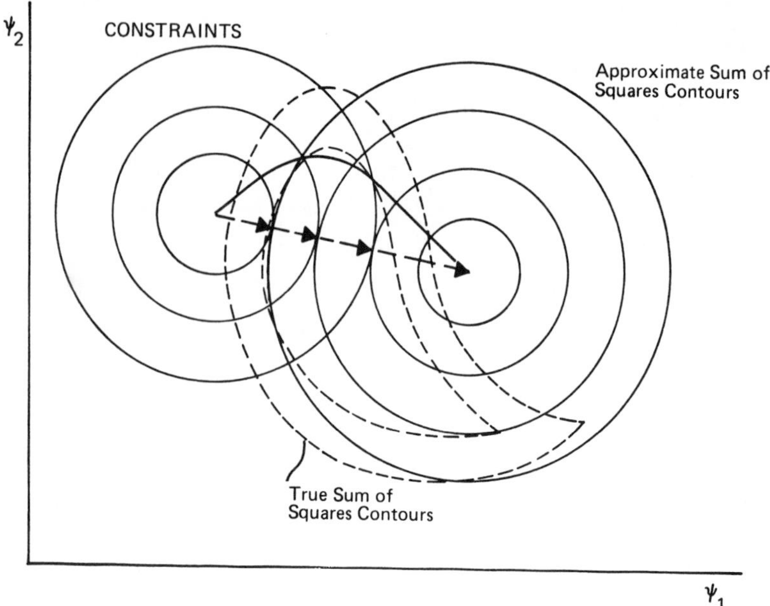

Figure 17.3 Constrained Minimization of the Sum of Squares in the Linearly Invariant Metric

The representation of the problem would now be that illustrated for p=2 in Figure 17.3 in which the contours for the linearized model are spheres. To the extent that the contours for the linearized model were like those for the true model, this would ensure that the contours of the true model were roughly spherical, enabling us to circumvent the difficulties arising in situations of a ridgy minimum.

If, in the linearly invariant metric, we apply appropriate spherical constraints we must minimize

$$F(\underset{\sim}{\psi}) = [\underset{\sim}{\varepsilon}_0 - Z_0(\underset{\sim}{\psi} - \underset{\sim}{\psi}^{(0)})]'[\underset{\sim}{\varepsilon}_0 - Z_0(\underset{\sim}{\psi} - \underset{\sim}{\psi}^{(0)})] \tag{5.11}$$

$$+ \lambda[(\underset{\sim}{\psi} - \underset{\sim}{\psi}^{(0)})'(\underset{\sim}{\psi} - \underset{\sim}{\psi}^{(0)}) - \gamma^2]$$

where $\underset{\sim}{\varepsilon}_0 = \underset{\sim}{y} - \underset{\sim}{f}_0$ and γ is the radius of the sphere. Setting the derivatives $\partial F(\underset{\sim}{\psi})/\partial\psi_i$ $(i = 1,\ldots,p)$ to zero,

$$-2Z_0'\underset{\sim}{\varepsilon}_0 + 2Z_0'Z_0(\underset{\sim}{\psi} - \underset{\sim}{\psi}^{(0)}) + 2\lambda(\underset{\sim}{\psi} - \underset{\sim}{\psi}^{(0)}) = 0 \tag{5.12}$$

Solving for $\underset{\sim}{\psi}$ we obtain

$$\underset{\sim}{\psi} - \underset{\sim}{\psi}^{(0)} = (Z_0'Z_0 + \lambda I_p)^{-1}Z_0'\underset{\sim}{\varepsilon}_0 \tag{5.13}$$

Because of (5.10), this is

$$\underset{\sim}{\psi} - \underset{\sim}{\psi}^{(0)} = \frac{1}{1+\lambda} Z_0'\underset{\sim}{\varepsilon}_0 \tag{5.14}$$

It can be shown that, as long as $(-\lambda)$ is chosen to be less than the minimum eigenvalue of $Z_0'Z_0$ (that is, less than unity), the solution (5.13) provides the point that minimizes in the space ψ the linearized model sum of squares on the sphere of radius $(1/1 + \lambda)^2\underset{\sim}{\varepsilon}_0'Z_0Z_0'\underset{\sim}{\varepsilon}_0$. (See Draper (1963) for the analysis of stationary points of the general quadratic response surface on spherical constraints.) We also note that this solution is the vector of steepest descent (except for the irrelevant scalar) since by differentiating the negative of the true sum of squares $S(\underset{\sim}{\psi}) = (\underset{\sim}{y} - \underset{\sim}{f}_\psi)'(\underset{\sim}{y} - \underset{\sim}{f}_\psi)$ with respect to ψ and evaluating the derivative at $\underset{\sim}{\psi} = \underset{\sim}{\psi}^{(0)}$ we obtain $2Z_0'\underset{\sim}{\varepsilon}_0$. Now transforming (5.14) back to the original metric $\underset{\sim}{\theta}$,

$$H(\underset{\sim}{\theta} - \underset{\sim}{\theta}^{(0)}) = \frac{1}{1+\lambda}(X_0H^{-1})'\underset{\sim}{\varepsilon}_0 \tag{5.15}$$

which gives

$$\underset{\sim}{\theta} - \underset{\sim}{\theta}^{(0)} = \frac{1}{1+\lambda} H^{-1}(X_0H^{-1})'\underset{\sim}{\varepsilon}_0 = \frac{1}{1+\lambda} (H'H)^{-1}X_0'\underset{\sim}{\varepsilon}_0 \tag{5.16}$$

Because of (5.5), we finally obtain

$$\underset{\sim}{\theta} - \underset{\sim}{\theta}^{(0)} = \frac{1}{1+\lambda} (X_0'X_0)^{-1}X_0'\underset{\sim}{\varepsilon}_0 \tag{5.17}$$

This is the same as (4.1), where the requirement $(-\lambda) < 1$ implies that $v = 1/(1 + \lambda)$ can range between 0 and ∞. Somewhat surprisingly then the Levenberg-Marquardt constrained minimization performed in the linearly invariant metric is exactly the modified Gauss method discussed earlier. It should be noted also that in the linearly invariant metric there is no question of a compromise between the two classical methods, because, using (5.10), the Gauss vector $(Z_0'Z_0)^{-1}Z_0'\underset{\sim}{\varepsilon}_0$ is equal to $Z_0'\underset{\sim}{\varepsilon}_0$ which is the steepest descent vector. (In Figure 17.3, the path that is the Gauss vector as well as the steepest descent vector is shown by the dotted straight line. The path which the Levenberg-Marquardt procedure would take is shown by the bold connected curve.)

So far as the problem of speedy convergence is concerned, there does not therefore seem to exist much concrete theoretical basis for interpolation between the two classical methods. One incidental advantage of the Levenberg-Marquardt procedure may be that the matrix $(X_0'X_0 + \lambda\Omega)$ can be inverted even when the matrix $X_0'X_0$ is singular or nearly singular, thus always giving a "solution". Practical experience, however, leads us to believe that the possibility of not having a singularity or near-singularity brought to one's attention is a disadvantage rather than an advantage. It has often been pointed out (for example, Box, 1954) that a minimum is often better approximated by a line, plane, or hyperplane than by a point. When this happens, it is important that it be brought to the investigator's notice. One method for ensuring this is by means of canonical analysis suggested, for example, by Box (1960).

VI. METHODS TO DETERMINE HOW FAR ONE SHOULD GO ALONG THE GAUSS SOLUTION VECTOR

In the preceding section, we have given theoretical support to the idea that we should explore the Gauss vector itself rather than the path followed by the Levenberg-Marquardt procedure. We have done this by demonstrating that the Gauss vector may be arrived at by applying Levenberg-Marquardt constrained minimization in the linearly invariant parameter metric.

We recall that the direction of the Gauss vector in the original parameter metric corresponds to that of the steepest descent considered in the linearly invariant metric. Also applying the Levenberg's result mentioned in Section 5.1 to the constrained minimization in the linearly invariant metric, we can be assured that the true sum of squares initially decreases as we start off from the current estimate $\underset{\sim}{\theta}^{(0)}$ on the Gauss vector provided $\underset{\sim}{\theta}^{(0)}$ is not already the stationary point.

The question of how far one should go along the Gauss vector still remains. The "halving and doubling" method already mentioned could be used. But its disadvantage is that, after the Gauss vector is determined for $\underset{\sim}{\theta}^{(0)}$, the function $f(\underset{\sim}{\xi}_u, \underset{\sim}{\theta})$ must be additionally evaluated for $u = 1,2,\ldots,n$ to calculate $S(\underset{\sim}{\theta})$ at each new "test point" in the parameter space. We also could apply the (λ,ν) algorithm of Marquardt to (5.17). Although no compromise between the original Gauss method and the steepest descent method is here involved, it may still make sense to gradually decrease λ so as to constrain the iteration less and less as the minimum is approached in successive iterations. Again, however, there is no way to know in advance how best to choose λ_0 and ν. Below we present two methods to determine the value of v in (4.1) at each iteration in such a way as to make use of only the information already available from the preceding computation.

Modification A

To obtain the Gauss solution vector it is necessary to compute the matrix X_0 of the partial derivatives $[\partial f(\underset{\sim}{\xi}_u, \underset{\sim}{\theta})/\partial\theta_j]_{\underset{\sim}{\theta}=\underset{\sim}{\theta}^{(0)}}$. Using this matrix, it is clearly possible to obtain the initial rate of change of the true sum of squares along the Gauss solution vector. In fact,

$$\left[\frac{dS}{dv}\right]_{v=0} = \sum_{i=1}^{p} \left[\frac{\partial S}{\partial \theta_i}\right]_{v=0} \left[\frac{d\theta_i}{dv}\right]_{v=0} = \left[\frac{\partial S}{\partial \underset{\sim}{\theta}}\right]'_{v=0} \left[\frac{d\underset{\sim}{\theta}}{dv}\right]_{v=0} \qquad (6.1)$$

where $\left[\frac{\partial S}{\partial \underset{\sim}{\theta}}\right]_{v=0}$ is the p×1 vector of $\left[\frac{\partial S}{\partial \theta_i}\right]_{v=0}$; i = 1,2,...,p, and $\left[\frac{d\underset{\sim}{\theta}}{dv}\right]_{v=0}$ is the p×1 vector of $\left[\frac{d\theta_i}{dv}\right]_{v=0}$; i = 1,2,...,p. However, since $S = (\underset{\sim}{y} - \underset{\sim}{f}_\theta)'(\underset{\sim}{y} - \underset{\sim}{f}_\theta)$ and $\underset{\sim}{\theta} - \underset{\sim}{\theta}^{(0)} = v(X_0'X_0)^{-1}X_0'\underset{\sim}{\varepsilon}_0$, we obtain

$$\left[\frac{dS}{dv}\right]_{v=0} = -2\underset{\sim}{\varepsilon}_0'X_0(X_0'X_0)^{-1}X_0'\underset{\sim}{\varepsilon}_0 = -2(\underset{\sim}{\theta}^{(g)} - \underset{\sim}{\theta}^{(0)})'X_0'\underset{\sim}{\varepsilon}_0 \tag{6.2}$$

Incidentally, equation (6.2) gives a direct proof of a corollary of Levenberg's result mentioned above that the true sum of squares initially decreases as we start moving from $\underset{\sim}{\theta}^{(0)}$ to $\underset{\sim}{\theta}^{(g)}$, because provided that $X_0'X_0$ is positive definite $\left[\frac{dS}{dv}\right]_{v=0}$ will be negative except when $X_0'\underset{\sim}{\varepsilon}_0 = -\frac{1}{2}\left[\frac{\partial S}{\partial \underset{\sim}{\theta}}\right]_{v=0}$ is $\underset{\sim}{0}$.

To locate a point along the Gauss solution vector at which the true sum of squares is approximately minimized, we first suppose that the true sum of squares follows, along this vector, a quadratic function $S = a + bv + cv^2$. We can determine the constants, a, b and c by setting $S = S_0$ for $v = 0$, $S = S_g$ for $v = 1$ and setting $\left[\frac{dS}{dv}\right]_{v=0}$ to that given by (6.2). Consequently, provided $c = S_g - S_0 + 2(\underset{\sim}{\theta}^{(g)} - \underset{\sim}{\theta}^{(0)})'X_0'\underset{\sim}{\varepsilon}_0 > 0$, the value of v for which the true sum of squares is approximately minimized is given by

$$v_{min} = \frac{(\underset{\sim}{\theta}^{(g)} - \underset{\sim}{\theta}^{(0)})'X_0'\underset{\sim}{\varepsilon}_0}{S_g - S_0 + 2(\underset{\sim}{\theta}^{(g)} - \underset{\sim}{\theta}^{(0)})'X_0'\underset{\sim}{\varepsilon}_0} \tag{6.3}$$

Thus, we may take our next estimate as $\underset{\sim}{\theta}^{(1)} = \underset{\sim}{\theta}^{(0)} + v_{min}(\underset{\sim}{\theta}^{(g)} - \underset{\sim}{\theta}^{(0)})$. If $c < 0$, we have that the actual sum of squares S_g at distance $v = 1$ is already smaller than that predicted by the initial slope since in this case $S_g < S_0 + [dS/dv]_{v=0}$. Therefore we may settle at $v = 1$, or double, or redouble the distance, checking at each point to see if the decreasing trend is continuing. Figure 17.4 illustrates various situations that could occur.

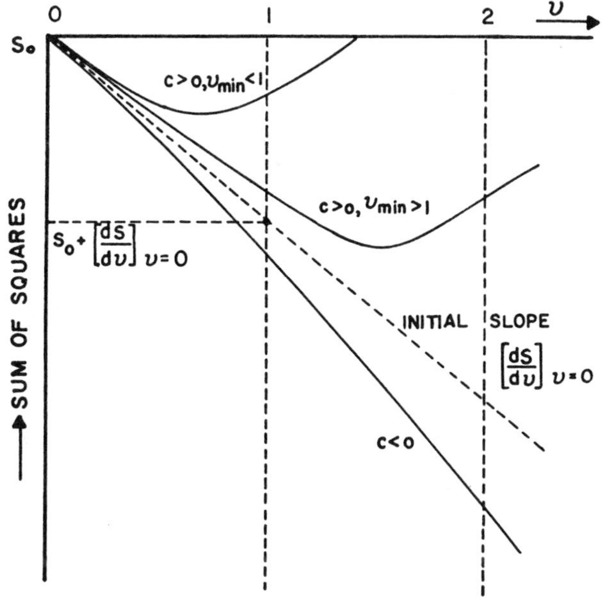

Figure 17.4 Various Quadratic Curves Approximating the Sum of Squares along Gauss Vector

Modification B

A procedure developed following a suggestion by Jack Draffen [personal communication (1972)] is another method that makes use of the existing information in an interesting manner. The quantities S_0 and S_g are obtained by computing, squaring, and summing the elements of the two residual vectors

$$\underset{\sim}{\varepsilon}'_0 = (y_1 - f(\underset{\sim}{\xi}_1,\underset{\sim}{\theta}^{(0)}),y_2 - f(\underset{\sim}{\xi}_2,\underset{\sim}{\theta}^{(0)}),\ldots,y_n - f(\underset{\sim}{\xi}_n,\underset{\sim}{\theta}^{(0)}))$$

and

$$\underset{\sim}{\varepsilon}'_g = (y_1 - f(\underset{\sim}{\xi}_1,\underset{\sim}{\theta}^{(g)}),y_2 - f(\underset{\sim}{\xi}_2,\underset{\sim}{\theta}^{(g)}),\ldots,y_n - f(\underset{\sim}{\xi}_n,\underset{\sim}{\theta}^{(g)})) \tag{6.4}$$

Consider a point along the Gauss solution vector with the distance $v\|\theta^{(g)} - \theta^{(0)}\|$ away from the origin $\theta^{(0)}$. By linear interpolation we can estimate the residuals at this point by

$$\varepsilon = (1 - v)\varepsilon_0 + v\varepsilon_g \tag{6.5}$$

which may be written

$$\varepsilon_0 = v(\varepsilon_0 - \varepsilon_g) + \varepsilon \tag{6.6}$$

Thus the value $\hat{v}$ of v for which the sum of squares of the estimated residuals will be as small as possible can be obtained by regressing ε_0 on $\varepsilon_0 - \varepsilon_g$ so that

$$\hat{v} = \frac{(\varepsilon_0 - \varepsilon_g)'\varepsilon_0}{(\varepsilon_0 - \varepsilon_g)'(\varepsilon_0 - \varepsilon_g)} \tag{6.7}$$

whence our estimate of parameters is obtained by $\theta^{(1)} = \theta^{(0)} + \hat{v}(\theta^{(g)} - \theta^{(0)})$. Again, computing $\hat{v}$ is very simple and makes use of information already available. [When the criterion to stop the iterative process is made too stringent, in the extreme neighborhood of the minimum both the numerator and the denominator of (6.3) and (6.7) may become too small for the particular arithmetic precision being used. A precautionary provision that checks this is desirable to avoid unnecessary use of resources.]

In the sample space, modification B corresponds to dropping a perpendicular line from y to $f_g - f_0$ whose foot gives the vector $(1 - \hat{v})f_0 + \hat{v}f_g$. This is illustrated in Figure 17.5 for a one parameter model. The relationship of this procedure to modification A can be seen as follows. The sum of squares of the estimated residuals can be written as

$$\varepsilon'\varepsilon = (y - f_0 - v(f_g - f_0))'(y - f_0 - v(f_g - f_0)) \tag{6.8}$$

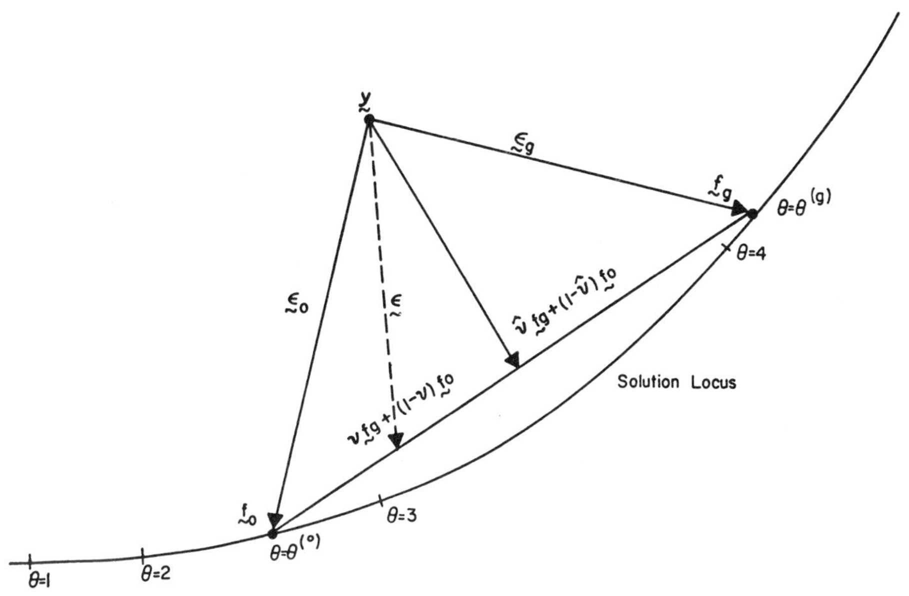

Figure 17.5 Sample Space Representation of Modification B for the Case of One Parameter Model

where $\underset{\sim}{f}_g$ is the n×1 vector of $f(\underset{\sim}{\xi}_u, \underset{\sim}{\theta}^{(g)})$, u = 1,2,...,n. This is quadratic in v and passing through the points $(0,S_0)$ and $(1,S_g)$. Furthermore, the initial slope of the sum of squares of the estimated residuals is given by

$$\left[\frac{d\underset{\sim}{\varepsilon}'\underset{\sim}{\varepsilon}}{dv}\right]_{v=0} = -2(\underset{\sim}{f}_g - \underset{\sim}{f}_0)'(\underset{\sim}{y} - \underset{\sim}{f}_0) = -2(\underset{\sim}{f}_g - \underset{\sim}{f}_0)'\underset{\sim}{\varepsilon}_0 \qquad (6.9)$$

which is identical to the initial slope used in the previous method provided that $\underset{\sim}{f}_g = \underset{\sim}{f}_0 + X_0(\underset{\sim}{\theta}^{(g)} - \underset{\sim}{\theta}^{(0)})$.

VII. EXAMPLE

In comparing various methods little in the way of general conclusions can be based on the performance of particular examples. The following simple example, taken from Box and Hunter (1965) in an abbreviated form, is provided only for the purpose of gaining some preliminary ideas of the behavior of the procedures we have discussed.

The model is $f(\underset{\sim}{\xi},\underset{\sim}{\theta}) = \theta_1\theta_2\xi_1/(1 + \theta_1\xi_1 + 5000\xi_2)$ and the observations (ξ_1,ξ_2,y) are (1,1,0.1165), (2,1,0.2114), (1,2,0.0684) and (2,2,0.1159). The sum of squares surface, plotted in Figure 17.6, is very curved and ridgy. The values $(\theta_1,\theta_2) = (300,6)$ corresponding to P_0 in the figure are the chosen initial estimates.

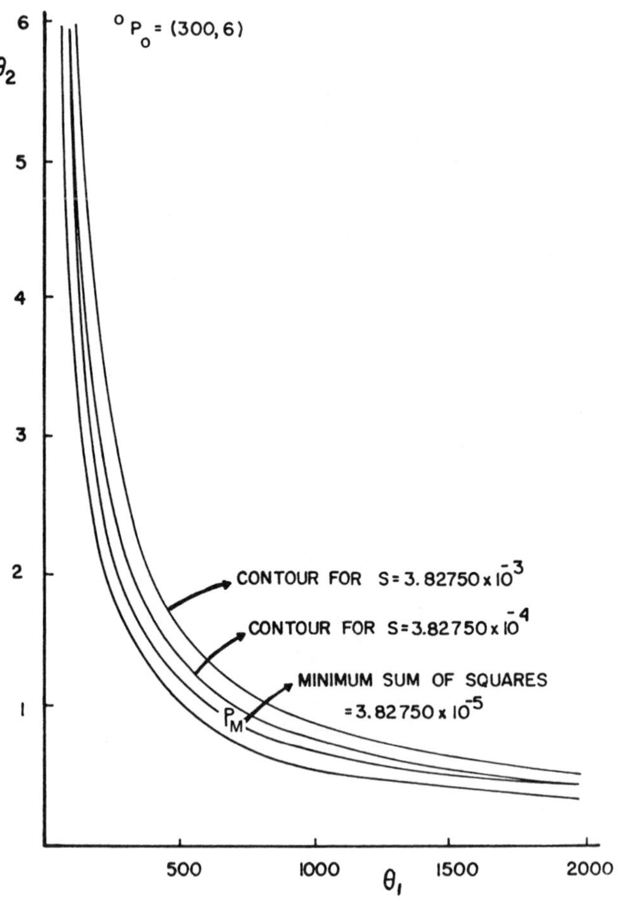

Figure 17.6 Sum of Squares Contours for the Example

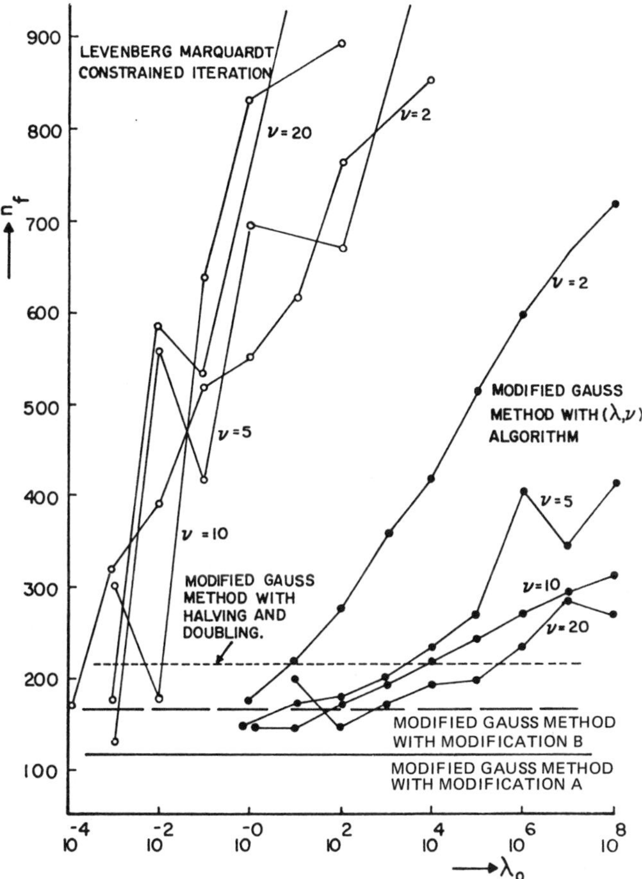

Figure 17.7 Comparison of Five Different Procedures in the Example

Figure 17.7 shows the result for five different methods. It shows the number of times, n_f, that $f(\underset{\sim}{\xi},\underset{\sim}{\theta})$ has to be evaluated before the iterative process reaches the point of minimum P_m (at which the sum of squares = 3.82750×10^{-5}). For the two methods that use the (λ,ν) algorithm n_f is shown for various choices of λ_0 and ν, while that for the methods which do not depend on any control parameters is indicated by a horizontal line.

Considering that the methods with the (λ,ν) algorithm require a suitable choice of λ_0 and ν in advance, the behavior of the modified Gauss methods with modifications A and B is certainly not discouraging. It will also be interesting that the modified Gauss method with the (λ,ν) algorithm appears stable over a wide range of λ_0, if we recall that the modified Gauss solution is the result of the constrained minimization in the space which requires no interpolation among the two classical approaches that often contradict each other.

VIII. CONCLUSION

The Levenberg-Marquardt's constrained iteration has been widely programmed and used. One apparent justification is that it provides a compromise between the steepest descent method and the Gauss method. However, if the parameters are transformed into the linearly invariant metric, the steepest descent vector and the Gauss solution vector are found to be identical and thus there is no need to compromise between directions given by these two vectors. It is also shown that the constrained minimization in the linearly invariant metric is equivalent to using in the original metric the modified Gauss method which was proposed earlier. Two methods are proposed to determine how far one should go along the Gauss vector. These both have the virtue that their computation only involves information that already exists.

ACKNOWLEDGMENTS

We wish to thank Professors William Hunter and Norman Draper of the University of Wisconsin for their encouragement and comments on this work.

Sponsored by the United States Army under Contract No. DAAG-80-C-0041.

REFERENCES

Booth, G.W., Box, G.E.P., Muller, M.E. and Peterson, T.I. (1959). Forecasting by Generalized Regression Methods, Nonlinear Estimation. *International Business Machines Corp., Mimeo.* (IBM SHARE Program No. 687.)

Box, G.E.P. (1954). The exploration and exploitation of response surface: some general considerations and examples. *Biometrics 10,* 16-60.

Box, G.E.P. (1956). Some notes on nonlinear estimation. Technical Report, Statistical Techniques Research Group, Princeton University.

Box, G.E.P. (1958). Use of statistical methods in the elucidation of physical mechanism. *Bull. Inst. Intern. De Statistique 36,* 215-255.

Box, G.E.P. (1960). Fitting empirical data. *Annals N.Y. Academy of Sciences 86,* 792-816.

Box, G.E.P. and Coutie, G.A. (1956). Application of digital computers in the exploration of functional relationship. *Proceedings of the Institute of Electrical Engineers 103, Part B, Supplement No. 1,* 100-107.

Box, G.E.P. and Hunter, W.G. (1962). A useful method for model building. *Technometrics 4,* 301-318.

Box, G.E.P. and Hunter, W.G. (1965). Sequential design of experiments for nonlinear models. *IBM Scientific Computing Symposium in Statistics 113.*

Box, G.E.P. and Wilson, K.B. (1951). On the experimental attainment of optimum conditions. *J. Roy. Stat. Soc. B 13,* 1-45.

Draper, N.R. (1963). "Ridge analysis" of response surfaces. *Technometrics 5,* 469-479.

Gauss, C.F. (1821). Theory of least squares. English translation by H.F. Trotter, *Statistical Techniques Research Group,* Technical Report No. 5, Princeton University (1957).

Hartley, H.O. (1961). The modified Gauss-Newton method for fitting of nonlinear regression functions by least squares. *Technometrics 3*, 269-280.

Koshal, R.S. (1933). Application of the method of maximum likelihood to the improvement of curves fitted by the method of moments. *J. Roy. Stat. Soc. 96*, 303.

Levenberg, K. (1944). A method for the solution of certain, nonlinear problems in least squares. *Quart. Appl. Math. 2*, 164-168.

Marquardt, D.W. (1963). An algorithm for least squares estimation of nonlinear parameters. *J. Soc. Ind. Appl. Math. 2*, 431-441.

Meeter, D.A. (1966). Nonlinear least squares (GAUSHAUS). University of Wisconsin Computing Center Users Manual, 4, Section 3, 22.

Morrison, D.D. (1960). Methods for nonlinear least squares problems and convergence proofs. Tracking Programs and Orbit Determination, *Proc. Jet Propulsion Laboratory Seminar, 1*.

PART IV

STATISTICAL AND POPULATION GENETICS

Chapter 18

EFFECTIVE BREEDING POPULATION SIZE FOR STRUCTURED RANDOM MATING WITH RANDOM OR DIRECTIONAL SELECTION*

Dewey L. Harris, Stephen S. Rich,† and Chuan-Ting Wang‡

Agricultural Research Service
U.S. Department of Agriculture and Purdue University
West Lafayette, Indiana

I. INTRODUCTION

Wright (1931) introduced the concept of effective population number to reflect the magnitude of expected random variation and fixation in gene frequency due to finite size of a population. In addition, he derived that the formula that the effective population number (N_e^W) for a population composed of N_m breeding males and N_f breeding females was

$$\frac{1}{N_e^W} = \frac{1}{4N_m} + \frac{1}{4N_f} \tag{1.1}$$

The effective population number for separate sexes indicates the size of a population with equal numbers of males and females which

---

*Journal paper No. 9434, Purdue Agricultural Experiment Station. Joint contribution from USDA-ARS-NCR and Department of Animal Sciences, Purdue University.

†Current affiliation: University of Minnesota, Minneapolis, Minnesota.

‡Current affiliation: Lincoln, Nebraska.

would result in equivalent inbreeding or variance due to genetic drift. Crow (1954) elaborated upon the concept and specified an *inbreeding effective number* reflecting expected inbreeding and a *variance effective number* reflecting expected sampling variance in gene frequency. In some cases, these two may differ but are often the same. When Kempthorne (1957) explained the derivation of Wright's formula in terms of Malécot (1948), he noted that the assumption was that of panmixia, where the two genes at a locus of each individual in a generation consist of a gene randomly drawn *with replacement* from the $2N_m$ genes possessed by males of the previous generation and another gene randomly drawn *with replacement* from the $2N_f$ genes carried by female parents. The assumption that mating is random *with replacement* is not satisfied in the breeding of most animal populations (mammals, in particular) since mating of a female to a certain male usually precludes (at least, for the period of gestation if the female conceives) mating to another male. In these and other cases (avian species, in particular), the experimental structure often dictates that only one male is mated to each female. Also, the specified number of female mates is usually the same for all males in an experiment. Thus, structured random mating of animals implies sampling *without replacement*. On the other hand, the assumptions of Wright would seem appropriate for open pollinated plant breeding populations when the number of pollen and seeds is large with bulk sampling of seed to be used.

Kimura and Crow (1963) developed formulae for effective population number for several situations where Wright's formula is not fully appropriate. In the present paper, the objective is to extend the formulae to include the situation, not fully considered before, of *structured* random mating where each of a specified number of males is mated to a specified number of females to produce families of offspring. From these offspring, males and females will be chosen for mating to produce the next generation. We will consider primarily the cases where offspring family sizes are equal with the choice of breeders either at random or by truncation directional selection. Variable offspring family sizes will also be considered.

II. GENETIC DRIFT

Our concern is for the variance effective breeding population size. A primary reason for this concern relates to computer simulation work by the authors to study the effects of small to moderate population size upon long-term response to selection. Previous studies (Gill, 1965a,b,c and Qureshi, Kempthorne and Hazel, 1968) simulated unstructured random mating with replacement. However, to be more consistent with actual practice, we have simulated structured random mating without replacement. Harris (1982) found a strong relationship between long-term response and N_e^W calculated as specified by Wright (1931) even though the conditions of simulation did not exactly match the theoretical assumptions. The question remains whether a more appropriate measure of population size would give more reliable results for later simulation studies and for numerous experimental studies.

The term, genetic drift, refers to the chance changes in frequency of alleles from one generation to the next. Of particular concern is the change which occurs from the breeding population of males and females through the male and female offspring produced to the breeding population of males and females chosen to produce the next generation. Thus, there are four pathways along which alleles pass between generations; male to male, male to female, female to male, and female to female with separate opportunities for chance changes along each path. Thus, we may determine the overall frequency, p, for a given generation by

$$p = \frac{1}{4}(p^{mm} + p^{mf} + p^{fm} + p^{ff})$$

where p^{yz} represents the allelic frequency transmitted from the $\underline{y}$

sex of parents to the z sex of offspring chosen to be used as parents with m = male and f = female. When the chance occurrences along these pathways are independent, we can represent V_p, the variance of the overall allelic frequency, by

$$V_p = \frac{1}{16}(V_p^{mm} + V_p^{mf} + V_p^{fm} + V_p^{ff})$$

where V_p^{yz} is the variance of the frequency along each path. In an idealized reference situation of random sampling with replacement of alleles, these variances would be binomial with

$$V_p^{yz} = \frac{p^{yz}(1 - p^{yz})}{2N_e^{yz}}$$

and

$$V_p = \frac{p(1 - p)}{2N_e}$$

where N_e^{yz} is the effective population number for a specific path, sex y to sex z, and N_e is the overall effective number. For the idealized reference situation, N_e^{yz} would be the actual number of parents, but becomes a measure of effective number for situations departing from this ideal. Since p is the average p^{yz} and since the four p^{yz} will be similar in magnitude, the approximation follows that

$$\frac{1}{N_e} \cong \frac{1}{16}\left(\frac{1}{N_e^{mm}} + \frac{1}{N_e^{mf}} + \frac{1}{N_e^{fm}} + \frac{1}{N_e^{ff}}\right) \tag{2.1}$$

Even though Kimura and Crow (1963) did not develop the specific formula needed, we can utilize their formula for monoecious populations to determine the effective number for each path for dioecious populations. Consider N_m males (sires) each mated to N_f different females (dams) with each mating producing K_m male offspring and K_f female offspring. The K_m and K_f might either be constant or variable. For generality, we will consider that ΣR_z individuals of sex *z* may be chosen by directional selection out of the ΣK_z available, so that variable R_z are chosen from the K_z available for a particular family. The symbol, Σ, represents summation over full-sib families for *fz* paths or over paternal half-sib families for *mz* paths. Then $N_z = \Sigma T_z$ are randomly chosen from the ΣR_z, so that a particular family has a variable T_z chosen to be sex *z* breeders for the next generation from the R_z previously selected for that family. With similar considerations for both males and females, the four pathways may be diagrammed as

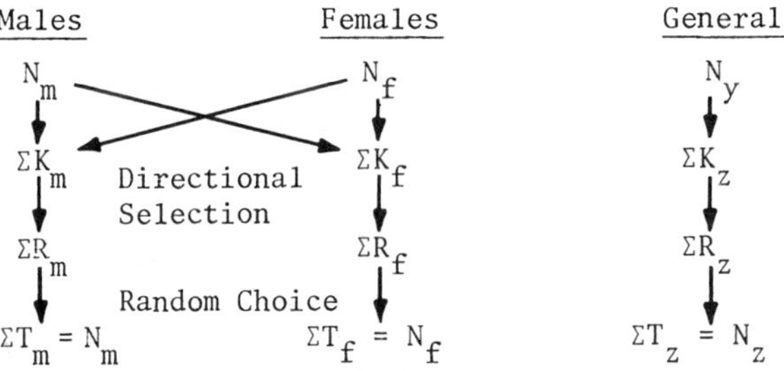

When the concern is for directional selection only, $R_z = T_z$ for all families; when the concern is for random selection only, $K_z = R_z$ for all families. In this context, Kimura and Crow's (1963) result for monoecious populations can be presented as

$$N_e^{yz} = \frac{(2N_y - 1)\overline{T}_z}{2(1 + V_T^{yz}/\overline{T}_z)} \tag{2.2}$$

where N_y is the number of parents of sex *y* in the previous generation, $\overline{T}_z$ is the average across families of T_z, the variable number of sex *z* offspring in a sex *y* family (and, thus, the number of gametes contributed by the sex *y* parent to the sex *z* offspring), and V_T^{yz} is the variance across families of T_z. Of course, *yz* may be mm, mf, fm, or ff. Note that $\overline{T}_z$ and V_T^{yz} may not be the same for $y = m$ as for $y = f$. Again, the families will be paternal half-sib groups when $y = m$ and full-sib groups when $y = f$.

III. STRUCTURED RANDOM MATING - EQUAL FAMILY SIZES

Consider now the structured random mating situation with each of s sires mated to d dams and each mating producing m male offspring and f female offspring, where d, m and f are constants. Relating the previous formulae to this situation, we have, for each of the four paths,

	male to male	male to female	female to male	female to female	
	yz = mm	yz = mf	yz = fm	yz = ff	
N_y	s	s	sd	sd	
ΣK_z	sdm	sdf	sdm	sdf	
$\overline{K}_z$	dm	df	m	f	
ΣT_z	s	sd	s	sd	
$\overline{T}_z$	1	d	1/d	1	
N_e^{yz}	$\dfrac{2s-1}{2(1+V_T^{mm})}$	$\dfrac{(2s-1)d}{2(1+V_T^{mf}/d)}$	$\dfrac{(2sd-1)/d}{2(1+dV_T^{fm})}$	$\dfrac{2sd-1}{2(1+V_T^{ff})}$	(3.1)

To complete the derivation, we need to develop the formulae for V_T^{yz} for each of three specific cases:

1. Random choice only $(V_{T|Rnd}^{yz})$
2. Truncation directional selection only $(V_{T|Trn}^{yz})$
3. Truncation directional selectional followed by random elimination of excess individuals selected but not to be used as breeders $(V_{T|TrR}^{yz})$

The latter case involving a combination of directional and random selection is of concern for some simulated situations of Harris (1982) and later studies. To illustrate combination selection, in one case in that paper, 6 sires were each mated to one dam to produce 3 male and 3 female progeny for a total of 18 of each sex. Selection of the top 2/3 yielded 12 males and 12 females, of which 6 males and 6 females were eliminated to obtain the required 6 males and 6 females to be parents of the next generation.

IV. STRUCTURED RANDOM MATING WITH RANDOM CHOICE OF BREEDERS

For the case where individuals to be used as breeders are chosen at random from the offspring ($K_z = R_z$) and with constant K_z across families (equal sizes of families produced), we symbolize the number of offspring of sex z chosen from family j to be breeders as T_{zj}. Thus, the situation is simply equal probability independent random sampling without replacement and the vector of numbers chosen per family $(T_{z1}, T_{z2}, \ldots, T_{zN_y})$ has a probability

$$P(T_{z1}, T_{z2}, \ldots, T_{zN_y}) = \frac{\prod_{j=1}^{N_y} C\begin{pmatrix} K_z \\ T_{zj} \end{pmatrix}}{C\begin{pmatrix} \Sigma K_z \\ \Sigma T_{zj} \end{pmatrix}} \tag{4.1}$$

where

$$C\begin{pmatrix} K_z \\ T_{zj} \end{pmatrix} = \frac{K_z!}{T_{zj}!(K_z - T_{zj})!}$$

In most situations, $\sum_{j=1}^{N_y} T_{zj}$ will be restricted to being equal to N_z, the number of breeders needed for that sex. Thus, for this situation, $V_{T|Rnd}^{yz}$ can be determined from

$$V_{T|Rnd}^{yz} = \sum_{\substack{\text{sets} \\ \text{of } T_z}} P(T_{z1}, T_{z2}, \ldots, T_{zN_y}) \sum_{j=1}^{N_y} \{T_{zj}^2 - N_y \bar{T}_z^2\}/N_y \tag{4.2}$$

with

$$\overline{T}_z = (\sum_{j=1}^{N_y} T_{zj})/N_y$$

and $\sum_{\text{sets of } T_z}$ indicates summation over the possible sets of T_{zj}, with the restrictions that $\sum_{\text{sets of } T_z} P(T_{z1}, T_{z2}, \ldots, T_{zN_y}) = 1$ and $\sum_{j=1}^{N_y} T_{zj} = N_y\overline{T}_z = N_z$ for the path *yz*. Thus, the variance of family sizes for each of the four paths may be determined with *y* = *m* or *f*, and with *z* = *m* or *f*, from formulae (4.1) and (4.2) with the appropriate values of N_y, $\Sigma T_z = N_z$, and $\overline{T}_z$. These four values of V_T^{yz} may then be used in (2.2) to obtain four values of N_e^{yz} which can be combined in (2.1) to obtain an overall N_e for random selection. Although tedious (especially when the number of families is large), the calculations involved are straight-forward and are readily programmable for computer calculation. We will symbolize the value of effective population number derived by these formulae as N_e^S to distinguish it from N_e^W as obtained from formula (1.1), with the superscript *S* indicating the *structured* mating system.

V. STRUCTURED RANDOM MATING WITH DIRECTIONAL SELECTION ONLY

When the choice of individuals to be used as breeders is not random, but by truncation directional selection, the formula for effective population number becomes more complex. Directional selection will cause the families to have differing probabilities of their members being chosen. Actually, individuals will have differing probabilities depending on their phenotypic values, but since we are concerned with how selection influences the distribution of family sizes, only the probabilities pertaining to families and dependent upon the average phenotypic values for the families are of concern.

Introducing the notation that a random member of sex *z* for family *j* has a probability P_{zj} of being chosen for breeding allows us to extend formulae (4.1) and (4.2) to include these differential probabilities for multivariate hypergeometric sampling from the families. Thus, formulae (4.1) and (4.2) for sex *z* offspring from sex *y* parents will extend to

$$P(T_{z1}, T_{z2}, \ldots, T_{zN_y}) = \frac{\prod_{j=1}^{N_y} C\binom{K_z}{T_{zj}} P_{zj}^{T_{zj}} (1-P_{zj})^{K_z - T_{zj}}}{C\binom{\Sigma K_z}{\Sigma T_{zj}} \overline{P}_z^{\Sigma T_{zj}} (1-\overline{P}_z)^{\Sigma K_z - \Sigma T_{zj}}} \tag{5.1}$$

where

$$\overline{P}_z = \frac{1}{N_y} \sum_{j=1}^{N_y} P_{zj} = \frac{\Sigma T_{zj}}{\Sigma K_{zj}} = \frac{\overline{T}_z}{\overline{K}_z}$$

is constrained by the chosen proportion of offspring to be selected. Equation (4.1) is a special case of (5.1) when $P_j = \overline{P}$ for all *j*.

Even though the value for $\overline{P}_z$ is directly specified from the selection procedure, the values for P_{zj} for specific family *j* are difficult to obtain. If selection was for all individuals above a fixed truncation point for sex *z*, say x_z, we would have

$$P_{zj} = pr(X_{zj} > x_z | \mu_{zj})$$

where μ_{zj} is the expected sex *z* family mean for the *jth* family for the selection criterion and X_{zj} is the phenotypic value for a sex *z* random member of the *jth* family with

$$E(X_{zj} | \mu_{zj}) = \mu_{zj}$$

$$E(\mu_{zj}) = \mu_z$$

$$V(X_{zj} | \mu_{zj}) = \sigma_W^2, \text{ variance within groups}$$

$$V(\mu_{zj}) = \sigma_B^2, \text{ variance between groups}$$

and

μ_z is the overall mean for sex z individuals

However, since the truncation point for selecting a specified proportion of a finite population depends upon the composition of the finite population, x_z is not constant. Harter (1961) has developed a complex algorithm to determine the expected value of the selected proportion of the total population. Extending this to determine the proportions selected in each family within the finite population seemed exceedingly difficult. However, it was anticipated that directional selection (with $P_{zj} \neq \overline{P}_z$) would increase the variance of number chosen per family, T_{zj}, relative to what it would be for random choice. This expectation occurs because superior families have greater probabilities of having several or all members selected and inferior families have greater probabilities of few or no family members being selected. An approximation can be developed for this increase following an approach suggested by Robertson (1961). Griffing (1960) noted that when the differences between groups were small relative to the variation within groups,

$$P_{zj} \cong \overline{P}_z\{1 + \frac{\overline{i}_z(\mu_{zj} - \mu_z)}{\sigma_W}\} \tag{5.2}$$

where $\overline{P}_z$ is the overall proportion selected in sex z, $\mu_{zj} - \mu_z$ is the deviation of the mean of the group for the criterion of selection, $\overline{i}_z$ is the selection differential for sex z in standard deviation units (related to $\overline{P}_z$), and σ_W is the standard deviation within groups for the criterion being selected. If normality of distribution of the selection criterion is assumed, the familiar result that $\overline{i}_z$ is equal to the ratio of the ordinate (Z_z) at the truncation point x_z to the proportion selected ($\overline{P}_z$) is obtained. Griffing's (1960) development was specific for the case where the groups were different genotypes; however it holds similarly for groups being different families. From approximation (5.2), we obtain the approximation

$$V(P_{zj}) \cong V\{\overline{P}_z(1 + \frac{\overline{i}_z(\mu_{zj} - \mu_z)}{\sigma_W}\} = \overline{P}_z^2\, \overline{i}_z^2\, \frac{\sigma_{By}^2}{\sigma_{Wy}^2} \qquad (5.3)$$

where $\sigma_{By}^2 = V(\mu_{zj} - \mu_z)$ is the variance among expected family means. When h^2 represents the heritability (the ratio of additive genetic to phenotypic variance) and r_y the intraclass genetic correlation between family members ($r_f = 0.5$ for full-sib families and $r_m = 0.25$ for paternal half-sib families), then

$$\frac{\sigma_{By1}^2}{\sigma_{Wy1}^2} = \frac{r_y h_0^2}{(1 - r_y h_0^2)} \qquad (5.4)$$

where h_0^2, σ_{By}^2, and σ_{Wy}^2, represent the initial heritability, initial variance between families, and initial within family variance, respectively. However, this formula pertains only to the first generation of selection. Usually, concern is for recurrent selection where selection is repeated each generation.

Previous generation selection will reduce the genetic variation in the next generation, especially variation between families, since selection reduces the variation among those chosen to become parents. Dickerson and Hazel (1944) and Cochran (1951) have shown that truncation directional selection reduces the initial additive genetic variance, $\sigma_{A_0}^2$, among chosen individuals by a fraction

$$h_{z0}^2 \overline{i}_z(\overline{i}_z - x_z)$$

to become

$$\sigma_{A_0}^2\{1 - h_{z0}^2 \overline{i}_z(\overline{i}_z - x_z)\}$$

for sex $\underline{z}$ individuals selected to become parents for the next generation. It is assumed here that $\sigma^2_{A_0}$ is the same for both sexes but the initial heritabilities, h^2_{m0} and h^2_{f0}, may differ due to different amounts of environmental variation. When these selected individuals are randomly mated to become the parents of the next generation, the additive genetic variance will become

$$\sigma^2_{A_1} = \frac{1}{4}\sigma^2_{A_0}\{1 - h^2_{m0}\bar{i}_m(\bar{i}_m - x_m)\}$$
$$+ \frac{1}{4}\sigma^2_{A_0}\{1 - h^2_{f0}\bar{i}_f(\bar{i}_f - x_f)\}$$
$$+ \frac{1}{2}\sigma^2_{A_0}$$

with the first term representing the reduced sire component of variance, the second term representing the reduced dam within sire component of variance and the third term representing the additive genetic variance within full-sib families, which is assumed to be unchanged by the previous generation selection. From this, we develop that

$$h^2_{m1} = \frac{h^2_{m0}\{1 - \frac{1}{4}h^2_{m0}\bar{i}_m(\bar{i}_m - x_m) - \frac{1}{4}h^2_{f0}\bar{i}_f(\bar{i}_f - x_f)\}}{1 - h^2_{m0}\{\frac{1}{4}h^2_{m0}\bar{i}_m(\bar{i}_m - x_m) + \frac{1}{4}h^2_{f0}\bar{i}_f(\bar{i}_f - x_f)\}}$$

and

$$h^2_{f1} = \frac{h^2_{f0}\{1 - \frac{1}{4}h^2_{m0}\bar{i}_m(\bar{i}_m - x_m) - \frac{1}{4}h^2_{f0}\bar{i}_f(\bar{i}_f - x_f)\}}{1 - h^2_{f0}\{\frac{1}{4}h^2_{m0}\bar{i}_m(\bar{i}_m - x_m) + \frac{1}{4}h^2_{f0}\bar{i}_f(\bar{i}_f - x_f)\}}$$

or, in general

$$h^2_{y1} = \frac{h^2_{y0}\{1 - \frac{1}{4}h^2_{m0}\bar{i}_m(\bar{i}_m - x_m) - \frac{1}{4}h^2_{f0}\bar{i}_f(\bar{i}_f - x_f)\}}{1 - h^2_{y0}\{\frac{1}{4}h^2_{m0}\bar{i}_m(\bar{i}_m - x_m) + \frac{1}{4}h^2_{f0}\bar{i}_f(\bar{i}_f - x_f)\}}$$

Then, for later generations (L), the ratio of between family variance to within family variance for paternal half-sib families would be

$$\frac{\sigma^2_{BmL}}{\sigma^2_{WmL}} = \frac{\frac{1}{4} h^2_{z1}\{1 - h^2_{m1}\bar{i}_m(\bar{i}_m - x_m)\}}{1 - \frac{1}{4} h^2_{z1}\{1 + h^2_{f1}\bar{i}_f(\bar{i}_f - x_f)\}} \tag{5.5}$$

for sex $\underline{z}$ offspring. For full-sib families, this ratio would be

$$\frac{\sigma^2_{BfL}}{\sigma^2_{WfL}} = \frac{\frac{1}{2} h^2_{z1}\{1 - \frac{1}{2} h^2_{m1}\bar{i}_m(\bar{i}_m - x_m) - \frac{1}{2} h^2_{f1}\bar{i}_f(\bar{i}_f - x_f)\}}{1 - \frac{1}{2} h^2_{z1}} \tag{5.6}$$

for sex $\underline{z}$ offspring. In order to approximate these formulae, the expectations for x_m and x_f can be determined as the truncation point for an infinite unit normal frequency distribution for the selection proportion $\overline{P}$. Various computer algorithms or tabulations are available for determining this value. Actual values of x_m and x_f for finite populations would fluctuate around this value but the inaccuracies of introducing this approximation do not seem serious. For K_{zj} equal to a constant, K_z (i.e., equal family sizes for number of offspring produced), it follows that since T_{zj} centers around $P_{zj}K_z$, the variance of number per family results both from the variation of P_{zj} and from the chance variation of how many actually are selected. Thus, we can approximate

$$V^{yz}_{T|Trn} \cong K^2_z V(P_{zj}) + V^{yz}_{T|Rnd} \tag{5.7}$$

where $V^{yz}_{T|Rnd}$ represents the value obtained by formula (4.2) for random selection. The additional approximation is that this latter value for random choice with $P_{zj} = \overline{P}_z$ may not quite be the same as the value for random variation around differing P_{zj} which average $\overline{P}_z$. Substituting (5.3) into (5.7), we have

$$V^{yz}_{T|Trn} \cong K^2_z \overline{P}^2_z \bar{i}^2_z \frac{\sigma^2_{By}}{\sigma^2_{Wy}} + V^{yz}_{T|Rnd} = \overline{T}^2_z \bar{i}^2_z \frac{\sigma^2_{By}}{\sigma^2_{Wy}} + V^{yz}_{T|Rnd} \tag{5.8}$$

which can be evaluated for early generations of selection with

$$\frac{\sigma^2_{By}}{\sigma^2_{Wy}} = \frac{\sigma^2_{By1}}{\sigma^2_{Wy1}} = \frac{r_y h^2_0}{1 - r_y h^2_0}$$

For later generations of recurrent selection for the cases of $\underline{y} = \underline{m}$, we can use $\frac{\sigma^2_{BmL}}{\sigma^2_{WmL}}$ of (5.5) to replace $\frac{\sigma^2_{By}}{\sigma^2_{Wy}}$ in (5.8). Using (5.6) in place of $\frac{\sigma^2_{By}}{\sigma^2_{Wy}}$ in (5.8), we obtain the appropriate formulae for later generations for $\underline{y} = \underline{f}$ cases. These values may be used in equation (2.2) to obtain N_e^{yz} values to combine in (2.1) to determine an over-all value of N_e for truncation directional selection. Since directional selection will increase the variance of number selected from families, the expected value for effective population number will be less than N_e^S, the value for <u>structured</u> random mating with random selection. The symbol for this <u>reduced</u> population number will be N_e^R and the approximation described above will be $\tilde{N}^R_{e1}$ for early generations [utilizing $\frac{\sigma^2_{By1}}{\sigma^2_{Wy1}}$ of (5.4) in (5.8)] and $\tilde{N}^R_{eL}$ for the later generations [utilizing (5.5) and (5.6) to replace $\frac{\sigma^2_{By}}{\sigma^2_{Wy}}$ in (5.8)]. The values of $\tilde{N}^R_{eL}$ will be slightly greater than those for $\tilde{N}^R_{e1}$.

VI. PAIRED MATINGS WITH EQUAL SELECTION PROPORTIONS IN TWO SEXES

For the special case where only one female is mated to each male (d=1), the number of offspring of each sex is the same for each mating, $K_m = m = K_f = f$, and the number selected in each sex is equal, $\Sigma R_m = \Sigma R_f$, the formulae simplify somewhat. When d=1, $N_m = \Sigma T_m = \Sigma T_f = \Sigma T_f = N_f = s$, $\overline{T}_z = 1$ and formula (2.2) becomes

$$N_e^{yz} = \frac{2s - 1}{2(1 + V_T^{yz})}$$

for all yz combinations with the appropriate alternative form of V_T^{yz}. In addition for these conditions of equality, $\overline{i}_m = \overline{i}_f = \overline{i}$ and $x_m = x_f = x$, thus, (5.5) simplifies to

$$\frac{\sigma^2_{BmL}}{\sigma^2_{WmL}} = \frac{\frac{1}{4} h^2_{z1}\{1 - h^2_{m1}\overline{i}(\overline{i} - x)\}}{1 - \frac{1}{4} h^2_{z1}\{1 + h^2_{f1}\overline{i}(\overline{i} - x)\}}$$

and (5.6) becomes

$$\frac{\sigma^2_{BfL}}{\sigma^2_{WfL}} = \frac{\frac{1}{2} h^2_{z1}\{1 - \frac{h^2_{m1} + h^2_{f1}}{2} \overline{i}(\overline{i} - x)\}}{1 - \frac{1}{2} h^2_{z1}}$$

for sex z offspring. If $h^2_{mL} = h^2_{fL}$, as would often be the case, then $V_T^{mm} = V_T^{mf}$ and $V_T^{fm} = V_T^{ff}$, so that $N_e^{mm} = N_e^{mf}$ and $N_e^{fm} = N_e^{ff}$ with resulting simplification of (2.1). These simplifications are indicated for $\tilde{N}^R_{eL}$ but similar simplifications hold for $\tilde{N}^R_{e1}$.

VII. COMBINED DIRECTIONAL AND RANDOM SELECTION

For two-stage combined selection with R_{zj} selected directionally from the K_z available in the jth family and with T_{zj} selected at random, from the R_{zj}, equation (5.8) would enlarge to become

$$V^{yz}_{T|TrR} \cong K_z^2 V(P_{zj}) + \sum_{\substack{\text{sets}\\ \text{of R}}} P(R_{z1},\ldots,R_{zN_y}) \sum_{\substack{\text{sets}\\ \text{of T}}} P(T_{z1},\ldots,T_{zN_y}|R_{z1},\ldots,R_{zN_y})V(T_{zj})$$

However, the calculations for the latter term would be quite difficult if N_y, R_{zj} and/or T_{zj} is at all large. Except in cases of very strong selection, the average variance of the T_{zj} does not seem to be greatly influenced by the sets of R_{zj} obtained, so that we can approximate

$$V^{yz}_{T|TrR} = \bar{R}_z^2 \bar{i}_z^2 \frac{\sigma^2_{By}}{\sigma^2_{Wy}} + V^{yz}_{T|Rnd} \tag{7.1}$$

for the initial generation of combined directional selection and random elimination. The approximation for later generations can be obtained by using (5.5) or (5.6) to replace $\frac{\sigma^2_{By}}{\sigma^2_{Wy}}$ in (7.1).

VIII. UNEQUAL OFFSPRING NUMBERS

When the sizes of family produced, the K_{zj}, also differ between families with a known variance, it seems that an appropriate approximation for combined directional selection and random elimination is

$$V^{yz}_{T|TrR} \cong V(K_{zj}) + \bar{R}_z^2 \bar{i}_z^2 \frac{\sigma^2_{By}}{\sigma^2_{Wy}} + V^{yz}_{T|Rnd} \tag{8.1}$$

In the last term, calculations are done as if all $K_{zj} = \bar{K}_z$ even though the K_{zj} will vary for this presumed situation. This approximation involves three potential sources of inaccuracy, 1) that $V(R|\text{sets of } K) \neq V(R|\bar{K})$ in the second term to the right of the equal sign, 2) that $V(T|\text{sets of } K) \neq V(T|\bar{K})$ in the last term, and 3) that $V(T|\bar{K})$ may not be an exact sum of $V(R|\bar{K}) + V(T|\bar{R})$. As pointed out by Latter (1959), with variation in number of offspring from a mating, a correlation may be induced between number of males, K_{mj}, and females, K_{fj}, produced in family *j*. This departure from independence along the four paths is also ignored in this approximation. It would be conceivable to calculate $V^{yz}_{T|TrR}$ more accurately without making these approximations, but the number of combinations involved in the summations would probably be prohibitive even on a computer, if the number of families and/or the size of the families are at all large. Formula (8.1) may be considered a general formula. For random elimination only, $R_{zj} = K_{zj}$, and, thus, $\bar{i}_z = 0$, and the second term goes to zero. For directional selection only, the formula remains symbolically the same, but since $R_{zj} = T_{zj}$, the second and third possible inaccuracies are not involved.

IX. NUMERICAL RESULTS

Table 18.1 presents various parameter combinations and the associated calculated values to show numerically the implications of various combinations of number of sires (*s*), dams per sire (*d*), number of male and female offspring per dam (*m* and *f*, respectively), selection proportion (P_m and P_f), number randomly eliminated ($R_m - T_m$ and $R_f - T_f$) and initial heritability (h_0^2) upon reduced heritability (h_1^2), standardized selection differentials ($\bar{i}_m$ and $\bar{i}_f$), and the alternative forms of effective population size (N_e^W for panmixia, N_e^S for structured random mating with random selection, $\tilde{N}_{e1}^R$ for the initial population with directional selection, and $\tilde{N}_{eL}^R$ for later generations with directional selection). Several trends are apparent. The values of N_e^S are always larger than N_e^W. The values of $\tilde{N}_{e1}^R$ are always smaller than N_e^S with the reduction greater for high heritability and small selection proportion. Thus, $\tilde{N}_{e1}^R$ is sometimes larger and sometimes smaller than N_e^W with the values of N_{e1}^R less than N_e^W occurring only for small selection proportion and high heritability. However, this type of combination also includes cases where the reduced heritability is considerably less than the initial heritability. This type of combination results in $\tilde{N}_{eL}^R$ being somewhat larger than $\tilde{N}_{e1}^R$ with $\tilde{N}_{eL}^R$ being only slightly larger than $\tilde{N}_{e1}^R$

TABLE 18.1

Various Parameter Combinations and Calculated Values for N_e^W, N_e^S, $\tilde{N}_{e1}^R$, and $\tilde{N}_{eL}^R$

(See text for definition of symbols)

s	d	m	f	$\overline{P}_m$	$\overline{P}_f$	R_m-T_m	R_f-T_f	h_0^2	h_1^2	$\overline{i}_m$	$\overline{i}_f$	N_e^W	N_e^S	$\tilde{N}_{e1}^R$	$\tilde{N}_{eL}^R$
3	1	3	3	.333	.333	0	0	.125	.121	.996	.996	6	6.67	6.46	6.48
				.333	.333	0	0	.500	.462	.996	.996	6	6.67	5.76	6.02
6	1	3	3	.667	.667	6	6	.125	.122	.521	.521	12	13.85	13.74	13.75
				.333	.333	0	0	.125	.121	1.042	1.041	12	13.85	13.40	13.45
				.667	.667	6	6	.500	.467	.521	.521	12	13.85	13.31	13.46
				.333	.333	0	0	.500	.457	1.042	1.042	12	13.85	11.91	12.55
4	3	3	3	.667	.667	20	12	.125	.121	.533	.533	12	12.83	12.74	12.75
				.333	.333	8	0	.125	.120	1.066	1.066	12	12.83	12.49	12.53
				.667	.667	20	12	.500	.466	.533	.533	12	12.83	12.44	12.55
				.333	.333	8	0	.500	.454	1.066	1.066	12	12.83	11.38	11.88

6	3	3	3	.667	.667	30	18	.125	.121	.537	.537	18	19.37	19.24	19.26
				.333	.333	12	0	.125	.120	1.074	1.074	18	19.37	18.87	18.92
				.667	.667	30	18	.500	.465	.537	.537	18	19.37	18.78	18.95
				.333	.333	12	0	.500	.453	1.074	1.074	18	19.37	17.21	17.97
12	1	3	3	.667	.667	12	12	.125	.121	.533	.533	24	28.25	28.00	28.03
				.333	.333	0	0	.125	.120	1.066	1.066	24	28.25	27.30	27.41
				.667	.667	12	12	.500	.466	.533	.533	24	28.25	27.12	27.45
				.333	.333	0	0	.500	.454	1.066	1.066	24	28.25	24.22	25.61
1	1	5	5	.800	.800	3	3	.03125	.03109	.291	.291	2	2	2.00	2.00
				.200	.200	0	0	.03125	.03107	1.163	1.163	2	2	1.97	1.97
				.800	.800	3	3	.500	.479	.291	.291	2	2	1.96	1.97
				.200	.200	0	0	.500	.475	1.163	1.163	2	2	1.51	1.60
4	1	5	5	.800	.800	12	12	.03125	.03107	.333	.333	8	8.58	8.57	8.57
				.200	.200	0	0	.03125	.03194	1.332	1.332	8	8.58	8.47	8.47
				.800	.800	12	12	.500	.474	.333	.333	8	8.58	8.44	8.48
				.200	.200	0	0	.500	.456	1.332	1.332	8	8.58	6.82	7.37
16	1	5	5	.800	.800	48	48	.03125	.03106	.346	.346	32	35.24	35.21	35.21
				.200	.200	0	0	.03125	.03090	1.382	1.382	32	35.24	34.79	34.80
				.800	.800	48	48	.500	.473	.346	.346	32	35.24	34.68	34.81
				.200	.200	0	0	.500	.488	1.382	1.382	32	35.24	28.00	30.59

TABLE 18.2

Parameter Combinations with N_e^W = 18 and 36 and Calculated Values for N_e^S, $\tilde{N}_{e1}^R$, and $\tilde{N}_{eL}^R$

(See text for definition of symbols)

s	d	m	f	$\overline{P}_m$	$\overline{P}_f$	R_m-T_m	R_f-T_f	h_0^2	h_1^2	$\bar{i}_m$	$\bar{i}_f$	N_e^W	N_e^S	$\tilde{N}_{e1}^R$	$\tilde{N}_{eL}^R$
9	1	27	27	.037	.037	0	0	.500	.444	2.161	2.151	18	18.28	11.44	13.70
		27	23	.037	.333	0	0	.500	.449	2.161	1.058	18	19.57	14.01	16.66
		27	27	.037	.333	0	72	.500	.447	2.161	1.087	18	18.28	13.30	15.69
		27	27	.037	.037	0	0	.125	.119	2.161	2.161	18	18.28	16.26	16.52
		3	3	.333	.333	0	0	.500	.455	1.058	1.058	18	21.05	18.07	19.07
		27	3	.037	.333	0	0	.125	.120	2.161	1.058	18	19.57	18.08	18.53
		27	27	.037	.333	0	72	.125	.120	2.161	1.087	18	18.28	16.96	17.36
		3	3	.333	.333	0	0	.125	.120	1.058	1.058	18	21.05	20.35	20.43
6	3	9	27	.037	.037	0	0	.500	.444	2.150	2.173	18	18.28	12.34	14.31
		27	27	.037	.037	12	0	.500	.441	2.173	2.173	18	18.14	12.24	14.30
		9	3	.037	.333	0	0	.500	.449	2.150	1.074	18	18.92	14.47	15.76
		9	9	.037	.333	0	36	.500	.449	2.150	1.085	18	18.43	14.17	15.42

		9	27	.037	.037	0	0	.125	.119	2.150	2.173	18	18.28	16.57	16.78
		27	27	.037	.037	12	0	.125	.119	2.173	2.173	18	18.14	16.44	16.66
		1	3	.333	.333	0	0	.500	.455	1.042	1.074	18	20.96	18.52	19.33
		3	3	.333	.333	12	0	.500	.453	1.074	1.074	18	19.37	17.21	17.97
		9	3	.037	.333	0	0	.125	.120	2.150	1.074	18	18.92	17.75	17.77
		9	9	.037	.333	0	36	.125	.120	2.150	1.085	18	18.43	17.31	17.33
		1	3	.333	.333	0	0	.125	.120	1.042	1.074	18	20.96	20.39	20.45
		3	3	.333	.333	12	0	.125	.120	1.074	1.074	18	19.37	18.87	18.92
5	9	3	27	.037	.037	0	0	.500	.444	2.143	2.180	18	18.26	12.87	14.65
		3	3	.037	.333	0	0	.500	.449	2.143	1.084	18	18.52	14.68	15.22
		3	27	.037	.037	0	0	.125	.119	2.143	2.180	18	18.26	16.72	16.92
		1	3	.333	.333	10	0	.500	.452	1.071	1.084	18	19.02	17.16	17.81
		3	3	.333	.333	40	0	.500	.451	1.084	1.084	18	18.52	16.73	17.36
		3	3	.037	.333	0	0	.125	.120	2.143	1.084	18	18.52	17.51	17.31
		1	3	.333	.333	10	0	.125	.120	1.071	1.084	18	19.02	18.58	18.63
		3	3	.333	.333	40	0	.125	.120	1.084	1.084	18	18.52	18.09	18.14
18	1	27	27	.037	.037	0	0	.500	.441	2.173	2.173	36	36.62	23.06	27.72
		27	27	.037	.037	0	0	.125	.119	2.173	2.173	36	36.62	32.64	33.15
		3	3	.333	.333	0	0	.500	.453	1.074	1.074	36	42.64	36.53	38.67
		3	3	.333	.333	0	0	.125	.120	1.074	1.074	36	42.64	41.21	41.38

for other combinations. All of this results in $\tilde{N}^R_{eL}$ being intermediate between $\tilde{N}^S_e$ and $\tilde{N}^R_{el}$ and closest to $\tilde{N}^R_{el}$ for weak selection and low heritability. Thus, $\tilde{N}^R_{eL}$ is usually larger than N^W_e for weak selection and/or low heritability but is often surprisingly similar to N^W_e for strong selection and/or high heritability.

Table 18.2 shows similar numerical values for numerous combinations most of which result in N^W_e being 18. The associated values of N^S_e vary from 18.28 up to 21.05 with the largest value associated with one dam per sire and weak selection ($P_m = P_f = .333$). The values of $\tilde{N}^R_{el}$ vary from 11.44 to 20.39 with small values for one dam per sire, high heritability ($h^2 = .5$) and strong selection ($P_m = P_f = .037$). The large value of $\tilde{N}^R_{el}$ is associated with 3 dams per sire, weak selection ($P_m = P_f = .333$) and low heritability ($h^2 = .125$). The values of $\tilde{N}^R_{eL}$ are less extreme from 13.70 to 20.45 but with the extremes occurring for the same combinations as for $\tilde{N}^R_{el}$.

Later simulation studies will study the reliability and accuracy of these calculations.

ACKNOWLEDGMENTS

This study was supported by Specific Cooperative Agreement No. 58-519B-0-931 between Purdue University and the Agricultural Research Service of the U.S. Department of Agriculture.

REFERENCES

Cochran, W.G. (1951). Improvement by means of selection. In: *Proc. Second Berkeley Symp. Math. Stat. and Prob.*, J. Neyman (ed.), 449-470.

Crow, J.F. (1954). Breeding structure of populations: II. Effective population number. Chapt. 43 of *Statistics and Mathematics in Biology*. O. Kempthorne, T.A. Bancroft, J.W. Gowan, and J.L. Lush (eds.), Ames: Iowa State College Press.

Dickerson, G.E. and Hazel, L.N. (1944). Effectiveness of selection on progeny performance as a supplement to earlier culling in livestock. *J. Agr. Res. 69*, 459-476.

Gill, J.L. (1965a). Effects of finite size on selection advance in simulated genetic populations. *Aust. J. Biol. Sci. 18*, 599-617.

Gill, J.L. (1965b). A Monte Carlo evaluation of predicted selection response, *Aust. J. Biol. Sci. 18*, 999-1007.

Gill, J.L. (1965c). Selection and linkage in simulated genetic populations. *Aust. J. Biol. Sci. 18*, 1171-1187.

Griffing, B. (1960). Theoretical consequences of truncation selection based on the individual phenotype. *Aust. J. Biol. Sci. 13*, 307-343.

Harris, Dewey L. (1982). Long-term response to selection. I. Relation to breeding population size, intensity, and accuracy with additive gene action. *Genetics 100*, 511-532.

Harter, H.L. (1961). Expected values of normal order statistics. *Biometrika 48*, 151-165.

Kempthorne, O. (1957). *An Introduction to Genetic Statistics*. New York: Wiley.

Kimura, M. and Crow, J.F. (1963). The measurement of effective population number. *Evolution 17*, 279-288.

Latter, B.D.H. (1959). Genetic sampling in a random mating population of constant size and sex ratio. *Aust. Journ. Biol. Sci. 12*, 500-505.

Malécot, G. (1948). *Les mathématiques de l'hérédité*. Paris: Masson.

Qureshi, A.W., Kempthorne, O. and Hazel, L.N. (1968). The role of finite population size and linkage in response to continued truncation selection. I. Additive gene action. *Theor. Appl. Genet. 38*, 256-263.

Robertson, A. (1961). Inbreeding in artificial selection programmes. *Genet. Res. 2*, 189-194.

Wright, S. (1931). Evolution in Mendelian populations. *Genetics 16*, 97-159.

Chapter 19

EFFECTS OF ERRORS IN PARAMETER ESTIMATES ON EFFICIENCY OF RESTRICTED GENETIC SELECTION INDICES

William G. Hill and Karin Meyer
University of Edinburgh
Edinburgh, Scotland

I. INTRODUCTION

In the standard selection index used in animal breeding (Hazel, 1943), a vector of weights $\underset{\sim}{b}$ is computed so as to maximize the correlation between the linear index, $I = \underset{\sim}{b}'\underset{\sim}{x}$ of the vector of observations, $\underset{\sim}{x}$, with total additive genetic merit or breeding value, H. The latter is customarily regarded as the sum, $H = \underset{\sim}{a}'\underset{\sim}{g}$, of economic weights $\underset{\sim}{a}$ and breeding values of individual traits $\underset{\sim}{g}$. Use of this index maximises rates of response in total merit, R_H, but no constraints are imposed on the changes occurring in any of the individual traits. In some computed indices the change in one of the traits might be in an undesired direction because of its unfavourable correlations with other traits featuring in the index; or, in other cases, any change in the trait might be undesirable because the population is assumed to be at some optimum value. To solve this problem Kempthorne and Nordskog (1959) proposed the "restricted selection index" in which changes in any traits or linear functions of traits can be restricted to zero. They gave the computing algorithm and used it in an example for poultry in which improvement in economic merit as a function of egg weight, age at first egg, egg production and egg quality (reduced blood spots) was required, but with body weight maintained constant.

Kempthorne and Nordskog's ideas have been developed subsequently and there is now a substantial literature. Tallis (1962) extended the methods to allow the relative changes in some of the traits or in linear combinations of them to take particular values, so as to achieve an "optimum genotype" using a "desired gains index". Some of the computational procedures have been generalized by Harville (1975) and Niebel (1979). For a recent review of index theory and the place of restricted indexes with further references, see James (1982).

In index calculations the variances and covariances among the observables ($\underset{\sim}{x}$) and their covariances with the breeding values (H, or its components $\underset{\sim}{g}$) are assumed to be known without error. Of course this is never the case in practice, and there has been an extensive discussion of the effects of errors in the parameter estimates on the efficiency of standard (unrestricted) indices (see James, 1982, for references), but not, as far as we know, on the restricted index. In this paper we shall discuss some of its properties when estimated from samples of data. The economic weights ($\underset{\sim}{a}$) and the restrictions required are assumed to be known without error; these may not always be realistic assumptions, but they reduce the number of variables to more manageable numbers.

II. THEORETICAL BACKGROUND

2.1 The Restricted Index

Consider a set of p traits, with known phenotypic covariance matrix (i.e., of observations $\underset{\sim}{x}$) $\underset{\sim}{P}$, known covariance matrix between observations and breeding values (i.e., $\underset{\sim}{g}$) $\underset{\sim}{G}$ and economic weights $\underset{\sim}{a}$. The optimal weights in the <u>unrestricted</u> index, $I = \underset{\sim}{b}'\underset{\sim}{x}$, are given by

$$\underset{\sim}{b} = \underset{\sim}{P}^{-1}\underset{\sim}{G}\underset{\sim}{a} \qquad (1)$$

and the response to selection in economic merit is, for one standard deviation of selection differential,

$$R_H = \rho_{IH}\sigma_H = \text{cov}(I,H)/\sigma_I = (\underset{\sim}{b}'\underset{\sim}{P}\underset{\sim}{b})^{\frac{1}{2}} \tag{2}$$

and in trait i the response is

$$R_i = \text{cov}(I,g_i)/\sigma_I = \underset{\sim}{b}'\underset{\sim}{G}_i(\underset{\sim}{b}'\underset{\sim}{P}\underset{\sim}{b})^{-\frac{1}{2}} \tag{3}$$

where $\underset{\sim}{G}_i$ is the ith column of $\underset{\sim}{G}$.

If a set of k < p <u>restrictions</u> $\underset{\sim}{C} = (\underset{\sim}{C}_1,\ldots,\underset{\sim}{C}_k)$ are imposed such that the responses $\Sigma_{i=1}^{p} c_{hi}R_i = 0$, $h = 1,\ldots,k$, then $\underset{\sim}{b}'\underset{\sim}{G}\underset{\sim}{C}_h = 0$. The index weights which satisfy these restrictions and maximise the residual correlation between I and H are given by

$$\underset{\sim}{b} = \{\underset{\sim}{I} - \underset{\sim}{P}^{-1}\underset{\sim}{G}\underset{\sim}{C}(\underset{\sim}{C}'\underset{\sim}{G}'\underset{\sim}{P}^{-1}\underset{\sim}{G}\underset{\sim}{C})^{-1}\underset{\sim}{C}'\underset{\sim}{G}'\}\underset{\sim}{P}^{-1}\underset{\sim}{G}\underset{\sim}{a} \tag{4}$$

(Kempthorne and Nordskog, 1959) where $\underset{\sim}{I}$ is the identity matrix. Cunningham, Moen and Gjedrem (1970) give an alternative derivation. The variance of index values equals their covariance with breeding value so the response in economic merit and in any individual trait are obtained by substituting (4) into (2) and (3), respectively.

2.2 Using Estimates of Parameters

If only estimates of the phenotypic and genetic covariances are available, then these have to be used to construct the index; but there will then be errors in predicting responses, in imposing the restricted responses, and a lower accuracy of the index. Let $\hat{\underset{\sim}{P}}$ and $\hat{\underset{\sim}{G}}$ be the estimates of $\underset{\sim}{P}$ and $\underset{\sim}{G}$ which are used in (1) or (4) to estimate the index weights $\hat{\underset{\sim}{b}}$. The predicted responses in economic merit or any trait are obtained by putting $\hat{\underset{\sim}{b}}$ into (2) or (3) and, of course, the <u>predicted</u> response is zero for any restricted trait or contrast. The actual responses, R_H^a and R_i^a, depend on the population parameters:

$$R_H^a = \text{Cov}(\hat{I},H)/\sigma_{\hat{I}} = \hat{\underset{\sim}{b}}'\underset{\sim}{G}\underset{\sim}{a}(\hat{\underset{\sim}{b}}'\underset{\sim}{P}\hat{\underset{\sim}{b}})^{-\frac{1}{2}} \tag{5}$$

$$R_i^a = \hat{\underset{\sim}{b}}'\underset{\sim}{G}_i(\hat{\underset{\sim}{b}}'\underset{\sim}{P}\hat{\underset{\sim}{b}})^{-\frac{1}{2}} \tag{6}$$

A Taylor series expansion (i.e. statistical differentiation) can be used to estimate the effects of sampling (e.g. Harris, 1964; Sales and Hill, 1976; Hayes and Hill, 1980). Let $\hat{\underset{\sim}{y}} = (\hat{y}_1,\ldots,\hat{y}_m) = (\hat{p}_{11},\hat{p}_{12},\ldots,\hat{p}_{nn}, \hat{g}_{11},\ldots,\hat{g}_{nn})$ denote the vector of estimates of elements of $\hat{\underset{\sim}{P}}$ and $\hat{\underset{\sim}{G}}$, with $\underset{\sim}{y}$ being the parameter values. Then, noting that the optimum response, R_H, equals R_H^a, when $\hat{\underset{\sim}{y}} = \underset{\sim}{y}$,

$$R_H^a \doteq R_H + \sum_k (\hat{y}_k - y_k) \left.\frac{\partial R_H^a}{\partial \hat{y}_k}\right|_{\hat{\underset{\sim}{y}}=\underset{\sim}{y}}$$

$$+ \frac{1}{2}\sum_{k\ell}\sum (\hat{y}_k - y_k)(\hat{y}_\ell - y_\ell) \left.\frac{\partial^2 R_H^a}{\partial \hat{y}_k \partial \hat{y}_\ell}\right|_{\hat{\underset{\sim}{y}}=\underset{\sim}{y}} \tag{7}$$

where higher order terms are ignored. Taking expectations over $\hat{\underset{\sim}{y}}$, and assuming $\hat{\underset{\sim}{y}}$ is unbiased, (7) reduces to

$$E(R_H^a) \doteq R_H + \frac{1}{2}\sum_{k\ell}\sum \operatorname{cov}(\hat{y}_k,\hat{y}_\ell) \left.\frac{\partial^2 R_H^a}{\partial \hat{y}_k \partial \hat{y}_\ell}\right|_{\hat{\underset{\sim}{y}}=\underset{\sim}{y}} \tag{8}$$

Similar expressions apply to $E(R_i^a)$. The derivatives in (7) and (8) can also be expressed in terms of the estimated index weights, for example $\partial R_H^a/\partial \hat{y}_k = (\partial R_H^a/\partial \hat{b}_i)(\partial \hat{b}_i/\partial \hat{y}_k)$.

The unrestricted index is computed by maximising the achieved response, so that $\partial R_H^a/\partial \hat{b}_i = 0$ for all i, when evaluated at the parameter values ($\hat{\underset{\sim}{y}}=\underset{\sim}{y}$). Thus the terms in first derivatives in (7) drop out and there is a quadratic relationship between the achieved response in the aggregate genotype (R_H^a) and the sample estimates ($\hat{\underset{\sim}{y}}$) about their parameter values. For the index weights, $\hat{b}_i$, and the achieved responses for individual traits, R_i^a, however, there is for most sets of parameters a linear as well as higher order relationship between them and the sample estimates, because $\partial \hat{b}_j/\partial \hat{y}_k \neq 0$ and $\partial R_i^a/\partial \hat{b}_j \neq 0$ (whether or not i = j).

The Taylor series expansion can also be used to find variances and covariances of index weights and responses. However $\mathrm{var}(R^a_H)$ in the unrestricted case is a function of fourth or higher moments (assuming third moments in $\hat{\underset{\sim}{y}}$ are negligible) because the first derivatives in (8) equal zero, and formulae are complicated. In the restricted index there is a constrained maximum to the economic response, so that $\partial R^a_H/\partial \hat{b}_i \neq 0$ for one or more values of i and $\mathrm{var}(R^a_H)$ generally depends on second moments. For index weights and thus for responses in any trait

$$\mathrm{cov}(\hat{b}_i,\hat{b}_j) \doteq \sum_{k\ell}\sum \mathrm{cov}(\hat{y}_k,\hat{y}_\ell) \left.\frac{\partial \hat{b}_i}{\partial \hat{y}_k}\frac{\partial \hat{b}_j}{\partial \hat{y}_\ell}\right|_{\hat{\underset{\sim}{y}}=\underset{\sim}{y}} \qquad (9)$$

and

$$\mathrm{cov}(R^a_i,R^a_j) \doteq \sum_{k\ell}\sum \mathrm{cov}(\hat{y}_k,\hat{y}_\ell) \left.\frac{\partial R^a_i}{\partial \hat{y}_k}\frac{\partial R^a_j}{\partial \hat{y}_\ell}\right|_{\hat{\underset{\sim}{y}}=\underset{\sim}{y}} \qquad (10)$$

$$\doteq \sum_{gh}\sum \mathrm{cov}(\hat{b}_g,\hat{b}_h) \left.\frac{\partial R^a_i}{\partial \hat{b}_g}\frac{\partial R^a_j}{\partial \hat{b}_h}\right|_{\hat{\underset{\sim}{b}}=\underset{\sim}{b}} \qquad (11)$$

using the chain rule for differentiation and ignoring higher moments.

2.3 Reparametrization

From standard multivariate theory it can be shown that any set of p observables $\underset{\sim}{x}$ and p breeding values $\underset{\sim}{g}$ can be linearly transformed to $\underset{\sim}{x}^* = \underset{\sim}{Q}\underset{\sim}{x}$, $\underset{\sim}{g}^* = \underset{\sim}{Q}\underset{\sim}{g}$ such that $\mathrm{var}(\underset{\sim}{x}^*) = \underset{\sim}{I}$ and $\mathrm{var}(\underset{\sim}{g}^*) = \mathrm{cov}(\underset{\sim}{x}^*,\underset{\sim}{g}^*) = \underset{\sim}{D}$, where $\underset{\sim}{I}$ is the identity matrix and $\underset{\sim}{D}$ a diagonal matrix (Hayes and Hill, 1980). Thus, in words, $\underset{\sim}{x}^*$ represents a set of reparametrized variables that are uncorrelated both phenotypically and genetically, each has unit phenotypic variance and with genetic variance and heritability given by the diagonal elements, d_i, of $\underset{\sim}{D}$. The corresponding economic weights are $\underset{\sim}{\alpha} = \underset{\sim}{Q}'^{-1}\underset{\sim}{a}$, and the optimum unrestricted weights are, from (1), $\underset{\sim}{\beta} = \underset{\sim}{D}\underset{\sim}{\alpha}$, or $\beta_i = d_i\alpha_i$, with variance of the index equal to $\Sigma d_i^2\alpha_i^2$. This reduces the number of parameters describing the model from the original p(p+2) to 2p (p each in $\underset{\sim}{D}$ and $\underset{\sim}{\alpha}$); and the sampling properties of the indices can be derived more easily because the variables are uncorrelated.

In the reparametrized model,

$$R_i^a = d_i\hat{\beta}_i(\Sigma\hat{\beta}_k^2)^{-\frac{1}{2}} \tag{12}$$

and the derivatives required in (11) are

$$\frac{\partial R_i^a}{\partial \hat{\beta}_i} = d_i \sum_{j\neq i} \hat{\beta}_j^2[(\Sigma\hat{\beta}_k^2)^{-3/2}]\Bigg|_{\hat{\underset{\sim}{\beta}}=\underset{\sim}{\beta}} \tag{13a}$$

and

$$\frac{\partial R_i^a}{\partial \hat{\beta}_j} = -d_i\hat{\beta}_i\hat{\beta}_j(\Sigma\hat{\beta}_k^2)^{-3/2}\Bigg|_{\hat{\underset{\sim}{\beta}}=\underset{\sim}{\beta}}, \quad i \neq j \tag{13b}$$

In the restricted case, the same reparametrization can, of course, be made. The restrictions $\underset{\sim}{b}'\underset{\sim}{G}\underset{\sim}{C} = \underset{\sim}{0}$ now become $\underset{\sim}{\beta}'\underset{\sim}{D}\underset{\sim}{Q}'^{-1}\underset{\sim}{C} = \underset{\sim}{0}$, so it is convenient to define a matrix, $\underset{\sim}{E} = \underset{\sim}{Q}'^{-1}\underset{\sim}{C}$, of the restrictions on the transformed variables. Thus the optimal weights are, from (4),

$$\underset{\sim}{\beta} = [\underset{\sim}{I} - \underset{\sim}{D}\underset{\sim}{E}(\underset{\sim}{E}'\underset{\sim}{D}^2\underset{\sim}{E})^{-1}\underset{\sim}{E}'\underset{\sim}{D}]\underset{\sim}{D}\underset{\sim}{\alpha}$$

Without loss of generality the elements of $\underset{\sim}{E}$ can be chosen such that $\sum_{j=1}^{p} e_{ij}^2 = 1$ for each i and $\sum_{j=1}^{p} e_{ij}e_{i'j} = 0$, $i\neq i'$, and the economic weight given to any restricted variable, i.e. $\sum_j e_{ij}d_j\alpha_j$, is irrelevant (Kempthorne and Nordskog, 1959). Hence, with p variables, the first restriction requires the additional definition of p-1 values of e_{1j} and allows the deletion of one value of α_j, i.e. p-2 parameters in all, the next restriction requires p-3 parameters, and so on, such that for p variables and k restrictions, the total number of parameters is 2p+k(2p-k-3)/2.

In the reparametrized model it follows from (12) that $\Sigma(R_i^a/d_i)^2 = 1$. Therefore the achieved responses fall on a p-dimensional ellipse with orthogonal axes or, if expressed relative to the heritabilities as R_i^a/d_i, on a p-dimensional sphere. With p=2 these response ellipses or circles give a useful representation of the alternative directions of response and have been used when parameter values were known (Moav and Hill, 1966). When these are estimated, it is useful to define the achieved direction of response

$$\hat{\Theta} = \tan^{-1}\{(R_2^a/d_2)/(R_1^a/d_1)\} = \tan^{-1}(\hat{\beta}_2/\hat{\beta}_1) \tag{14}$$

2.4 Sampling of Data

In the following examples observations are assumed to be multivariate normally distributed with covariance matrices as for the reparametrized model. Variances and covariances are assumed to be estimated from a multivariate analysis of variance with a single classification of s half-sib families with n individuals in each family. Monte Carlo simulation was carried out as described by Hill and Thompson (1978), and formulae for variances of mean squares in the analysis of variance are given by Hayes and Hill (1980) in terms of the intra-class correlation of half-sibs, $t_i = d_i/4$. For simplicity s and n will be assumed to be sufficiently large and t_i sufficiently small so that $s-1 \sim s$, $1-t_i \sim 1$ and $1+(n-1)t_i \sim 1+nt_i$ for all i. Thus for the unrestricted index, Hayes and Hill (1980) use the Taylor series expansion (9) to give

$$\text{cov}(\hat{\beta}_i,\hat{\beta}_j) \doteq \frac{16}{n^2}\alpha_i\alpha_j(1-t_i)(1-t_j)\{1+(n-1)t_i\}$$
$$\times \{1+(n-1)t_j\}\{\frac{1}{s-1} + \frac{1}{s(n-1)}\}$$

which reduces to, setting $d_i = 4t_i$,

$$\mathrm{cov}(\hat{\beta}_i,\hat{\beta}_j) \doteq \alpha_i\alpha_j(4/n+d_i)(4/n+d_j)/s,\ i\neq j \tag{15a}$$

and similarly, from Hayes and Hill (1980),

$$\mathrm{var}(\hat{\beta}_i) \doteq \{2\alpha_i^2(4/n+d_i)^2 + \sum_{j\neq i} \alpha_j^2(4/n+d_i)(4/n+d_j)\}/s \tag{15b}$$

To put these in perspective, note that the variance of the heritability estimate for a single trait is, to this level of approximation,

$$\mathrm{var}(\hat{d}_i) = \mathrm{var}(\hat{\beta}_i|p=1,\ \alpha_i=1) \doteq 2(4/n+d_i)^2/s$$

III. RESULTS

3.1 Unrestricted Index for Two Traits

These results are needed for subsequent comparison with the restricted index. At the optimum, $\beta_i = d_i\alpha_i$ and variances of responses can be obtained using (11), (12) and (13). Letting

$$K_u = \frac{32\alpha_1^2\alpha_2^2(d_1-d_2)^2/n^2 + (\alpha_1^2 d_1 + \alpha_2^2 d_2)^2(4/n+d_1)(4/n+d_2)}{s(\alpha_1^2 d_1^2 + \alpha_2^2 d_2^2)^3} \tag{16}$$

it can be shown that

$$\mathrm{var}(R_1^a) \doteq \alpha_2^2 d_1^2 d_2^2 K_u,\ \mathrm{var}(R_2^a) \doteq \alpha_1^2 d_1^2 d_2^2 K_u \tag{17a}$$

$$\mathrm{cov}(R_1^a,R_2^a) \doteq -\alpha_1\alpha_2 d_1^2 d_2^2 K_u,\ \mathrm{corr}(R_1^a,R_2^a) \doteq -1 \tag{17b}$$

$$\mathrm{var}(\hat{\Theta}) = (\alpha_1^2 d_1^2 + \alpha_2^2 d_2^2)K_u \tag{17c}$$

$$E(R_H^a) \doteq R_H\{1 - \tfrac{1}{2}(\alpha_1^2 d_1^2 + \alpha_2^2 d_2^2)K_u\} \tag{17d}$$

and it follows that

$$\text{var}(R_H^a) = \text{var}(\alpha_1 R_1^a + \alpha_2 R_2^a) \doteq 0 \tag{17e}$$

Note that $\text{var}(R_H^a)$ and also, for example, $\text{var}(R_1^a)$ when $\alpha_2 = 0$ depend on higher than second moments and do not strictly equal zero. Some simple examples help for illustration and for contrast with the restricted case.

3.1.1 $\underline{\alpha_1 = \alpha_2 = \alpha,\ d_1 = d_2 = d}$. Equation (16) reduces to $K_u = (4/n+d)^2/(2s\alpha^2 d^4)$ and (17a) to $\text{var}(R_i^a) = (4/n+d)^2/(2s)$. More generally, it can be shown that, for any number p of identically distributed and equally important traits: $\text{var}(R_i^a) \doteq (4/n+d)^2(1-1/p)/s$, $\text{cov}(R_i^a, R_j^a) \doteq -(4/n+d)^2/(ps)$ and $\text{corr}(R_i^a, R_j^a) \doteq -1/(p-1)$.

3.1.2 $\underline{d_1 = d_2 = d}$. If, for example, $\alpha_1 > \alpha_2$, eq. (17a) shows that $\text{var}(R_2^a) > \text{var}(R_1^a)$, i.e. the less important trait varies more. However, the variance in the direction of response, $\hat{\Theta}$ (see (14) and (17c)), is given by $\text{var}(\hat{\Theta}) \doteq (4/n+d)^2/(sd^2)$ for <u>all</u> values of α_1 and α_2.

3.1.3 $\underline{\alpha_2 = 0}$. From (16) and (17), it follows that

$$\text{var}(R_2^a) \doteq (4/n+d_1)(4/n+d_2)(d_2^2/d_1^2)/s \tag{18}$$

and that $\text{var}(R_1^a)$ and $\text{var}(R_H^a)$ depend on higher moments and

$$E(R_H^a) \doteq R_H\{1 - \text{var}(R_2^a)/(2d_2^2)\} \tag{19}$$

where $R_H = \alpha_1 d_1$. Note that the loss in expected economic response, $E(R_H^a) - R_H$, is high when the heritability of the important trait is low, as pointed out by Sales and Hill (1976).

3.2 Restricted Index for Two Traits

Let us assume the restraint to be applied is

$$e_1R_1 + e_2R_2 = 0$$

so the estimated index weights must satisfy

$$e_1(\hat{\beta}_1\hat{g}_{11} + \hat{\beta}_2\hat{g}_{12}) + e_2(\hat{\beta}_1\hat{g}_{12} + \hat{\beta}_2\hat{g}_{22}) = 0$$

where $\hat{g}_{ij}$ are estimates of the genetic covariances (on the transformed scale). It is convenient to take $\hat{\beta}_1 = e_1\hat{g}_{12} + e_2\hat{g}_{22}$ and $\hat{\beta}_2 = -e_1\hat{g}_{11} - e_2\hat{g}_{12}$ which reduce to $\beta_1 = e_2d_2$, $\beta_2 = -e_1d_1$ if the parameters are known. Using the same approximations as for (15), the variances and covariances of the index weights are

$$\text{var}(\hat{\beta}_1) \doteq \{e_1^2(4/n+d_1)(4/n+d_2) + 2e_2^2(4/n+d_2)\}/s$$

$$\text{cov}(\hat{\beta}_1,\hat{\beta}_2) \doteq -e_1e_2(4/n+d_1)(4/n+d_2)/s$$

From (11), letting

$$K_r = \frac{32e_1^2e_2^2(d_1-d_2)^2/n^2 + (e_1^2d_1+e_2^2d_2)^2(4/n+d_1)(4/n+d_2)}{s(e_1^2d_1^2 + e_2^2d_2^2)^3} \qquad (20)$$

formulas analogous to (17) are obtained:

$$\text{var}(R_1^a) \doteq e_1^2d_1^4K_r,\ \text{var}(R_2^a) \doteq e_2^2d_2^4K_r \qquad (21a)$$

$$\text{cov}(R_1^a,R_2^a) \doteq e_1e_2d_1^2d_2^2K_r,\ \text{corr}(R_1^a,R_2^a) \doteq 1 \qquad (21b)$$

$$\text{var}(\hat{\Theta}) \doteq (e_1^2d_1^2 + e_2^2d_2^2)K_r \qquad (21c)$$

and the variances of the "restrained" quantity and of economic response are

$$\text{var}(e_1R_1^a + e_2R_2^a) \doteq (e_1^2d_1^2 + e_2^2d_2^2)^2K_r \tag{21d}$$

$$\text{var}(\alpha_1R_1^a + \alpha_2R_2^a) \doteq (\alpha_1e_1d_1^2 + \alpha_2e_2d_2^2)^2K_r \tag{21e}$$

Some special cases can be considered for illustration.

3.2.1 $\underline{d_1 = d_2 = d.}$ In this case $\text{var}(\hat{\Theta}) \doteq (4/n+d)^2/(sd^2)$ as in the unrestricted index, and taking $e_1^2 + e_2^2 = 1$, $\text{var}(e_1R_1^a + e_2R_2^a) \doteq (4/n+d)^2/s$, i.e. the variance in direction and in the restricted quantity is independent of the relative magnitude of the restrictions applied to the two traits.

3.2.2 $\underline{e_1 = 0,\ e_2 = 1.}$ This corresponds to the unrestricted index with $\alpha_1 = 1$, $\alpha_2 = 0$ in that $R_2 = 0$ if the parameters are known. From (21a)

$$\text{var}(R_2^a) \doteq (4/n+d_1)(4/n+d_2)/s \tag{22}$$

and $E(R_H^a)$ is given by substitution of (22) into (19). If the restricted trait, trait 2, has a low heritability relative to trait 1, i.e. $d_2 < d_1$, the variance of the restricted trait is <u>greater</u> using the <u>restricted</u> index than using the unrestricted index, and vice versa.

3.2.3 <u>Correspondence to Unrestricted Index.</u> In the unrestricted index, if the parameters are known, $\alpha_1R_1 + \alpha_2R_2$ is maximised and the direction of response is such that $\alpha_2d_2^2R_1 - \alpha_1d_1^2R_2 = 0$. The equivalent restricted index is therefore given by $e_1 = \alpha_2d_2^2$ and $e_2 = -\alpha_1d_1^2$. Substituting these into (20) gives

$$K_{ru} = \frac{32\alpha_1^2\alpha_2^2(d_1-d_2)^2/n^2 + (\alpha_1^2d_1^2/d_2+\alpha_2^2d_2^2/d_1)^2(4/n+d_1)(4/n+d_2)}{s(\alpha_1^2d_1^2 + \alpha_2^2d_2^2)^3}$$

which can be used in (17). The special case of $\alpha_2 = 0$ (cf. $e_1 = 0$) has been mentioned previously; in general K_{ru} exceeds K_u if the trait of more economic importance has high heritability in which case the unrestricted index is more robust.

3.3 Simulation Results

In Figure 19.1 results are given for an unrestricted index with traits of equal heritability and economic weight. This illustrates the relationship between the variability of achieved responses in the individual traits and the loss of economic merit. If the economic weights were changed, the same graph would apply but with the axes rotated; e.g. for $\alpha_1 = 1$, $\alpha_2 = 0$ the axes need to be rotated 45° anticlockwise. The same pattern of results is obtained for the equivalent restricted index.

In Figure 19.2 results are given for the case of unequal heritabilities for four separate cases, placed around the ellipse: the unrestricted and restricted index with the important trait of relatively low (d = 0.2) or high (d = 0.4) heritability (see caption). This illustrates the behaviour of the two alternatives derived previously and provides, perhaps, a visual "explanation" of the results. If the trait to be changed has the higher heritability, the unrestricted index does well because changes in the direction of selection have a large effect on economic change, whereas the restricted index does poorly because changes in the direction of selection have a small effect on the response in the restricted trait.

Simulation results are given in Table 19.1 for the case of equal heritabilities and a range of restricted indices (values of e_1 for fixed e_2). These are compared with predictions in (21) based on the Taylor series expansion and it is seen that, except where

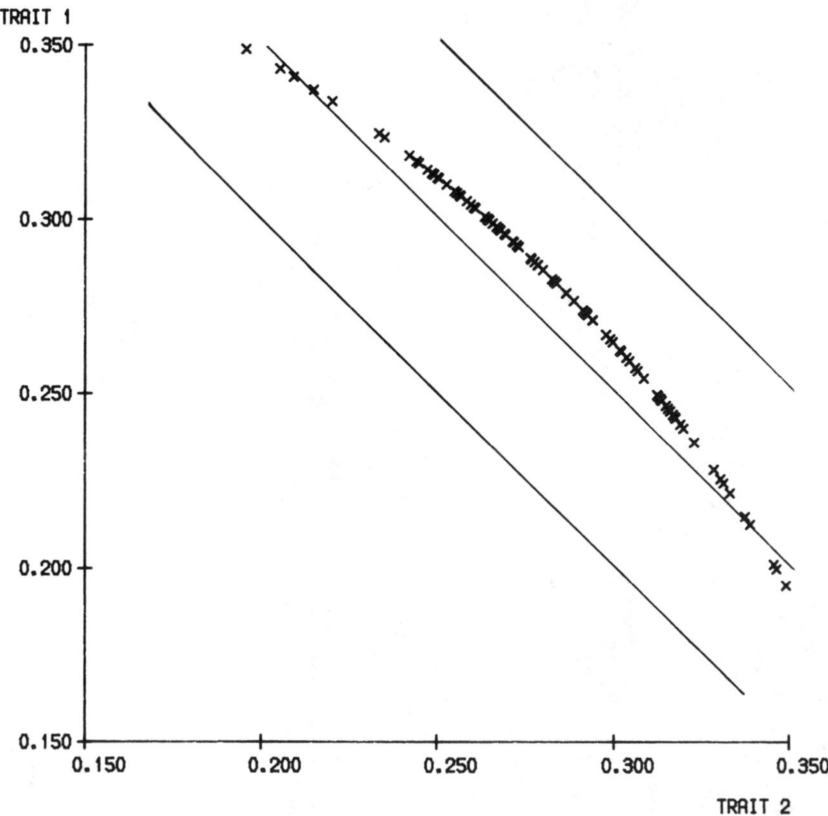

Figure 19.1 Sample Values of Achieved Responses in Trait 1 (R_1^a), Trait 2 (R_2^a) and in Economic Merit (R_H^a) Shown as Contours 0.05 Units Apart, Using Monte Carlo Simulation with $d_1 = d_2 = 0.4$, $\alpha_1 = \alpha_2 = 1$, $s = 100$, $n = 20$ and 100 Replicates

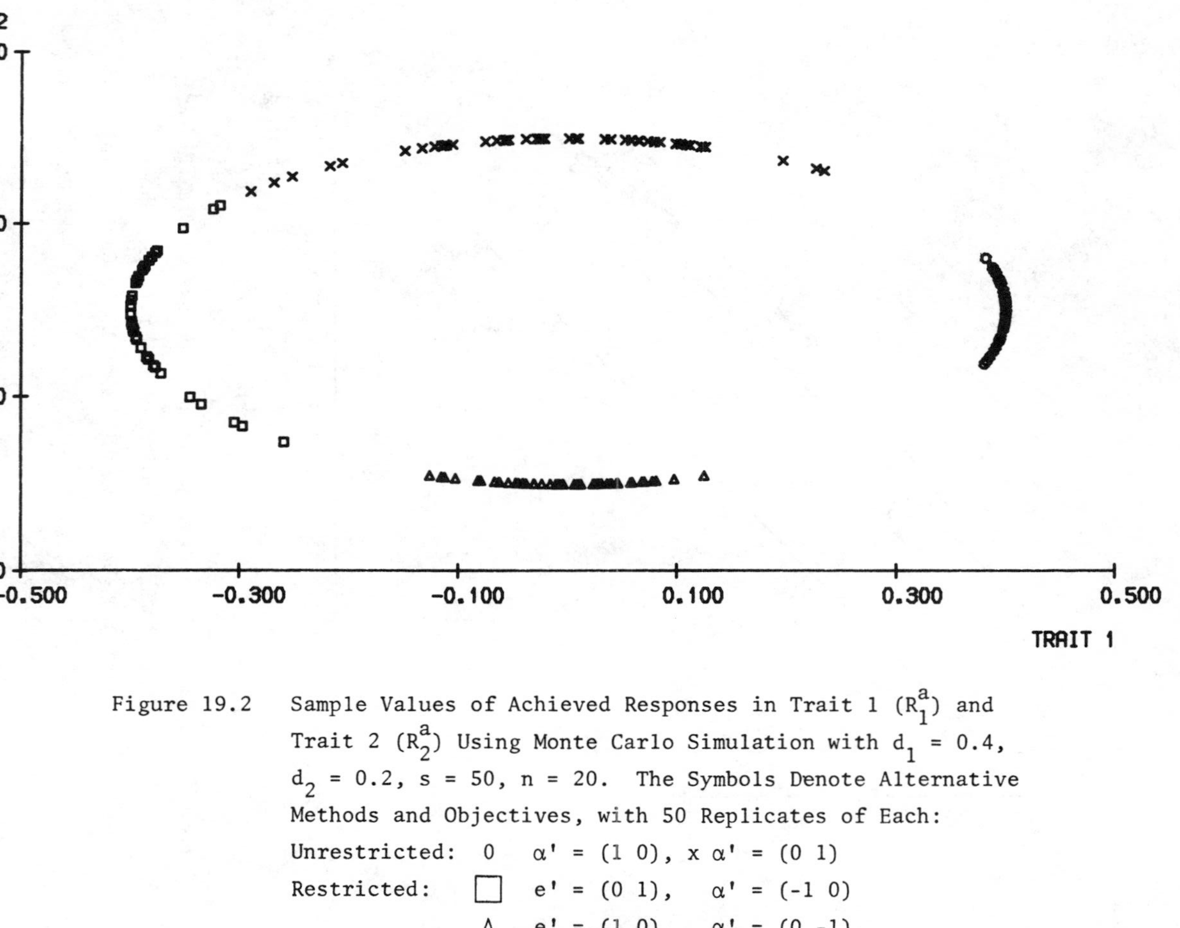

Figure 19.2 Sample Values of Achieved Responses in Trait 1 (R_1^a) and Trait 2 (R_2^a) Using Monte Carlo Simulation with $d_1 = 0.4$, $d_2 = 0.2$, $s = 50$, $n = 20$. The Symbols Denote Alternative Methods and Objectives, with 50 Replicates of Each:

Unrestricted: 0 $\alpha' = (1\ 0)$, x $\alpha' = (0\ 1)$

Restricted: □ $e' = (0\ 1)$, $\alpha' = (-1\ 0)$

Δ $e' = (1\ 0)$. $\alpha' = (0\ -1)$

TABLE 19.1

Mean (E) and Standard Deviation (SD) of Achieved Response for a Two Trait Restricted Index after Reparametrization. The Restricted Function Is $e_1R_1 + e_2R_2$, with $e_2 = 1$ Throughout. Achieved Responses Are Also Shown for Economic Response, and for the Restricted Function (See Footnote 1). $d_1 = d_2 = 0.4$. Monte Carlo Simulation, 2000 Reps, s = 100, n = 20, and Predicted Values from Taylor Series Expansion.

	R_1^a (see footnote 2)			R_2^a			R_H^a $\alpha' = (1\ 1)$		
e_1	E Sim.	SD Sim.	SD Pred.	E Sim.	SD Sim.	SD Pred.	E Sim.	SD Sim.	SD Pred.
.0	.396	.0064	.0000	-.001	.0562	.0600	.395	.0586	.0600
.05	.395	.0070	.0030	-.021	.0579	.0599	.374	.0611	.0629
.2	.389	.0129	.0118	-.079	.0565	.0588	.309	.0679	.0706
.4	.367	.0221	.0223	-.148	.0535	.0557	.219	.0749	.0800
.6	.339	.0304	.0309	-.205	.0499	.0514	.134	.0797	.0823
.8	.308	.0370	.0375	-.248	.0459	.0468	.060	.0825	.0843
.95	.286	.0410	.0413	-.273	.0429	.0435	.013	.0833	.0848

1. Simulated $SD(e_1R_1^a + e_2R_2^a) \doteq 0.058\ (1 + e_1^2)^{1/2}$ for all e_1, predicted SD = $0.06\ (1 + e_1^2)^{1/2}$.

2. $R_1^a = R_H^a$ for $\alpha' = (1\ 0)$.

TABLE

Mean (E) and Standard Deviation (SD) of Achieved Response

tion as a Function of Sample Size, $\alpha_1 = 1$, $\alpha_2 = 0$,

No. of sires (s)	25			50		
	R_1^a		R_2^a †	R_1^a		R_2^a
No. of progeny (n)	E	SD	SD	E	SD	SD
						$d_1 = d_2 =$
5	.1778	.2861	.2158	.2963	.2043	.1747
10	.3268	.1628	.1635	.3806	.0458	.1141
20	.3791	.0432	.1201	.3909	.0130	.0840
40	.3870	.0181	.0996	.3937	.0088	.0703
						$d_1 = 0.4$, $d_2 =$
5	.1638	.2633	.1264	.2402	.2157	.1181
10	.2679	.1906	.1140	.3376	.1033	.0940
20	.3435	.0888	.0923	.3749	.0373	.0672
40	.3700	.0416	.0731	.3853	.0199	.0529
						$d_1 = 0.2$, $d_2 =$
5	.0351	.1665	.2104	.0820	.1617	.1688
10	.1115	.1477	.1519	.1710	.0900	.1031
20	.1798	.0708	.1032	.1964	.0133	.0704
40	.1955	.0135	.0802	.1980	.0028	.0556

† $E(R_2^a) = 0$

19.2
for a Two Trait Restricted Index after Reparametriza-
$e_1 = 0$, $e_2 = 1$ Monte Carlo Simulation 2000 Reps.

100			200		
R_1^a		R_2^a	R_1^a		R_2^a
E	SD	SD	E	SD	SD
0.4 (R_1 = 0.4)					
.3702	.0832	.1267	.3898	.0233	.0867
.3919	.0118	.0791	.3961	.0055	.0553
.3956	.0062	.0591	.3978	.0030	.0416
.3969	.0043	.0496	.3985	.0021	.0350
0.2 (R_1 = 0.4)					
.3117	.1425	.1031	.3574	.0748	.0817
.3721	.0461	.0697	.3874	.0188	.0488
.3880	.0169	.0478	.3941	.0082	.0340
.3926	.0101	.0379	.3963	.0051	.0270
0.4 (R_1 = 0.2)					
.1379	.1371	.1210	.1832	.0689	.0818
.1943	.0317	.0703	.1985	.0021	.0489
.1985	.0022	.0490	.1993	.0010	.0343
.1990	.0013	.0389	.1995	.0007	.0273

second moment terms in the latter become very small, e.g. for var(R_1^a) when $e_1 = 0$, there is good agreement.

The restricted index with $e_1 = 0$ is investigated further in Table 19.2 for a range of values of number (s) and size (n) of families sampled in the original data set. Comparison of the cases of s = 100 and 200 show that, as predicted from (22) and (19), SD(R_a^2) and E(R_1^a) - R_1 change closely in proportion to $[(4/n+d_1)(4/n+d_2)/s]^{\frac{1}{2}}$. For smaller values of s and particularly when n is small, the agreement is much poorer. Note that values of SD(R_1^a) decline at a much more rapid rate than SD(R_2^a) as n and s increase. Table 19.2 can also be used to compare efficiencies for alternative arrangements of s and n for fixed total resources (ns). For equal heritabilities of d = 0.4 the optimal family size is n = 10, i.e. the inverse of the intra-class correlation, as for heritability estimation (Robertson, 1959). If the heritabilities are unequal, family sizes should be close to those optimal for the more lowly heritable trait.

IV. DISCUSSION

Our analysis has been in terms of the reparametrized model to permit simple formulation and simulation with limited numbers of parameters. Hill and Thompson (1978) give examples of the correspondence between values of d_1 and d_2 (there shown as $t_i = d_i/4$) and values of heritability and correlations on the untransformed scale If $d_1 = d_2$, for example, one possibility is to take $h_1^2 = h_2^2 = d_1 = d_2$ and $r_P = r_G$, with these phenotypic (r_P) and genetic (r_G) correlations taking any value $-1 < r < 1$. The restriction $R_2 = 0$, for example, on the untransformed scale corresponds to some more complicated restriction with $e_1 \neq 0$, $e_2 \neq 0$ after reparametrization.

In Table 19.3 results are presented for a model in which the traits are correlated, specifically with phenotypic equal to genetic correlations, and with restriction of no change in the second trait. Thus with, say, $r_P = r_G = 0.2$, $h_1^2 = h_2^2 = 0.4$ and $\tilde{a}' = (1,0)$, i.e. only the first trait is of economic importance, in the unrestricted index the expected response in trait 1 ± SD over replicates of data obtained with 100 sires and 20 progeny each is $(0.396 \pm 0.006)i\sigma_{P1}$ and in trait 2 it is $(0.078 \pm 0.055)i\sigma_{P2}$. In the restricted index the equivalent figures are $(0.388 \pm 0.013)i\sigma_{P1}$ and $(0.000 \pm 0.059)i\sigma_{P2}$; so although the actual change in the second trait is very variable, it will almost always be much less than if the unrestricted index is applied. If selection were practised solely for trait 1, i.e. $\tilde{b}' = (1,0)$, the actual change in trait 2 would be, in these terms, $(0,080 \pm 0.000)i\sigma_{P2}$ since no sampling of index weights would be involved. If the original data collection were based on more or less families, the SD of responses in the second trait would change by $s^{-1/2}$, approximately. With higher correlations between the traits, applying the restrictions reduces the economic response rather more, but the standard deviation of the restricted trait is little affected.

We have not yet investigated the possibilities of improving the estimates of the parameters by, for example, "bending" (Hayes and Hill, 1981). Since this can greatly improve the efficiency in the unrestricted case, it seems likely that it will do so in the restricted case. The basic assumption being made is that covariances within families are, approximately, a scalar multiple of those between families - for example that genetic and phenotypic correlations are similar.

In conclusion, we find that Kempthorne and Nordskog's proposed restricted index can be an effective procedure if based on reliable estimates of parameters. If the sample of data is small, the restricted index, like any other, should be applied with caution. In particular, application of any restriction, other than one orthogonal to the direction of maximum improvement, leads to a reduction of economic response when the genetic parameters are known. With poor

TABLE 19.3

Mean (E) and Standard Deviation (SD) of Achieved Response for Two Trait Unrestricted and Restricted Indices without Reparametrization. In each Case $\sigma_P = 1$ (Phenotypic Standard Deviation), $r_G = r_P$ (Genetic and Phenotypic Correlations). Monte Carlo Simulation, 2000 Reps, s = 100, n = 20.

	Unrestricted a' = (1 1)					Unrestricted a' = (1 0)					Restricted a' = (1 0)			
r_P	R_H	R_H^a		R_1^a (see footnote 1)		R_H	$R_H^a = R_1^a$		R_2^a		R_H	$R_H^a = R_1^a$		R_2^a (see footnote 2)
	$h_1^2 = h_2^2 = 0.4$													
		E	SD	E	SD		E	SD	E	SD		E	SD	SD
.0	.566	.560	.0076	.280	.0386	.400	.396	.0056	-.001	.0556	.400	.396	.0063	.0593
.1	.593	.588	.0081	.294	.0366	.400	.396	.0056	.039	.0553	.398	.394	.0084	.0592
.2	.620	.614	.0086	.307	.0345	.400	.396	.0056	.078	.0545	.392	.388	.0130	.0590
.4	.669	.663	.0095	.332	.0300	.400	.396	.0056	.157	.0511	.367	.363	.0236	.0586
.8	.759	.752	.0108	.376	.0183	.400	.396	.0056	.316	.0339	.240	.237	.0468	.0583

						$h_1^2 = 0.4$ $h_2^2 = 0.2$								
.0	.447	.443	.0071	.354	.0260	.400	.397	.0040	-.000	.0233	.400	.388	.0174	.0482
.1	.467	.462	.0075	.359	.0250	.400	.397	.0040	.022	.0231	.398	.386	.0195	.0482
.2	.487	.482	.0080	.363	.0239	.401	.398	.0039	.044	.0227	.391	.380	.0254	.0479
.4	.526	.521	.0087	.375	.0211	.403	.401	.0039	.090	.0210	.361	.352	.0420	.0475
.8	.639	.633	.0085	.419	.0114	.429	.427	.0036	.198	.0120	.210	.206	.0740	.0457
						$h_1^2 = 0.2$ $h_2^2 = 0.4$								
.0	.447	.443	.0067	.088	.0255	.200	.194	.0103	-.002	.0952	.200	.198	.0091	.0492
.1	.467	.462	.0071	.104	.0244	.200	.194	.0101	.042	.0947	.199	.197	.0095	.0492
.2	.487	.482	.0074	.118	.0232	.201	.195	.0099	.085	.0929	.196	.194	.0105	.0490
.4	.526	.521	.0081	.146	.0200	.203	.197	.0095	.173	.0853	.182	.181	.0138	.0486
.8	.639	.633	.0084	.214	.0102	.228	.224	.0078	.366	.0459	.112	.111	.0232	.0479

1. Values for R_2^a are the same as for the appropriate set of R_1^a

2. $E(R_2^a) \doteq 0$ in restricted index

estimation of the parameters, the actual response in the restricted variable may differ quite markedly from that intended.

REFERENCES

Cunningham, E.P., Moen, R.A. and Gjedrem, T. (1970). Restriction of selection indexes. *Biometrics 26*, 67-74.

Harris, D.L. (1964). Expected and predicted progress from index selection involving estimates of population parameters. *Biometrics 20*, 46-72.

Harville, D.A. (1975). Index selection with proportionality constraints. *Biometrics 31*, 223-225.

Hayes, J.F. and Hill, W.G. (1980). A reparametrization of a genetic selection index to locate its sampling properties. *Biometrics 36*, 237-248.

Hayes, J.F. and Hill, W.G. (1981). Modification of estimates of parameters in the construction of genetic selection indices ('Bending'). *Biometrics 37*, 483-493.

Hazel, L.N. (1943). The genetic basis for constructing selection indexes. *Genetics 28*, 476-490.

Hill, W.G. and Thompson, R. (1978). Probabilities of nonpositive definite between-group or genetic covariance matrices. *Biometrics 34*, 429-439.

James, J.W. (1982). Construction, uses and problems of multitrait selection indices. *Proc. 2nd Wld. Congr. Genet. Appl. Livestock Prodn. 5*, 130-139.

Kempthorne, O. and Nordskog, A.W. (1959). Restricted selection indices. *Biometrics 15*, 10-19.

Moav, R. and Hill, W.G. (1966). Specialised sire and dam lines. IV. Selection within lines. *Animal Production 8*, 375-390.

Niebel, E. (1979). Optimale Restriktion von Leistungsmerkmalen bei gleichzeitiger Verwendung mehrerer Selektionsindices in einer Population. *Zeitschrift für Tierzüchtung und Züchtungsbiologie 95*, 211-221.

Robertson, A. (1959). Experimental design in the estimation of genetic parameters. *Biometrics 15,* 219-226.

Sales, J. and Hill, W.G. (1976). Effect of sampling errors on efficiency of selection indices. 2. Use of information on associated traits for improvement of a single important trait. *Animal Production 23,* 1-14.

Tallis, G.M. (1962). A selection index for optimum genotype. *Biometrics 18,* 120-122.

Chapter 20

CONDITIONAL RECURRENCE RISK TO DISEASE GIVEN THE DISEASE STATUS OF BOTH PARENTS

H.J. Khamis
Wright State University
Dayton, Ohio

Klaus Hinkelmann
Virginia Polytechnic Institute
and State University
Blacksburg, Virginia

I. INTRODUCTION

In human populations, the disease status of both parents is usually known. Hence, it is useful to analyze the recurrence risk to disease of a child given the disease status of both parents. Based on information from a semi-symmetric intraclass contingency table, Norwood and Hinkelmann (1978) derived the recurrence risks to disease of a child, given that (i) both parents are diseased, (ii) one parent is diseased, and (iii) neither parent is diseased.

Khamis and Hinkelmann (1984) applied a modification of the log-linear model (e.g. Bishop, Fienberg and Holland, 1975) to the aforementioned semi-symmetric intraclass contingency table to show that for data corresponding to certain structures of association among the genetic variables, the bias in the recurrence risks is reduced by conditioning the recurrence risks on some genotypic information about the relative as well as the phenotypic information. Such "conditional recurrence risks" were derived for a propositus and a single relative by Khamis (1980).

In general, there is a wide range of choices for the added genotypic information about the two parents that one may incorporate into the conditional recurrence risk. Two special cases are considered here: (i) one of the alleles for each parent at the locus under study is known, and (ii) the genotypes of both parents at the locus under study are known. This leads to a general formula for the conditional recurrence risk to disease of a child.

Following other studies of the disease genotype association problem (see, e.g. Norwood and Hinkelmann, 1978, Banjević and Tucić, 1979, and Khamis and Hinkelmann, 1984) it is assumed, since we only study the genetic component of incidence of the disease, that an individual's disease status is not influenced by a relative's disease status or genotype when that individual's genotype is given. That is, concerning only the genetic aspect of the problem, any information about the phenotype or genotype of an individual's relative is nugatory in the presence of genotypic information about the individual in determining the likelihood that the individual is diseased. The mathematical formulation and interpretation of this assumption is given in the Appendix.

Additional genetic assumptions are as follows: (i) the disease is associated with a single locus, (ii) random mating with regard to the locus under consideration, (iii) no inbreeding, and (iv) no selection prior to onset of the disease, so that genotypic proportions are the Hardy-Weinberg proportions. As a practical matter, it is also necessary that the possible genotypes be distinguishable, and that it can be completely ascertained as to whether a person has the disease or not. Finally, we shall assume that the parents are genetically unrelated and that no environmental factors influence the incidence of disease in such a way as to cause a correlation between spouses with regard to disease incidence.

II. CONDITIONAL RECURRENCE RISKS: SPECIAL CASES

In the following derivations we shall use the notation of Norwood and Hinkelmann (1978) for allele (p_i), genotype (p_{ij}) and disease ($P_D, P_{D,ij}, P_{D(ij)}$) frequencies.

For the conditional recurrence risk (CRR) to disease of a child given the disease status of both parents, some genotypic information, I, about the parent will be included in the conditional part of the probability. Letting Y = 1 denote the event that the child is diseased and, similarly,

$$X = \begin{cases} (0,0): & \text{neither parent is diseased} \\ (0,1): & \text{one parent is diseased} \\ (1,1): & \text{both parents are diseased} \end{cases}$$

we are then interested in obtaining $\mathrm{pr}(Y=1|X=(\delta_1,\delta_2),I)$ for (δ_1,δ_2) = (1,1), (0,1), (0,0) and some specified I.

2.1 One Allele for each Parent Known:

Let $I = G_{kk}$ represent the event that each parent has one A_k-allele at the locus under study. Also, let $Y = A_iA_j$ denote the event that the child has the ordered genotype A_iA_j and $X = (A_mA_n,A_uA_v)$ that the ordered pair of parents has the ordered genotypes A_mA_n and A_uA_v. Using conditional probabilities and the assumptions of Section I we can then write the CRR for the case in which both parents are diseased as follows:

$$\begin{aligned}
&\mathrm{pr}\{Y=1|X=(1,1),G_{kk}\} \\
&\quad = \Sigma_{i,j}\ \Sigma_{m,n}\ \Sigma_{u,v}\ \mathrm{pr}\{Y=1|Y=A_iA_j,X=(1,1),X=(A_mA_n,A_uA_v),G_{kk}\} \\
&\quad \times \mathrm{pr}\{Y=A_iA_j|X=(A_mA_n,A_uA_v),X=(1,1),G_{kk}\} \\
&\quad \times \mathrm{pr}\{X=(A_mA_n,A_uA_v)|X=(1,1),G_{kk}\} \\
&\quad = \Sigma_{i,j}\ \Sigma_{m,n}\ \Sigma_{u,v}\ \mathrm{pr}(Y=1|Y=A_iA_j) \\
&\quad \times \mathrm{pr}\{Y=A_iA_j|X=(A_mA_n,A_uA_v),G_{kk}\} \\
&\quad \times \mathrm{pr}\{X=(A_mA_n,A_uA_v)|X=(1,1),G_{kk}\} \qquad (2.1)
\end{aligned}$$

Now,

$$\text{pr}\{X=(A_mA_n,A_uA_v)\,|\,X=(1,1),G_{kk}\}$$

$$= \text{pr}\{X=(1,1),G_{kk}\,|\,X=(A_mA_n,A_uA_v)\}$$

$$\times\ \text{pr}\{X=(A_mA_n,A_uA_v)\}/\text{pr}\{X=(1,1),G_{kk}\} \qquad (2.2)$$

Further,

$$\text{pr}\{X=(1,1),G_{kk}\} = P^2_{D,k\cdot} \qquad (2.3)$$

where $P_{D,k\cdot} = \Sigma_\ell\, P_{D,k\ell}$ is the probability that an individual is diseased and has the allele A_k,

$$\text{pr}\{X=(A_mA_n,A_uA_v)\} = p_mp_np_up_v \qquad (2.4)$$

and

$$\text{pr}\{X=(1,1),G_{kk}\,|\,X=(A_mA_n,A_uA_v)\}$$

$$= \begin{cases} P_{D(mn)}P_{D(uv)}\ , & (m,n,u,v)\ \varepsilon\ A^{kk} \\ 0\ , & \text{otherwise} \end{cases} \qquad (2.5)$$

where $A^{kk} = \{(m,n,u,v): m = u = k \text{ or } m = v = k \text{ or } n = u = k \text{ or } n = v = k\}$. Substituting (2.3) - (2.5) in (2.2) and then in (2.1) yields

$$\text{pr}\{Y=1\,|\,X=(1,1),G_{kk}\} = \Sigma_{i,j}\ \Sigma_{A^{kk}}\ p_mp_np_up_v\ P_{D(mn)}P_{D(uv)}P_{D(ij)}$$

$$\times\ \text{pr}\{Y=A_iA_j\,|\,X=(A_mA_n,A_uA_v\}/P^2_{D,k\cdot}$$

$$= \Sigma_{A^{kk}}\ p_mp_np_up_v\ P_{D(mn)}P_{D(uv)}P_{D(mu)}/P^2_{D,k\cdot} \qquad (2.6)$$

Note that if disease is independent of genotype, i.e., if $P_{D(ij)} = P_D$ for all i,j, then (2.6) reduces to P_D. That is, if disease is independent of genotype then knowledge of the parents' phenotypes and genotypes does not alter the recurrence risk to disease of the child from the overall probability of disease in the population. If disease is independent of allele (Norwood and Hinkelmann, 1978), however, then (2.6) can be written as follows (Khamis, 1980):

$$\mathrm{pr}\{Y=1|X=(1,1),G_{kk}\}$$

$$= \frac{1}{4p_k^2\, P_D^2}\left(P_{D,kk}\, P_D^2 + 2P_D\, \Sigma_i P_{D,ki}^2/p_i\right.$$

$$\left. + \Sigma_{i,j}\, P_{D,ki}\, P_{D,kj}\, P_{D,ij}/p_{ij}\right) \tag{2.7}$$

The reason that independencies between disease and allele is not a sufficient condition to reduce (2.6) to P_D is discussed by Khamis (1980). As a generalization of (2.6), let $G_{k\ell}$ be the event that one of the alleles for one parent is A_k and one of the alleles for the other parent is A_ℓ. Then

$$\mathrm{pr}\{Y=1|X=(1,1),G_{k\ell}\}$$

$$= \Sigma_{A^{k\ell}}\, p_m p_n p_u p_v\, P_{D(mn)}\, P_{D(uv)}\, P_{D(mu)}/P_{D,k\cdot}\, P_{D,\ell\cdot} \tag{2.8}$$

where

$A^{k\ell}$ = {(m,n,u,v): m=k and u=ℓ, or m=k and v=ℓ, or n=k and u=ℓ, or n=k and v=ℓ}

For the case in which just one parent is diseased we have

$$\mathrm{pr}\{Y=1|X=(0,1),G_{k\ell}\} = \Sigma_{i,j}\, \Sigma_{m,n}\, \Sigma_{u,v}\, \mathrm{pr}(Y=1|Y=A_iA_j)$$

$$\times\, \mathrm{pr}\{Y=A_iA_j\,|X=(A_mA_n,A_uA_v),G_{k\ell}\}$$

$$\times\, \mathrm{pr}\{X=(A_mA_n,A_uA_v)\,|X=(0,1),G_{k\ell}\} \tag{2.9}$$

Following an argument similar to that leading from (2.1) to (2.6), using the fact that

$$\mathrm{pr}\{X=(0,1),G_{k\ell}\} = P_{D,k\cdot}\,\overline{P}_{D,\ell\cdot} + P_{D,\ell\cdot}\,\overline{P}_{D,k\cdot}$$

where

$$\overline{P}_{D,k\cdot} = \Sigma_m\, p_m p_k (1-P_{D(mk)}) = p_k - P_{D,k\cdot}$$

is the probability that an individual has an A_k-allele and is not affected by the disease, and

$$\mathrm{pr}\{X=(0,1),G_{k\ell}|X=(A_mA_n,A_uA_v)\}$$

$$= \begin{cases} (1-P_{D(mn)})P_{D(uv)} + (1-P_{D(uv)})P_{D(mn)} & \text{if } (m,n,u,v) \in A^{k\ell} \\ 0 & \text{otherwise} \end{cases}$$

we obtain

$$\mathrm{pr}\{Y=1|X=(0,1),G_{k\ell}\} = 2\,\Sigma_{A^{k\ell}}\, p_m p_n p_u p_v\,(1-P_{D(mn)})$$

$$\times\, P_{D(uv)}\,P_{D(mu)}/(P_{D,k\cdot}\,\overline{P}_{D,\ell\cdot} + P_{D,\ell\cdot}\,\overline{P}_{D,k\cdot}) \qquad (2.10)$$

Finally, for the case in which neither parent is affected by the disease we derive, analogously to (2.6) and (2.10)

$$\mathrm{pr}\{Y=1|X=(0,0),G_{k\ell}\} = \Sigma_{A^{k\ell}}\, p_m p_n p_u p_v$$

$$\times\, (1-P_{D(mn)})(1-P_{D(uv)})P_{D(mu)}/\overline{P}_{D,k\cdot}\,\overline{P}_{D,\ell\cdot} \qquad (2.11)$$

2.2 Genotype for each Parent Known:

If the genotype of both parents is known then the genotype array of the offspring is known. With the assumption given in Section I that the disease status of the child is not influenced by the parents' disease status when the child's genotype is known it follows then that the CRR is identical for all three cases considered in Section 2.1. It is easy to see that, in fact, we have for $I = G_{mn,uv}$ = {the first parent has genotype A_mA_n and the second parent has genotype A_uA_v}

$$\mathrm{pr}\{Y=1|X=(\delta_1,\delta_2),G_{mn,uv}\}$$

$$= \frac{1}{4}(P_{D(mu)} + P_{D(mv)} + P_{D(nu)} + P_{D(nv)}) \qquad (2.12)$$

for $(\delta_1,\delta_2) = (1,1),\ (0,1),\ (0,0)$.

III. CONDITIONAL RECURRENCE RISKS: GENERAL CASE

The CRR formulas of Section II can be rewritten in a somewhat more condensed form which, at the same time, leads to CRR formulas for the general case of genotypic information I on the parents. Let $I = G^*$ denote the event of interest that relates to the parent genotype combination, e.g., $G^* = G_{k\ell}$ or $G^* = G_{mn,uv}$. Let $A^* = \{(m,n,u,v)$: $\{X = (A_mA_n,A_uA_v)\}$ is an occurrence of $G^*\}$ and

$$Q(m,n;u,v) = \frac{1}{4}\, p_m p_n p_u p_v\ (P_{D(mu)} + P_{D(mv)} + P_{D(nu)} + P_{D(nv)})$$

We can then write the CRR for the three cases of phenotypic information on the parents as follows:

$$\mathrm{pr}\{Y=1|X=(1,1),G^*\}$$

$$= \Sigma_{A^*}\ Q(m,n;u,v)P_{D(mn)}\ P_{D(uv)}/\mathrm{pr}\{X=(1,1),G^*\} \qquad (3.1)$$

$$\mathrm{pr}\{Y=1|X=(0,1),G^*\} = \Sigma_{A^*}\ Q(m,n;u,v)$$

$$\times\ \{P_{D(mn)}(1-P_{D(uv)}) + P_{D(uv)}(1-P_{D(mn)})\}/\mathrm{pr}\{X=(0,1),G^*\} \qquad (3.2)$$

and

$$\text{pr}\{Y=1|X=(0,0),G^*\}$$

$$= \Sigma_{A^*} Q(m,n;u,v)(1-P_{D(mn)})(1-P_{D(uv)})/\text{pr}\{X=(0,0),G^*\} \tag{3.3}$$

If G* is symmetric with respect to the genotypes of the two parents; i.e., if

$$\Sigma_{A^*} Q(m,n;u,v)\ P_{D(mn)}(1-P_{D(uv)}) = \Sigma_{A^*} Q(m,n;u,v)\ P_{D(uv)}(1-P_{D(mn)})$$

we then have, for $(\delta_1,\delta_2) = (1,1),\ (0,1),\ (0,0)$,

$$\text{pr}\{Y=1|X=(\delta_1,\delta_2),G^*\} = \Sigma_{A^*} Q(m,n;u,v)$$

$$\times \{(3\ P_{D(uv)} - 2)P_{D(mn)}{}^{\delta_1} - (1 - P_{D(uv)})(1 - 3\ P_{D(mn)})^{\delta_2}$$

$$+ (1 - P_{D(mn)})(1 - P_{D(uv)})\}/\text{pr}\{X=(\delta_1,\delta_2),G^*\} \tag{3.4}$$

IV. COMPARISON OF RECURRENCE RISK AND CONDITIONAL RECURRENCE RISK

For the case of one relative and a propositus Khamis and Hinkelmann (1984) have shown that for highly structured data it is more appropriate to consider CRR rather than recurrence risk (RR). The same can be said for the situation considered in this paper. It is informative to compare the expressions for CRR and RR.

Formulas for RR when the disease status of both parents is known were given by Norwood and Hinkelmann (1978) as

$$\text{pr}\{Y=1|X=(1,1)\} = B/P_D^2 \tag{4.1}$$

$$\text{pr}\{Y=1|X=(0,1)\}=\{\rho_A'\ P_D(1 - P_D) + 2\ P_D^2 - 2\ B\}/2\ P_D(1 - P_D) \tag{4.2}$$

and

$$\mathrm{pr}\{Y=1|X=(0,0)\}=\{P_D(1 - 2P_D) - \rho_A' \, P_D(1 - P_D) + B\}/(1 - P_D)^2 \qquad (4.3)$$

where

$$B = \Sigma_{i,j} \, \Sigma_{k,\ell} \, p_i p_j p_k p_\ell \, P_{D(ij)} \, P_{D(k\ell)} \, P_{D(ik)}$$

and ρ_A' is the coefficient of association between disease and allele (Norwood and Hinkelmann, 1978). We note that if I = G* is the universe of parent genotype combinations then (3.1), (3.2) and (3.3) yield (4.1), (4.2) and (4.3), respectively.

If there is no association between disease and genotype (ρ_G = 0, Norwood and Hinkelmann, 1978), or between disease and allele (ρ_A' = 0), then the probability that the child has the disease becomes the overall probability of disease, P_D, regardless of the disease status of the parents, i.e. all three RR (4.1), (4.2) and (4.3) reduce to P_D. We also have the monotonic relationship (Norwood, 1974).

$$\mathrm{pr}\{Y=1|X=(0,0)\} \leq \mathrm{pr}\{Y=1|X=(0,1)\} \leq \mathrm{pr}\{Y=1|X=(1,1)\} \qquad (4.4)$$

Computations show that these inequalities do not necessarily hold for all G* and other parameters for the CRR. Counter examples are given in Table 20.1 together with a comparison of RR and CRR, with I = G_{11}, for the two-allele case and selected values of $P_{D(11)}$, $P_{D(12)}$, $P_{D(22)}$ and p_1. The monotonicity property does hold when G* is a singleton, i.e. when the genotypes of both parents are known.

As an example of how a nontrivial event, G*, interacts with the disease status of the parents to influence the CRR of a child, Y, we consider the two-allele case with G* representing the event that the parent genotype combination is (A_1A_1,A_1A_1) or (A_1A_1,A_1A_2); i.e., G* = $\{(A_1A_1,A_1A_1)$ or $(A_1A_1,A_1A_2)\}$. Then

TABLE 20.1

Recurrence Risk and Conditional Recurrence Risk for $I = G_{11}$ When the Disease Status of Both Parents Is Given

$P_{D(11)}$	$P_{D(12)}$	$P_{D(22)}$	P_1	P_D	RR† X = (1,1)	RR† X = (0,1)	RR† X = (0,0)	CRR X = (1,1)	CRR X = (0,1)	CRR X = (0,0)
** 0.01	0.07	0.13	0.3	0.094	0.110	0.101	0.092	0.067	0.059	0.051
**			0.5	0.070	0.096	0.082	0.068	0.063	0.051	0.039
**			0.7	0.046	0.079	0.062	0.044	0.055	0.041	0.027
** 0.04	0.07	0.10	0.3	0.082	0.087	0.084	0.082	0.064	0.062	0.061
**			0.5	0.070	0.076	0.073	0.070	0.059	0.057	0.055
**			0.7	0.058	0.065	0.061	0.058	0.053	0.051	0.049
0.01	0.07	0.01	0.3	0.035	0.040	0.039	0.035	0.040	0.039	0.037
*			0.5	0.040	0.040	0.040	0.040	0.040	0.038	0.032
			0.7	0.035	0.040	0.039	0.035	0.038	0.035	0.025
0.01	0.13	0.01	0.3	0.060	0.070	0.068	0.059	0.070	0.069	0.064
*			0.5	0.070	0.070	0.070	0.070	0.070	0.068	0.053
			0.7	0.060	0.070	0.068	0.059	0.069	0.063	0.038
0.01	0.04	0.16	0.3	0.096	0.138	0.113	0.092	0.055	0.048	0.042
			0.5	0.063	0.116	0.083	0.060	0.048	0.038	0.030
			0.7	0.036	0.084	0.054	0.035	0.038	0.028	0.021
0.04	0.04	0.13	0.3	0.084	0.106	0.093	0.082	0.051	0.051	0.051
			0.5	0.063	0.082	0.070	0.061	0.046	0.046	0.046
			0.7	0.048	0.056	0.051	0.048	0.042	0.042	0.042

	0.07	0.13	0.10	0.3	0.110	0.110	0.110	0.110	0.104	0.102	0.101
				0.5	0.108	0.108	0.108	0.107	0.099	0.097	0.094
				0.7	0.098	0.102	0.100	0.097	0.092	0.089	0.085
***	0.07	0.01	0.13	0.3	0.074	0.105	0.085	0.072	0.058	0.049	0.050
				0.5	0.055	0.067	0.059	0.055	0.063	0.054	0.051
				0.7	0.050	0.050	0.050	0.050	0.067	0.060	0.056
***	0.04	0.01	0.10	0.3	0.057	0.083	0.067	0.056	0.033	0.032	0.034
***				0.5	0.040	0.055	0.045	0.040	0.035	0.032	0.033
				0.7	0.033	0.033	0.033	0.033	0.037	0.035	0.034
	0.07	0.04	0.10	0.3	0.072	0.079	0.075	0.072	0.060	0.060	0.060
				0.5	0.063	0.065	0.063	0.062	0.062	0.061	0.061
				0.7	0.060	0.060	0.060	0.060	0.065	0.064	0.063
	0.10	0.10	0.13	0.3	0.115	0.116	0.115	0.114	0.104	0.140	0.104
				0.5	0.107	0.109	0.108	0.107	0.102	0.102	0.102
				0.7	0.103	0.103	0.103	0.103	0.101	0.101	0.101
	0.01	0.01	0.07	0.3	0.039	0.061	0.048	0.039	0.017	0.017	0.017
				0.5	0.025	0.048	0.034	0.025	0.014	0.014	0.014
				0.7	0.015	0.028	0.020	0.015	0.011	0.011	0.011

† Norwood (1974)

* $\rho'_A = 0$

** $\rho'_I = 0$ ($\rho'_I = \rho_G - \rho'_A$, see Norwood and Hinkelmann, 1978)

*** violation of monotonicity property in CRR

$$\mathrm{pr}\{Y=1|X=(\delta_1,\delta_2),G^*\} = \frac{\mathrm{pr}\{X=(\delta_1,\delta_2),X=(A_1A_1,A_1A_1)\}}{\mathrm{pr}\{X=(\delta_1,\delta_2),G^*\}} P_{D(11)}$$

$$+ \frac{\mathrm{pr}\{X=(\delta_1,\delta_2),X=(A_1A_1,A_1A_2)\}}{\mathrm{pr}\{X=(\delta_1,\delta_2),G^*\}} (P_{D(11)}+P_{D(12)})/2 \qquad (4.5)$$

On the other hand,

$$\mathrm{pr}\{Y=1|G^*\} = \frac{\mathrm{pr}\{X=(A_1A_1,A_1A_1)\}}{\mathrm{pr}(G^*)} P_{D(11)}$$

$$+ \frac{\mathrm{pr}\{X=(A_1A_1,A_1A_2)\}}{\mathrm{pr}(G^*)} (P_{D(11)}+P_{D(12)})/2 \qquad (4.6)$$

In both cases the conditional probabilities (4.5) and (4.6) are convex linear combinations of $P_{D(11)}$ and $P_{D(11)} + P_{D(12)}$. Whereas the weights in (4.5) involve the disease status of the parents as well as the genotypic information about the parents, the weights in (4.6) involve only the genotypic information about the parents. Also, the RR (4.1) - (4.3) are quite different from (4.5) and (4.6); in fact, they include terms involving $P_{D(22)}$ not contained in (4.5) and (4.6).

V. REMARKS

(i) There is a great deal of flexibility in the proposed CRR formula. The event G* may be any event relating to the genotypes of the two parents, just so long as it is commensurate with the complexity of the data, as indicated in Khamis and Hinkelmann (1984). Hence, if the structure of association among the genetic variables is simple, we may set G* = U, where U is the universe of genotype combinations for the two parents -- thereby reducing the CRR formulas to the RR formulas. If the structure of association among the genetic variables is complex, then we use $G^* = G_{k\ell}$ for various values of k and

ℓ, or use G* = {$X=(A_mA_n, A_uA_v)$} for various m, n, u, and v. Other choices of G* are possible; e.g., G* = {both parents are homozygous}, G* = {one parent is homozygous while the other is heterozygous}, etc. In general, the choice of G* should be based on the complexity of the structure of association among the genetic variables, the experimenter's research interests, and computational convenience.

(ii) The CRR formulas presented here can be used for any number of alleles at a locus, $s \geq 2$.

(iii) The CRR formulas can also be extended to accommodate a range of disease categories, say $k = 1,2,\ldots,\ell$. In this case we would have

$$(\delta_1,\delta_2) = (1,1),(1,2),\ldots,(1,\ell),\ (2,2),(2,3),\ldots,(2,\ell),\ldots,(\ell,\ell)$$

(iv) The case in which the disease is related to more than one locus (multiple-loci case) will be the subject of a future paper.

REFERENCES

Banjević, D. and Tucić, N. (1979). The monofactorial model for inheritance of liability to disease and its implications for relatives at risk. *Genetika 11 (3):* 221-230.

Bishop, Y.M.M., Fienberg, S.E. and Holland, P.W. (1975). *Discrete Multivariate Analysis.* Cambridge: The MIT Press.

Khamis, H.J. (1980). Log-linear Model Analysis of the Association between Disease and Genotype. Unpublished Ph.D. dissertation. Virginia Polytechnic Institute and State University.

Khamis, H.J. and Hinkelmann, K.H. (1984). Log-linear model analysis of the association between disease and genotype. *Biometrics.* To appear.

Norwood, P.K. (1974). Statistical Analysis of Association between Disease and Genotype. Unpublished Ph.D. dissertation. Virginia Polytechnic Institute and State University

Norwood, P.K. and Hinkelmann, K.H. (1978). Measures of association between disease and genotype. *Biometrics 34:* 593-602.

APPENDIX

We consider here the case of a single relative, X, of the propositus, Y, and use the notation $\{X=k\}$ to represent the event that X has disease status k, $k=1,2,\ldots,\ell$. The following theorem can be extended to accommodate the disease status of both parents.

Let M_{ijk} denote the total number of propositi, Y, having genotype A_iA_j and having a relative, X, with disease status k and let m_{ijk} denote the number of diseased propositi having genotype A_iA_j and having relative, X, with disease status k $(i,j = 1,2,\ldots,s;$ $k = 1,2,\ldots,\ell)$. We then state

<u>Theorem</u>. In the disease-genotype problem, in order for

$$\mathrm{pr}\{Y=1|Y=A_iA_j,X=k\} = \mathrm{pr}\{Y=1|Y=A_iA_j\} \tag{A1}$$

it is necessary and sufficient that m_{ijk}/M_{ijk} is independent of k for all $i,j = 1,2,\ldots,s$; $k = 1,2,\ldots,\ell$.

<u>Proof</u>. By the definition of conditional probability,

$$\mathrm{pr}\{Y=1|Y=A_iA_j,X=k\} = \frac{m_{ijk}}{M_{ijk}} \tag{A2}$$

and

$$\mathrm{pr}\{Y=1|Y=A_iA_j\} = \frac{m_{ij\cdot}}{M_{ij\cdot}} \tag{A3}$$

where $m_{ij:} = \Sigma_k\, m_{ijk}$, $M_{ij\cdot} = \Sigma_k M_{ijk}$.

(i) If (A1) holds then it follows from (A2) and (A3) that $m_{ijk}/M_{ijk} = m_{ij\cdot}/M_{ij\cdot}$ and m_{ijk}/M_{ijk} is independent of k.

(ii) If m_{ijk}/M_{ijk} is independent of k then

$$\frac{m_{ij1}}{M_{ij1}} = \frac{m_{ij2}}{M_{ij2}} = \cdots = \frac{m_{ij\ell}}{M_{ij\ell}} \tag{A4}$$

For $\ell = 2$, $m_{ij1}/M_{ij1} = m_{ij2}/M_{ij2}$ implies $m_{ijk}/M_{ijk} = m_{ij\cdot}/M_{ij\cdot}$ $(k = 1,2,)$. Using an inductive proof assume then that

$$\frac{m_{ij1}}{M_{ij1}} = \frac{m_{ij2}}{M_{ij2}} = \cdots = \frac{m_{ij,\ell-1}}{M_{ij,\ell-1}}$$

implies that

$$\frac{m_{ijk}}{M_{ijk}} = \frac{\Sigma_{r=1}^{\ell-1}\, m_{ijr}}{\Sigma_{r=1}^{\ell-1}\, M_{ijr}} \quad (k = 1,2,\ldots,\ell-1) \tag{A5}$$

Now suppose (A4) holds. From (A5) we have

$$\frac{\Sigma_{r=1}^{\ell-1}\, m_{ijr}}{\Sigma_{r=1}^{\ell-1}\, M_{ijr}} = \frac{m_{ij\ell}}{M_{ij\ell}}$$

or

$$M_{ij\ell}\, \Sigma_{r=1}^{\ell-1}\, m_{ijr} = m_{ij\ell}\, \Sigma_{r=1}^{\ell-1}\, M_{ijr} \tag{A6}$$

Adding $m_{ij\ell}\, M_{ij\ell}$ on both sides of (A6) yields

$$M_{ij\ell}\, m_{ij\cdot} = m_{ij\ell}\, M_{ij\cdot}$$

and hence $m_{ijk}/M_{ijk} = m_{ij\cdot}/M_{ij\cdot}$ for $k = 1,2,\ldots,\ell$, i.e. (A1) is true.

The condition that m_{ijk}/M_{ijk} is independent of k is consistent with the assumptions of the disease-genotype association problem. The problem under study only concerns the genetic component of disease, or the genetic contribution to the conditional probability that an individual is diseased. Hence, it is appropriate that the ratio, m_{ijk}/M_{ijk}, of diseased individuals be a function only of the genotype, A_iA_j, not of the disease status, k, of the relative.

Chapter 21

THE EWENS SAMPLING FORMULA IN A POPULATION THAT VARIES IN SIZE*

Edward Pollak
Iowa State University
Ames, Iowa

I. INTRODUCTION

During the last two decades electrophoretic analysis of many loci in several organisms has revealed an enormous amount of genetic variability. It is not surprising that there should be much potential variability at a locus coding for a complex protein, which is made up of many amino acids, because each of the codons that transmit a message to insert one amino acid could possibly mutate.

In view of these facts, a sensible and relatively simple way to model the situation mathematically is to assume that there is an infinite number of possible alleles at a locus. In addition, it is generally assumed, as a way to make the mathematics manageable, that u is the probability that any of these alleles mutates every time a generation of adults produces gametes.

Now the potential for a great deal of variability does not necessarily imply that a particular small sample from a population

*Journal Paper No. J-10990 of the Iowa Agriculture and Home Economics Experiment Station, Ames, Iowa. Project 2588.

will contain several alleles at substantial frequencies. If a sample of such a structure is found, a question arises about whether the presence of some fairly frequent alleles is due to selection or is merely what is to be expected from random sampling. Kimura (1968) and King and Jukes (1969) have argued that the bulk of variability at the molecular level is associated with alleles that are neutral or nearly so. I refer the reader to Kimura and Ohta (1971), Ohta and Kimura (1971), Crow (1972) and Stebbins and Lewontin (1972) for the presentation of various points of view associated with the ensuing controversy.

Ewens (1972) has originated a theory that can serve as a foundation for testing whether observed patterns of variability are consistent with neutrality. Previous discussions of this subject have been based upon the assumption of fixed finite population sizes. My purpose in this paper is to indicate how the theory may be extended to populations that vary in size.

II. EWENS SAMPLING FORMULA

Data on variability in samples are of a type that one might call configurational. That is to say, we can envisage that a sample of r genes is taken from among the 2N that were present in the zygotes and that it contains k alleles, which are respectively represented $n_1, n_2, \ldots, n_k$ times in the sample. In the remainder of this paper I shall refer to the collection $\{n_1, n_2, \ldots, n_k\}$ as a configuration, where it is to be assumed that $n_1 \geq n_2 \geq \ldots \geq n_k$.

Ewens (1972) made a very important contribution to an understanding of the resulting sampling theory for populations at equilibrium. He showed that the probability that a random sample of size r contains the configuration $\{n_1, n_2, \ldots, n_k\}$ is equal to

$$P(r;k;n_1,\ldots,n_k) = \frac{r!}{n_1 n_2 \ldots n_k \alpha_1! \ldots \alpha_k!} \frac{\theta^k}{\theta(\theta+1)\ldots(\theta+r-1)} \tag{1}$$

where α_j is the number of frequencies $n_1, \ldots, n_k$ that are equal to j,

$\theta = 4Nu$ and N is the population size. Kingman (1980) calls this expression the Ewens sampling formula. The derivation of Ewens (1972) was partially heuristic. Karlin and McGregor (1972) were the first to give a rigorous proof of (1), by using a combinatorial argument.

III. OBJECT OF THIS PAPER

To the best of my knowledge, all previous authors who have written on (1) and consequences of it have only given proofs that apply to populations that do not change in size from generation to generation. However, Watterson (1976) has given a derivation of (1) that is based upon the diffusion approximation. Thus, in view of the increasing evidence of the robustness of this approximation, as, for example, that provided by Ethier and Nagylaki (1980), we are led to hope that the Ewens sampling formula is very generally true.

However, this still leaves open the question of what is the proper effective population number N_e to substitute for N in (1). The object of this paper is to answer this question if there is a monoecious population of varying size that reproduces in each generation according to the Wright-Fisher model. Following Karlin (1968) and Chia and Pollak (1974), we shall assume that the possible population sizes are the finite numbers $N^{(1)},\ldots,N^{(s)}$, and that the sequence of sizes $\{N_t,\ t=0,1,\ldots\}$ is a finite irreducible Markov chain. The associated transition matrix is $C = [c_{iv}]$, where

$$c_{iv} = P[N_{t+1} = N^{(v)} | N_t = N^{(i)}] \tag{2}$$

Our method of proof will be to generalize a combinatorial argument that is sketched by Ewens (1979). It is of interest to note that if u is set equal to zero, this argument generates probabilities of configurations in which n_i alleles are identical by descent rather than identical in allelic state. Moreover, the same combinatorial argument has been sketched by Kempthorne (1967) in this special case.

The method relies on the derivation of two equations relating configuration probabilities. One of them relates such probabilities at time t+1 to their analogues at time t. This is also part of the reasoning of Karlin and McGregor (1972), who do not explicitly present this expression, but state that its derivation is similar to, but simpler than, that of another one needed in their proof. They explicitly derive the latter.

The second equation applies to a set of configuration probabilities for one generation. It is derived by the device of considering a sample of size r+1 to be made up of two parts, the first being a subsample of size r and the second being a single observation that follows it.

IV. DERIVATION OF THE TWO FUNDAMENTAL EQUATIONS

We shall first derive a recurrence equation that relates the probabilities of configurations in generations t and t+1. We assume that the Wright-Fisher model holds. Thus if $N_t = N^{(i)}$ and $N_{t+1} = N^{(v)}$, the $2N^{(v)}$ genes among the zygotes of the offspring generation are obtained by repeated sampling with replacement of the $2N^{(i)}$ genes of the parent generation. We also assume that as each gene is passed from the parent to the offspring it has a probability u of being a mutant to a type that did not previously exist in the population. Then if a sample of r+1 genes is taken from among the offspring and we denote by $q(r+1, m|N_t = N^{(i)}, N_{t+1} = N^{(v)})$ the probability that this sample was transmitted from exactly m distinct parent genes,

$$q(r+1, r+1|N_t = N^{(i)}, N_{t+1} = N^{(v)})$$

$$= \frac{2N^{(i)}(2N^{(i)}-1)\ldots(2N^{(i)}-r)}{[2N^{(i)}]^{r+1}} = 1 - \frac{r(r+1)}{4N^{(i)}} + O((N^{(i)})^{-2}) \quad (3)$$

and

$$q(r+1,r|N_t = N^{(i)}, N_{t+1} = N^{(v)})$$

$$= {}_{r+1}C_2 \frac{2N^{(i)}(2N^{(i)}-1)\ldots(2N^{(i)}-r+1)}{[2N^{(i)}]^{r+1}}$$

$$= \frac{r(r+1)}{4N^{(i)}} + O((N^{(i)})^{-2}) \tag{4}$$

as $N^{(i)} \to \infty$. We obtain expression (4) because ${}_{r+1}C_2$ is the number of ways to choose 2 genes from among r+1 that were derived from the same parent and $2N^{(i)}(2N^{(i)}-1)\ldots(2N^{(i)}-r+1)$ is the number of ways to select r distinct parental genes and a specified repeated parent from among $2N^{(i)}$.

Now let $P_{t+1}(r+1;k;n_1,\ldots,n_k|m,N^{(i)},N^{(v)})$ and $P_{t+1}(r+1;k;n_1,\ldots,n_k|N^{(i)},N^{(v)})$ respectively denote conditional probabilities of the offspring configuration $\{n_1,\ldots,n_k\}$, given $m,N^{(i)},N^{(v)}$ and $N^{(i)},N^{(v)}$. It then follows from (3) and (4) that

$$P_{t+1}(r+1;k;n_1,\ldots,n_k|N^{(i)},N^{(v)})$$

$$\doteq \{1 - \frac{r(r+1)}{4N^{(i)}}\} P_{t+1}(r+1;k;n_1,\ldots,n_k|r+1,N^{(i)},N^{(v)})$$

$$+ \frac{r(r+1)}{4N^{(i)}} P_{t+1}(r+1;k;n_1,\ldots,n_k|r,N^{(i)},N^{(v)}) \tag{5}$$

if the possible population sizes are all large.

We shall now calculate the probabilities on the right side of (5) if $\alpha_1 = 0$. In this case the r+1 offspring genes must be unmutated copies of the parental genes. Thus

$$P_{t+1}(r+1;k;n_1,\ldots,n_k|r+1,N^{(i)},N^{(v)})$$

$$= (1-u)^{r+1}P_t(r+1;k;n_1,\ldots,n_k|N^{(i)}) \tag{6}$$

where the configuration probability on the right-hand side is conditional on $N^{(i)}$ alone. The second conditional configuration probability of $\{n_1,\ldots,n_k\}$ in (5) is the probability of an event that can occur in several mutually exclusive ways. Each of these is associated with r parental genes, one of which is used as a parent twice. One way is to have the parental configuration $\{n_1,\ldots,n_{j-1}, n_j-1, n_{j+1},\ldots,n_k\}$. The probability of this is $(\alpha(n_j))^{-1}$ times as large as $P(r;k;n_1,\ldots,n_{j-1}, n_j-1, n_{j+1},\ldots,n_k)$ if there are $\alpha(n_j)$ offspring genes in a sample of size r+1 that are represented n_j times. Now if the offspring genes represented n_j-1 times are $\alpha(n_j-1)$ in number, there are $\alpha(n_j-1)+1$ of such alleles among the parents. Hence some parental allele represented n_j-1 times is chosen to produce two copies with the probability

$$\frac{(n_j-1)[\alpha(n_j-1) + 1]}{r}$$

if parental genes are randomly chosen to be replicated twice. Hence

$$P_{t+1}(r+1;k;n_1,\ldots,n_k|r,N^{(i)},N^{(v)}) = \sum_{j=1}^{k} \frac{(n_j-1)[\alpha(n_j-1)+1]}{r\alpha(n_j)}$$

$$\times P_t(r;k;n_1,\ldots,n_{j-1},n_j-1,n_{j+1},\ldots,n_k|N^{(i)}) \tag{7}$$

Therefore, if we combine (2), (5), (6) and (7) we obtain

$$\sum_{i=1}^{s} P_{t+1}(r+1;k;n_1,\ldots,n_k|N^{(i)},N^{(v)})P(N_t = N^{(i)})c_{iv}$$

$$= P_{t+1}(r+1;k;n_1,\ldots,n_k|N^{(v)})P(N_{t+1} = N^{(v)})$$

$$\doteq \sum_{i=1}^{s} P(N_t = N^{(i)})\ c_{iv}\ [1-(r+1)(\frac{r}{4N^{(i)}} + u)]$$

$$\times P_t(r+1;k;n_1,\ldots,n_k|N^{(i)}) + \frac{r(r+1)}{4N^{(i)}} \sum_{j=1}^{k} \frac{(n_j-1)[\alpha(n_j-1)+1]}{r\alpha(n_j)}$$

$$\times P_t(r;k;n_1,\ldots,n_{j-1},n_j-1,n_{j+1},\ldots,n_k|N^{(i)}) \tag{8}$$

which is one of our two fundamental equations. It applies if $\alpha_1 = 0$ among the offspring.

We shall now derive the second of the fundamental equations, which only refers to configurations within one generation. To do this, a sample of r+1 genes in generation t will be looked upon as consisting of two parts. First r genes are drawn, and next the resulting subsample is supplemented by the drawing of one more gene.

With random sampling, the configuration $\{n_1,\ldots,n_k\}$ among the first r genes is $1/(r+1)$ times as probable as all sets of r+1 genes containing this as a subset of size r. Also, one of the ways to have $\{n_1,\ldots,n_k\}$ among the first r genes is to have it followed by a gene represented n_j+1 times among r+1 genes and n_j times among r. Thus, if $\alpha(n_j)$ is equal to the number of alleles represented n_j times among the r genes, each of these ways is $1/\alpha(n_j)$ times as probable as all configurations of the type $\{n_1,n_2,\ldots,n_{j-1},n_j+1,n_{j+1},\ldots,n_k\}$. Finally, there are $(n_j+1)[\alpha(n_j+1)+1]$ ways to pick a gene to be in the r+1th position in the sample if it is represented n_j+1 times in the sample. Therefore

$$P_t(r;k;n_1,\ldots,n_k) = \sum_{j=1}^{k} \frac{[\alpha(n_j+1)+1](n_j+1)}{\alpha(n_j)(r+1)}$$

$$\times\, P_t(r+1;k;n_1,\ldots,n_{j-1},n_j+1,n_{j+1},\ldots,n_k)$$

$$+\, \frac{\alpha_1+1}{r+1}\, P_t(r+1;k+1;n_1,\ldots,n_k,1) \tag{9}$$

where α_1 is the number of alleles represented once among the first r genes. This is the second of the fundamental equations.

In the special case in which s = 1, equations (8) and (9) may be used to derive probabilities for all configurations in the following manner:

(i) Since one gene must certainly be identical in state with itself, the probability of the configuration {1} is equal to 1.

(ii) One may then use (8) to compute probabilities for all configurations $\{n_1\}$.

(iii) One may then use (9) to compute the probabilities of configurations $\{n_1,1\}$.

(iv) Then (8) may be used again to compute probabilities for $\{n_1,n_2\}$, $n_1,n_2 > 1$.

(v) Next (9) may be used again to compute probabilities for $\{n_1,n_2,1\}$, $n_1,n_2 > 1$.

(vi) By continuing this reasoning (1) may be derived in its full generality.

V. PROBABILITIES FOR SOME SMALL SAMPLES

If we consider small samples from a population of size N, the notation can be simplified without loss of clarity by eliminating r and k from terms given by (1). Thus, for example, we may write $P_t(4)$ in place of $P_t(4;1;4)$ and $P_t(2,1)$ in place of $P_t(3;2;2,1)$. Since $P_t(1) = 1$ for all t it follows from (8) that

$$
\begin{aligned}
P_{t+1}(2) &= \left(1 - \frac{2\theta+2}{4N}\right)P_t(2) + \frac{2}{4N} \\
P_{t+1}(3) &= \left(1 - \frac{3\theta+6}{4N}\right)P_t(3) + \frac{6}{4N} P_t(2) \\
P_{t+1}(4) &= \left(1 - \frac{4\theta+12}{4N}\right)P_t(4) + \frac{12}{4N} P_t(3) \\
P_{t+1}(2,2) &= \left(1 - \frac{4\theta+12}{4N}\right)P_t(2,2) + \frac{4}{4N} P_t(2,1)
\end{aligned}
\tag{10a}
$$

and from (9) that

$$
P_t(2,1) = 3[P_t(2)-P_t(3)] \tag{10b}
$$

The equilibrium solutions of (10a,b) are

$$P_t(2) = P(2) = \frac{1}{\theta+1}$$

$$P_t(3) = P(3) = \frac{2}{(\theta+1)(\theta+2)}$$

$$P_t(4) = P(4) = \frac{6}{(\theta+1)(\theta+2)(\theta+3)} \tag{11}$$

$$P_t(2,1) = P(2,1) = \frac{3\theta}{(\theta+1)(\theta+2)}$$

$$P_t(2,2) = P(2,2) = \frac{3\theta}{(\theta+1)(\theta+2)(\theta+3)}$$

These equations are not only special cases of (1), but are of interest in their own right. Thus, with random mating, $P_t(2)$ is equal to the probability of homozygotes in generation t and $P_t(4)$ and $P_t(2,2)$ may be used to calculate the between population variance of the sample homozygosity. For if x_w is equal to the frequency of allele A_w in generation t the probability of homozygosity is $J_t = \sum_w x_w^2$. Hence

$$E(J_t) = E(\sum_w x_w^2) = P_t(2)$$

$$E(J_t^2) = E(\sum_w x_w^4 + \sum_w \sum_{w' \neq w} x_w^2 x_{w'}^2)$$

while

$$P_t(2,2) = E\{\sum_w \sum_{w' \neq w} (x_w^2 x_{w'}^2 + x_w x_{w'} x_w x_{w'} + x_w x_{w'} x_{w'} x_w)\}$$

$$= 3E(\sum_w \sum_{w' \neq w} x_w^2 x_{w'}^2)$$

Therefore,

$$\mathrm{Var}(J_t) = P_t(4) + \frac{1}{3} P_t(2,2) - [E(J_t)]^2 \tag{12}$$

At equilibrium equations (11) and (12) imply that

$$\mathrm{Var}(J) = \frac{6+\theta}{(\theta+1)(\theta+2)(\theta+3)} - \frac{1}{(\theta+1)^2} = \frac{2\theta}{(\theta+1)^2(\theta+2)(\theta+3)} \tag{13}$$

as found by Lewis and Pollak (1982), after arguing in the manner just described.

The first of equations (11) was given first by Kimura and Crow (1964). Expression (13) was derived in somewhat different ways by Li and Nei (1975) and Stewart (1976). It can also be shown that if (12) is solved for finite t, the result is consistent with that given by Li and Nei (1975).

VI. GENERAL SAMPLING FORMULA

Now let us suppose that the transition matrix C is periodic with period d. Then there exist numbers π_{bi} such that

$$\lim_{t\to\infty} P[N_{td+b} = N^{(i)}] = \pi_{bi}$$

and

$$\pi_{b+1,v} = \sum_i \pi_{bi}\, c_{iv}$$

Hence

$$c'_{vi,b,b+1} = \frac{\pi_{bi} c_{iv}}{\pi_{b+1,v}} = \lim_{t\to\infty} P(N_{td+b} = N^{(i)} \,|\, N_{td+b+1} = N^{(v)}) \tag{14}$$

is a transition probability associated with what Feller (1968) calls a reversed Markov chain. It can easily be shown that if $c_{iv}^{(p)}$ is a p-step transition probability corresponding to the original Markov chain,

$$c'^{(p)}_{vi;b,b+p} = \frac{\pi_{bi} c_{iv}^{(p)}}{\pi_{b+p,v}}$$

is a p-step transition probability corresponding to the reversed Markov chain and that the latter is also periodic with period d.

As $t \to \infty$, we may use (14) to recast (8) as

$$
\begin{aligned}
&P_{td+b+1}(r+1;k;n_1,\ldots,n_k|N^{(v)}) \\
&\quad \doteq \sum_{i=1}^{s} c'_{vi;b,b+1}[1-(r+1)(\frac{r}{4N^{(i)}}+u)] \\
&\quad \times P_{td+b}(r+1;k;n_1,\ldots,n_k|N^{(i)}) \\
&\quad + \frac{r(r+1)}{4N^{(i)}} \sum_{j=1}^{k} \frac{(n_j-1)[\alpha(n_j-1)+1]}{r\alpha(n_j)} \\
&\quad \times P_{td+b}(r;k;n_1,\ldots,n_{j-1},n_j-1,n_{j+1},\ldots,n_k|N^{(i)})
\end{aligned} \tag{15}
$$

At equilibrium the subscripts td+b+1 and td+b in (15) can be replaced by b+1 and b. Hence, if d is small and all the $N^{(i)}$ are large, (15) implies that

$$
\begin{aligned}
&P_{b+1}(r+1;k;n_1,\ldots,n_k|N^{(v)}) \\
&\quad \doteq \sum_{i} c'_{vi;b,b+1}\ P_b(r+1;k;n_1,\ldots,n_k|N^{(i)}) \\
&\quad \doteq \sum_{i} c'^{(d)}_{vi;b,b+d}\ P_{b+1}(r+1;k;n_1,\ldots,n_k|N^{(i)})
\end{aligned}
$$

Therefore, $P_{b+1}(r+1;k;n_1,\ldots,n_k|N^{(v)})$ is equal to, say, $P_{b+1}(r+1;k;n_1,\ldots,n_k)$ for all v because it is an element of a right eigenvector associated with an eigenvalue 1 that corresponds to one of the d classes into which the Markov chain can be decomposed. In class b+1 all such eigenvectors are multiples of a vector of ones. It is also clear from this derivation that

$$P_{b+1}(r+1;k;n_1,\ldots,n_k) = P_b(r+1;k;n_1,\ldots,n_k)$$

Because, in addition,

$$\frac{1}{d}\sum_{b=0}^{d-1} \pi_{b+1,v} = \frac{1}{d}\sum_{b=0}^{d-1} \sum_i \pi_{bi}\, c_{iv} = 1$$

it therefore follows from (15) that

$$P(r+1;k;n_1,\ldots,n_k) = \frac{1}{d}\sum_{b=0}^{d-1} \sum_v \pi_{b+1,v} P_{b+1}(r+1;k;n_1,\ldots,n_k)$$

$$\doteq [1-(r+1)(\frac{r}{4N_e} + u)]P(r+1;k;n_1,\ldots,n_k)$$

$$+ \frac{r(r+1)}{rN_e} \sum_{j=1}^{k} \frac{(n_j-1)[\alpha(n_j-1)+1]}{r\alpha(n_j)}$$

$$\times P(r;k;n_1,\ldots,n_{j-1},n_j-1,n_{j+1},\ldots,n_k) \qquad (16)$$

where

$$N_e = [\frac{1}{d} \sum_{b=0}^{d-1} \sum_{i=1}^{s} \pi_{bi}/N^{(i)}]^{-1} \qquad (17)$$

It is evident that (16) is the same as the system of equations for equilibrium configuration probabilities that is applicable when $s = d = 1$ and the population does not vary in size. Equation (9) also holds because all configuration probabilities in it refer to the same generation and population size. Therefore we obtain (1), if N in that equation is replaced by N_e, which is calculated from (17).

It can be shown by induction that the equilibrium configuration probabilities in (16) are approached as $t \to \infty$. For we know that if

$$P_{td+b}(r;k;n_1,\ldots,n_k|N^{(v)})$$

$$= P(r;k;n_1,\ldots,n_k|N^{(v)}) + \delta^{(v)}_{td+b}(r;k;n_1,\ldots,n_k)$$

then $\delta^{(v)}_{td+b}(1;1;1) = 0$ for all t. Let us next assume as an induction hypothesis that

$$\delta^{(v)}_{td+b}(r;1;r) \to 0, \quad t \to \infty$$

for $r = 1,2,\ldots,m$. Then we see from (15) that

$$\delta^{(v)}_{td+b+1}(m+1;1;m+1)$$

$$\doteq \sum_{i=1}^{s} c'_{vi;b,b+1}[1-(m+1)(\frac{m}{4N^{(i)}} + u)]\delta^{(i)}_{td+b}(m+1;1;m+1) \quad (18)$$

as $t \to \infty$. Now the matrix $C_1 = (c'_{vi;b,b+1})$ has d distinct characteristic roots that are of absolute value 1 while the other roots are smaller in absolute value. Hence the T-th power of the matrix of coefficients in (18), which has elements that are smaller than those of C_1, tends to 0 as $T \to \infty$. It then follows from the method of calculation of the P values, discussed in Section IV, that all δ values tend in the long run to 0.

VII. CONCLUDING REMARKS

Equation (17) generalizes earlier results in the literature. Particular cases are what Karlin (1968) calls uniform variation of population number and cyclic variation in population size, which was considered by Wright (1938, 1939). I refer the reader to Chia and Pollak (1974) for details. Chia and Pollak gave more general expressions for N_e than (17), which apply to dioecious populations as well as monoecious ones and allow as well for more general methods of reproduction than in the Wright-Fisher model. Their results are consistent with expressions for N_e obtained by Kimura and Crow (1963) in the special case in which s = d = 1. However, Chia and Pollak (1974) only calculated probabilities of homozygosity and did not derive an expression like (17) that is applicable to all transition matrices C. But by making full use of the properties of reversed Markov chains, as in this paper, their results can be shown to hold for all C. This encourages us to hope that it may be possible to generalize (1) in a similar way. This is consistent with remarks of Kingman (1980), who expressed the view that we may have confidence in the reliability of the Ewens sampling formula when the sample size is small in comparison with that of the population, selection plays a negligible role, mutation is nonrecurrent and the population is at equilibrium under mutation and random drift.

REFERENCES

Chia, A.B. and Pollak, E. (1974). The inbreeding effective number and the effective number of alleles in a population that varies in size. *Theor. Pop. Biol. 6,* 149-172.

Crow, J.F. (1972). Darwinian and non-Darwinian evolution. *Proceedings of the 6th Berkeley Symposium on Mathematical Statistics and Probability, Vol. V,* 1-22. Berkeley: University of California Press.

Ethier, S.N. and Nagylaki, T. (1980). Diffusion approximations of Markov chains with two time scales and applications to population genetics. *Advan. Appl. Prob. 12,* 14-49.

Ewens, W.J. (1972). The sampling theory of selectively neutral alleles. *Theor. Pop. Biol. 3,* 87-112.

Ewens, W.J. (1979). *Mathematical Population Genetics*. Berlin: Springer.

Feller, W. (1968). *An Introduction to Probability Theory and its Applications, Vol. 1,* Third Edition. New York: John Wiley.

Karlin, S. (1968). Rates of approach to homozygosity for finite stochastic models with variable population size. *Amer. Natur. 102,* 443-455.

Karlin, S. and McGregor, J. (1972). Addendum to a paper of W. Ewens. *Theor. Pop. Biol. 3,* 113-116.

Kempthorne, O. (1967). The concept of identity of genes by descent. *Proceedings of the 5th Berkeley Symposium on Mathematical Statistics and Probability, Vol. IV,* pp. 333-348. Berkeley: University of California Press.

Kimura, M. (1968). Evolutionary rate at the molecular level. *Nature 217,* 624-626.

Kimura, M. and Crow, J.F. (1963). The measurement of effective population number. *Evolution 17,* 279-288.

Kimura, M. and Crow, J.F. (1964). The number of alleles that can be maintained in a finite population. *Genetics 49,* 725-738.

Kimura, M. and Ohta, T. (1971). On the rate of molecular evolution. *Journal of Molecular Evolution 1,* 1-17.

King, J.L. and Jukes, T.H. (1969). Non-Darwinian evolution. *Science 164,* 788-798.

Kingman, J.F.C. (1980). *The Mathematics of Genetic Diversity.* CBMS-NSF Regional Conf. Ser. No. 34. Society for Industrial and Applied Mathematics. Philadelphia.

Lewis, J.W. and Pollak, E. (1982). Genetic identity in subdivided populations. 1. Two equal-sized subpopulations. *Theor. Pop. Biol. 22,* 218-240.

Li, W.H. and Nei, M. (1975). Drift variances of heterozygosity and genetic distance in transient states. *Genet. Res. 25,* 229-248.

Ohta, T. and Kimura, M. (1971). On the constancy of the evolutionary rate of cistrons. *Journal of Molecular Evolution 1,* 18-25.

Stebbins, G.L. and Lewontin, R.C. (1972). Comparative evolution at the levels of molecules, organisms, and populations. *Proceedings of the 6th Berkeley Symposium on Mathematical Statistics and Probability, Vol. V,* pp. 23-42. Berkeley: University of California Press.

Stewart, F.M. (1976). Variability in the amount of heterozygosity maintained by neutral mutations. *Theor. Pop. Biol. 9,* 188-201.

Watterson, G.A. (1976). The stationary distribution of the infinitely-many neutral alleles diffusion model. *J. Appl. Prob. 13,* 639-651.

Wright, S. (1938). Size of population and breeding structure in relation to evolution. *Science 87*, 430-431.

Wright, S. (1939). Statistical genetics in relation to evolution. In *"Actualités Scientifiques et Industrielles", No. 802*, 5-64. Exposés de Biometrie et de la Statistique Biologique XIII. Paris: Hermann.

INDEX